New Developments in Coatings Technology

ACS SYMPOSIUM SERIES **962**

New Developments in Coatings Technology

Peter Zarras, Editor
U.S. Department of Navy

Timothy Wood, Editor
Rohm and Haas Company

Brough Richey, Editor
Rohm and Haas Company

Brian C. Benicewicz, Editor
Rensselaer Polytechnic Institute

Sponsored by the
ACS Divisions of Polymer Chemistry, Inc. and
Polymeric Materials: Science and Engineering, Inc.

American Chemical Society, Washington, DC

Library of Congress Cataloging-in-Publication Data

New developments in coatings technology / Peter Zarras, editor,...[et al.] ; sponsored by the ACS Divisions of Polymer Chemistry, Inc., and Polymeric Materials: Science and Engineering, Inc.

    p. cm.—(ACS symposium series ; 962)

Includes bibliographical references and index.

ISBN 978–0–8412–3963–0 (alk. paper)

    1. Coating processes—Congresses. 2. Coatings—Congresses.

    I. Zarras, Peter, 1961- II. American Chemical Society. Division of Polymer Chemistry, Inc. III. American Chemical Society. Division of Polymeric Materials: Science and Engineers, Inc.

TP156.C37N49   2006
667'.9—dc22

                                                  2006052611

The paper used in this publication meets the minimum requirements of American National Standard for Information Sciences—Permanence of Paper for Printed Library Materials, ANSI Z39.48–1984.

Copyright © 2007 American Chemical Society

Distributed by Oxford University Press

All Rights Reserved. Reprographic copying beyond that permitted by Sections 107 or 108 of the U.S. Copyright Act is allowed for internal use only, provided that a per-chapter fee of $36.50 plus $0.75 per page is paid to the Copyright Clearance Center, Inc., 222 Rosewood Drive, Danvers, MA 01923, USA. Republication or reproduction for sale of pages in this book is permitted only under license from ACS. Direct these and other permission requests to ACS Copyright Office, Publications Division, 1155 16th Street, N.W., Washington, DC 20036.

The citation of trade names and/or names of manufacturers in this publication is not to be construed as an endorsement or as approval by ACS of the commercial products or services referenced herein; nor should the mere reference herein to any drawing, specification, chemical process, or other data be regarded as a license or as a conveyance of any right or permission to the holder, reader, or any other person or corporation, to manufacture, reproduce, use, or sell any patented invention or copyrighted work that may in any way be related thereto. Registered names, trademarks, etc., used in this publication, even without specific indication thereof, are not to be considered unprotected by law.

PRINTED IN THE UNITED STATES OF AMERICA

Foreword

The ACS Symposium Series was first published in 1974 to provide a mechanism for publishing symposia quickly in book form. The purpose of the series is to publish timely, comprehensive books developed from ACS sponsored symposia based on current scientific research. Occasionally, books are developed from symposia sponsored by other organizations when the topic is of keen interest to the chemistry audience.

Before agreeing to publish a book, the proposed table of contents is reviewed for appropriate and comprehensive coverage and for interest to the audience. Some papers may be excluded to better focus the book; others may be added to provide comprehensiveness. When appropriate, overview or introductory chapters are added. Drafts of chapters are peer-reviewed prior to final acceptance or rejection, and manuscripts are prepared in camera-ready format.

As a rule, only original research papers and original review papers are included in the volumes. Verbatim reproductions of previously published papers are not accepted.

ACS Books Department

Contents

1. Overview of Coatings for Advanced Applications............1
   P. Zarras, T. Wood, B. Richey, and B. C. Benicewicz

Anticorrosion and Antifouling Coatings

2. Scanning Probe Studies of Active Coatings for Corrosion Control of Al Alloys............8
   Jie He, Dante Battocchi, Alda M. Simões, Dennis E. Tallman, and Gordon P. Bierwagen

3. Optimization of One-Step, Low-VOC, Chromate-Free Novel Primer Coatings Using a Taguchi Method Approach............24
   Anuj Seth and Wim J. van Ooij

4. Investigation of Electroactive Polymers and Other Pretreatments as Replacements for Chromate Conversion Coatings: A Neutral Salt Fog and Electrochemical Impedance Spectroscopy Study............40
   P. Zarras, N. Prokopuk, N. Anderson, and J. D. Stenger-Smith

5. Reversible Addition–Fragmentation Chain Transfer Polymerization of 4-Anilinophenyl(meth)acrylates............54
   Ru Chen, Chunzhao Li, and Brian C. Benicewicz

6. New Technologies for the Analysis of Marine Coatings............69
   Thomas E. Ready, Johnson Thomas, Seok-bong Choi, and Philip Boudjouk

7. Laboratory Methods to Access the Antifouling and Foul-Release Properties of Novel, Non-Biocidal Marine Coatings............91
   M. E. Callow, J. A. Callow, and D. E. Wendt

8. Ion Exchange Compounds for Corrosion Inhibiting Pigments in Organic Coatings..........108
   R. G. Buchheit and S. P. V. Mahajanam

9. Polyurethane–Polysiloxane Ceramer Coatings..........135
   Hai Ni, William J. Simonsick, and Mark D. Soucek

Specialty Coatings

10. Effect of Selected Processing Parameters on the Key Properties of Solvent-Cast Polyimides for Optical Waveguides..........146
    Andrew J. Guenthner, K. R. Davis, L. Steinmetz, and J. M. Pentony

11. Sensor Coatings: Responsive Coatings for the Detection, Identification, and Removal of Surfaceborne Plutonium and Uranium..........162
    H. Neil Gray and Betty Jorgensen

12. High-Performance UV-Cured Acrylic Coatings..........176
    Christian Decker

13. Polyamide-11 Powder Coatings: Exceptional Resistance to Cavitation Erosion..........190
    T. Page McAndrew, Marc Audenaert, Jerry Petersheim, Dana Garcia, and Thomas Richards

14. Using VOC-Exempt Solvents in Coatings: Performance, Productivity, and Lower Environmental Impact..........201
    Daniel B. Pourreau

15. Development of UV-Curable Waterborne Polyurethane Dispersion for Soft Feel Application..........225
    Peter D. Schmitt, Lyuba Gindin, Sr., Aaron Lockhart, and Phil Lunney

Advanced Techniques for Measuring Coating Performance

16. Combinatorial and High-Throughput Development of Polymer Sensor Coatings for Resonant and Optical Sensors..........240
    Radislav A. Potyrailo

17. **Using Quartz Crystal Microbalance–Heat Conduction Calorimetry to Monitor the Drying and Curing of an Alkyd Spray Enamel**..................261
    Allan L. Smith

18. **Predicting Service Life Performance: Our Analytical Toolbox**..........279
    Karlis Adamsons

19. **Evaluation of the Protective Properties of Novel Chromate-Free Polymer Coatings Using Electrochemical Impedance Spectroscopy**..................297
    E. Kus, M. Grunlan, W. P. Weber, N. Anderson, C. Webber, J. D. Stenger-Smith, P. Zarras, and F. Mansfeld

Indexes

Author Index..................325

Subject Index..................326

New Developments in Coatings Technology

Chapter 1

Overview of Coatings for Advanced Applications

P. Zarras[1], T. Wood[2], B. Richey[2], and B. C. Benicewicz[3]

[1]Polymer Science and Engineering Branch (Code 498200D), Naval Air Warfare Center Weapons Division, Department of the Navy, 1900 North Knox Road (Stop 6303), China Lake, CA 93555–6106
[2]Rohm and Haas Company, 727 Norristown Road, Springhouse, PA 19477–0904
[3]Department of Chemistry, Rensselaer Polytechnic Institute, 110 8th Street, Troy, NY 12180–3590

This American Chemical Society book will provide a detailed overview of the many advances in coatings technology over the past decade. The book "New Developments in Coatings Technology" is based on the Fall National American Chemical Society Meeting symposium held in Philadelphia, Pennsylvania, August, 22-26, 2004. The book will cover such diverse topics as electroactive polymers (EAPs), meta-rich primers and organofunctional silanes for replacements of hexavalent chromium (CrVI) based pretreatments and primers. Anti-fouling and foul release coatings are discussed in this volume. Laboratory methods are examined in this section to evaluate the best performing anti-fouling and foul release coatings. Specialty coatings with an emphasis on sensing based applications and environmentally benign coatings are also presented. The final section of this book will examine recent developments in monitoring a coating's performance. Electrochemical impedance spectroscopy (EIS) and calorimetry methods are a select few of the techniques that will be discussed in this volume. This book will give the reader a sense of the many varied applications that coatings can be used for as well as the methods needed to evaluate them.

© 2007 American Chemical Society

Introduction

The earliest coating materials were natural resins that were used by the ancients. In the 12th century, several reports are found on combining natural resins with chemically hardening natural oils (1). The last century saw the development of the production line which facilitated the onset of industrial scale painting. Quick drying paints and faster coating processes allowed for the growth of the "coatings industry."

Coatings are used for a variety of applications. These applications include protection from environmental factors, beautifying structures by changing their surface properties and coatings that "react to" and "sense" their environment, alternatively referred to as "intelligent coatings."

In this volume series the term "coating" will apply to both surface coatings and paints. In the strictest sense, the term "surface coating" refers to any material that is applied as a thin continuous layer, whereas, the term "paint" refers to any pigmented materials used in a film layer(s) (2).

Anticorrosion Coatings

Corrosion is the destructive result of chemical reactions that occur between a metal or a metal alloy and its environment (3). Corrosion impacts many aspects of our daily lives, and various estimates put the costs to the US economy between 100-300 billion dollars annually (4,5). The most striking features of the corrosion process are the immense variety of conditions under which it occurs and the large number of forms in which it appears (6). Corrosion affects all structural materials and infrastructure of society which can cause in many cases grave economic consequences or life-threatening situations. There are numerous infrastructure items that can be significantly damaged and eventually destroyed by corrosion. These structures include pipelines, bridges, automobiles, storage tanks, airplanes and ships (both military and commercial). The most common environments for corrosion to occur are in natural waters, atmospheric moisture, rain and man-made solutions (such as storage tanks). In this section several recent developments in corrosion inhibiting pretreatments and primers will be discussed.

Electroactive polymers (EAPs) have been investigated for the past two decades for their corrosion-inhibiting properties (7-9). Chapters 2, 4 and 5 present recent developments of EAPs for corrosion-inhibition using polypyrrole and magnesium rich primers on Al 2024-T3, poly(phenylene vinylene) (PPV) derivatives as alternative pretreatment coatings for replacing chromate conversion coatings (CCC) and poly(acrylates)/poly(acrylamides) containing oligoaniline side chains are presented. Additional chapters on corrosion-inhibiting coatings encompass organofunctional silanes as "super primers", organic/inorganic hybrid based adhesion promoter/corrosion inhibiting coatings and synthetic inorganic ion-exchange compounds as pigments in organic coatings that can store and release non-Cr(VI) inhibitors.

Antifouling Coatings

In addition to corrosion-inhibiting coatings, anti-fouling/foul-release coatings are examined in this section. Marine biofouling is a serious and costly problem resulting in loss of operating efficiency of marine vessels (*10*). Due to increased environmental regulations, coatings derived from copper and tin are banned due to their toxic effects on marine life (*11*). The regulation of anti-fouling paints has drastically increased over the past decade. More stringent rules are expected across the world requiring developers of anti-fouling coatings to comply with future regulations in order to minimize the environmental impact on marine life (*12*).

Alternative coatings are therefore presented in this section that addresses this ongoing problem. Chapter 6 presents recent developments on silicone based coatings that possess non-leaching/non-metallic biocidal and foul-release components. A high-throughput method is described in this chapter to maximize efficiency in the design and evaluation of these coatings. Chapter 7 also examines laboratory-based methods to evaluate the effectiveness of foul-release and anti-fouling coatings.

Specialty Coatings

In this section, several chapters will be presented that focus on "intelligent coatings" that can respond to their environment and coatings that are "environmentally friendly." These environmentally friendly coatings are characterized by their low volatile organic content (VOC) and compliance with environmental regulations. A contaminant-sensing coating is presented in Chapter 11. This coating is water-based, plutonium and uranium sensing and can decontaminate surfaces. This coating exhibits responsive behavior by indicating areas of contamination.

Several additional chapters presented in this section focus on environmentally friendly coatings, metal replacement using polyamides, UV-cured coatings for improved exterior durability of metals, UV-curable waterborne PUDs and polyimide coatings for optical devices.

Techniques for Measuring a Coatings Performance

The final section of this book examines several techniques that can evaluate a coatings performance. UV-visible, IR, DSC, MS are several analytical techniques that are used to measure the surface and depth profiling of films. Simulated atmospheric, salt-spray and immersion tests are used to evaluate coatings for their corrosion-inhibiting performance (*13*). Electrochemical testing methods are effective tools for understanding the mechanisms of corrosion. Specifically, electrochemical impedance spectroscopy (EIS) has been extensively used to evaluate the properties of coatings in a corrosive environment (*14*).

A combinatorial screening process flow is described in Chapter 16. This method was employed to design and optimize sensor materials. The use of quartz crystal microbalance /heat conduction calorimetry is presented in Chapter 17. This analytical technique measures small mass changes in films which allowed the author to monitor the curing of commercial alkyd spray enamel. The final chapter (Chapter 19) examines chromate-free coatings using EIS. EAP and nonfluorinated pentosiloxane coatings were examined for corrosion-inhibition and anti-fouling performance respectively.

References

1. Goldschmidt, A., and Streitberger, H-J., BASF Handbook of Coating Technology, Vincentz Network, Germany, Chapter 1, pp. 15-25, 2003.
2. Lambourne, R.; In *Paint Composition and Applications-A General Introduction*, Lambourne, R. and Strivens, T. A., Paint and Surface Coatings -Theory and Practice, 2nd Edition, William Andrew Publishing, 1999, Chapter 1, p. 3.
3. Marek, M. I., *Thermodynamics of Aqueous Corrosion*, in ASM Handbook, ed., J. R. Davis., et. al., Vol.13, Corrosion, ASM International, 1987, p. 18.
4. Lu, W-K., Basak, S., and Elsenbaumer, R. L., *Corrosion Inhibition of Metals by Conductive Polymers*, in Handbook of Conducting Polymers, eds., T. A. Skotheim, R. L. Elsenbaumer and J. R. Reynolds, Marcel Dekker, New York, 1998, p. 881.
5. Brumbaugh, D., *AMPITIAC Newsletter*, **1999**, 3(1), 1
6. Uhling, H. H., and Reive, R. W., *Corrosion and Corrosion Control*, Wiley, New York, 1985.
7. Mengoli, G., Munari, M. T., Bianco, P., nd Musiana, M.M., *J. Appl. Polym. Sci.*, **1981**, 26, 4247.
8. DeBerry, D. W., *J. Electrochem. Soc.*, **1985**, 132, 1022.

9. Ahmand, N., and MacDiarmid, A. G., *Bull. Am. Phys. Soc.*, **1987**, 32, 548.
10. Efimenko, K., and Genzer, J., *Functional-Siloxane-based Foul-Release Coatings: Preparation and Properties*, Pacific Polymer Federation Proceedings, December 11-14, 2005, The Westin Maui, American Chemical Society, Division of Polymer Chemistry, Session F#12.
11. Alzieu, C. L., *Mar. Poll. Bull.*, **1986**, 17, 494.
12. IMO (2001), Antifouling systems-International Convention on the Control of Harmful Anti-fouling Systems on Ships. Resolution 3-Approval and test methodologies for anti-fouling systems on ships. p. 29.
13. Davis, J. R. Corrosion: Undestanding the Basics, ASM International, Ohio, 2000, Chapter 11, pp.427-487.
14. Mansfeld, F., *J. Appl. Electrochem.*, **1995**, 25, 187.

Anticorrosion and Antifouling Coatings

Chapter 2

Scanning Probe Studies of Active Coatings for Corrosion Control of Al Alloys

Jie He[1], Dante Battocchi[2], Alda M. Simões[3], Dennis E. Tallman[1,2,*], and Gordon P. Bierwagen[2]

Departments of [1]Chemistry and Molecular Biology and [2]Coatings and Polymeric Materials, North Dakota State University, Fargo, ND 58105
[3]ICEMS/Chemical Engineering Department, Instituto Superio Tecnico, Ave. Rovisco Pais, 1049–001 Lisboa, Portugal

Active corrosion protection coatings such as conducting polymer coatings and metal-rich primers interact chemically, electrically and/or electrochemically with the underlying active metal, thereby altering its corrosion behavior. In this report, the scanning vibrating electrode technique, the scanning ion electrode technique and scanning electrochemical microscopy have been used to study a polypyrrole coating and also a newly developed Mg-rich primer on AA 2024-T3 aluminum alloy. Using these techniques, current density and either pH or oxygen concentration were mapped across the coating and/or alloy surfaces as a function of time, providing new spatial and temporal information on the interactions between these active coatings and the Al alloy. The polypyrrole coating functions to mediate oxygen reduction leading to alloy oxidation/ passivation, whereas the Mg-rich primer functions by a cathodic protection (sacrificial) mechanism.

© 2007 American Chemical Society

Introduction

Active corrosion protection coatings are defined as coatings that not only provide barrier properties but also contain or consist of a material that can interact chemically, electrically and/or electrochemically with the metal, altering its corrosion behaviour. Based on this definition, two coating systems that fall into this category are electroactive conjugated polymers (1, 2) and metal-rich coatings (3-5). In this paper, polypyrrole (Ppy) doped with 4,5-dihydroxy-1,3-benzenedisulfonic acid disodium salt (DBA) and a newly developed Mg-rich primer are evaluated for their corrosion protection of the Al alloy AA 2024-T3.

The Ppy/DBA film (Figure 1) is directly electrodeposited on the metal by an electron-transfer mediation mechanism (6). The current efficiency for the electrodeposited film formation was nearly 100% and the uniform and adherent film showed quite high electronic conductivity (typically 2.5 S/cm).

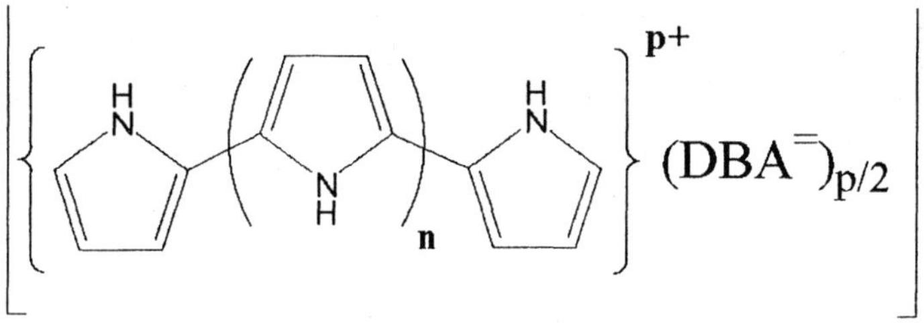

Figure 1. Polypyrrole doped with DBA. The doping level $p/(n+2) = 0.37$ from XPS measurement (6).

The metal-rich primer examined in this article is a Mg-rich primer system, a coating system that was recently developed in our laboratory to protect Al alloys (7). The Mg-rich primer is based on the concept of sacrificial protection in analogy to the usage of the zinc-rich primer to protect steel. This novel coating system may potentially replace the chromate pigments and pre-treatments currently used in the aerospace industry to protect Al alloys. Chromate has been proven to be a health and an environmental hazard and its use as a corrosion inhibitor will eventually be phased out (8, 9).

In this chapter, a series of scanning probe techniques, including the scanning vibrating electrode technique (SVET), the scanning ion electrode technique (SIET), and scanning electrochemical microscopy (SECM) were used to study the above two forms of active coatings on AA 2024-T3 alloy. The electrical and electrochemical interactions of the coatings with the Al alloy were investigated using SVET. The SIET was employed to map the pH distribution on the Ppy/DBA coated alloy surface while the SECM (10, 11) was used to generate oxygen concentration profiles resulting from the galvanic interaction between Mg and AA 2024-T3.

Experimental

Sample Preparation

The AA 2024-T3 (Q panel, Cleveland OH) substrates for Ppy/DBA coated samples were first polished using 600 grit silicon carbide, washed with hexane, and air-dried. The immersion solution was dilute Harrison's solution (DHS), which is an aqueous electrolyte of 0.35 wt.% $(NH_4)_2SO_4$ and 0.05 wt.% NaCl.

Electrochemically synthesized Ppy/DBA was prepared in a one-compartment, 100 mL 3-electrode cell including a working electrode (AA 2024-T3), a platinum plate counter electrode, and a saturated Ag/AgCl reference electrode. The working and counter electrodes were arranged parallel to one another to ensure uniform current distribution. The aqueous electrodeposition solution contained 0.10 M pyrrole monomer (freshly distilled) and 0.10 M DBA. The polypyrrole films were electrodeposited using the galvanostatic mode of the EG&G Princeton Applied Research 273A at a current density of 1 mA/cm^2. The electrodeposition time was 1000 seconds, at which time a continuous and uniform film was obtained. The film thickness of the Ppy/DBA averaged 2.6 ± 0.3 μm (*6*).

The AA 2024-T3 substrates for Mg-rich primer coated samples were brushed with a wire brush and degreased with hexane. Mg granules of average particle size ~50μm (Non Ferrum Metalpullver-Ecka, GmbH, St Georgen, Austria) are used as pigments and dispersed in an organic binder (*7*). Electrical contact between the pigment particles and the substrate was established by formulating the primer near the critical pigment volume concentration (CPVC) of the system, ~50%.

The Mg-rich primer was then spray applied at a thickness of approximately 70 μm. The primer was allowed to dry for three days at room temperature before an epoxy topcoat (for the topcoated experiments) was applied using a drawdown bar at a 3 mil thickness. This clear topcoat offers low barrier properties but was used only to inhibit the bubbling that otherwise occurred at the surface due to hydrogen evolution when exposed Mg particles from the very high pigment volume concentration primer came into direct contact with electrolyte.

Electrochemical Cell and Instrumentation

The SVET and SECM were used to study the local electrochemistry that occurred at the sample surface. Two electrochemical cell configurations described previously (*12*) were used in this work. One was a "single-substrate" configuration in which only one sample substrate was mounted in a Teflon sample cell, and a 2×2 mm area of the coated surface was directly exposed to electrolyte, which determined the scanning area of the SVET. An artificial defect (0.1-0.3 mm^2) was made by scribing through the coating with an awl. All the Ppy/DBA coated samples and some of the Mg-rich coated samples shown in this article were studied using this single-substrate configuration. The other sample cell configuration was based on our "two-substrate" configuration (*12*). Basically, two different substrates (one a bare AA 2024-T3 substrate, the other a Mg-rich primer coated AA 2024-T3 substrate) were electronically connected by a mechanical switch (Figure 2). The vibrating probe of the SVET then scanned either of the two substrate surfaces with the switch open (substrates electrically isolated from each other) or closed (substrates electrically connected).

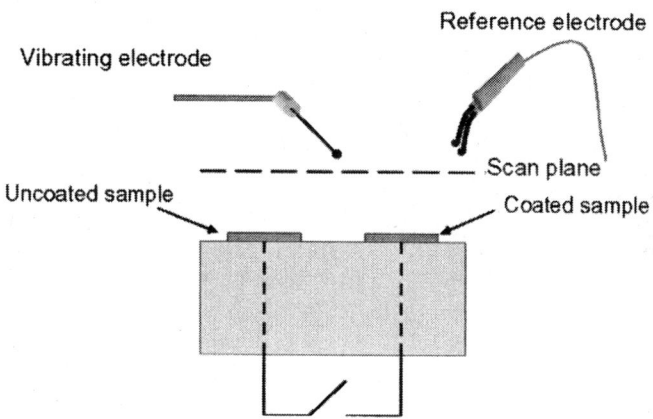

Figure 2. The two-substrate cell setup for SVET.

The SVET instrumentation used in these experiments was from Applicable Electronics (Forestdale, MA) and is described in detail elsewhere (*13*). The vibrating probe with a 20 µm platinum-black-coated tip was carefully placed above the surface of the substrate at a height of about 200 µm. All SVET measurements were performed in approximately 5 mL DHS at the open circuit potential (OCP). Scans were initiated within 5 min of immersion and were collected every 30 min for the duration of the experiment. Each scan consisted of 400 data points obtained on a 20 x 20 grid. An integration time of 1-second per scanning point was used. As described in our previous publication (*14*), the current density maps of SVET are displayed using two methods. One method displays the current density as a 3-dimensional surface, with the z-component of the measured current density plotted as a function of the x, y position in the probe scan plane. In this format, positive and negative current densities represent anodic and cathodic regions, respectively. The other method displays current density vectors (representing current density magnitude and direction) superimposed onto an optical micrograph of the immersed sample. In all cases, the bottom edge of the optical micrograph corresponds to the x-axis of the 3-dimensional plot. This latter display permits correlation of anodic and cathodic current processes with visual features on the substrate surface.

The SIET from Applicable Electronics (Forestdale, MA) was used to map the pH over the substrate surface area. For these measurements, ion-selective microelectrodes sensitive to protons were prepared using a commercial micropipette puller (Model P-97, Flaming/Brown Micropipette Puller). Single-barreled glass tubes (World Precision Instruments, Inc., Sarasota, FL) with an outer diameter of 1.5 mm were pulled to make micropipettes with a tip diameter of 3 μm. The micropipette tips were silanized to make them hydrophobic and then filled with a hydrogen ionophore (Fluka I-cocktail B). The body of each micropipette was filled with 50 mM KCl and 50 mM KH_2PO_4 and a AgCl coated silver wire was inserted to form an inner Ag/AgCl reference electrode. The micro-pH electrode showed a linear response to pH changes over the range of 3 to 10. The Nernst factor was 57.5 ± 1.0 mV. The outer reference electrode for the SIET measurements was a Ag/AgCl (3M NaCl) electrode. The micro-pH electrode was manually placed approximately 100 μm above the Ppy coated Al 2024 surface.

A scanning electrochemical microscope (SECM 900B, CH instruments, Inc. Austin, TX) was used to evaluate the dissolved oxygen concentration profile resulting from galvanic interaction between Al alloy and Mg metal. In this experiment, a modified form of the two-substrate cell was used. The Al alloy and Mg electrodes were embedded in epoxy resin and electrically connected on the back side through a mechanical switch. The exposed electrode areas were 5 mm x 0.75 mm for the Al alloy and 5 mm x 0.25 mm for Mg. The surface of each of the two metal substrates was mirror polished and the reservoir for the electrolyte solution was formed using plastic tape; DHS was used as the working electrolyte. A 10 μm planar disk Pt microelectrode (probe) was used to detect the oxygen concentration above the surface of both electrodes. The potential of the probe was set -0.75 V (vs. 3 M Ag/AgCl reference electrode) and the substrates were at open circuit potential. The separation distances for the probe-Al alloy and the probe-Mg metal were 500 μm and 200 μm, respectively. One dimensional linear scans were performed across the epoxy and metal surfaces. The probe current was sampled at 0.017 s/μm intervals.

Results and Discussion

Ppy/DBA coating on AA 2024-T3

In order to study the electricaland/or electrochemical interactions between the electrodepositedPpy/DBA and AA 2024-T3 during the initiation period of a corrosion process, an artificialscribe extending through the Ppy film to the AA 2024-T3 surface was introduced. A typical SVET result for the current density mapping of the Ppy/DBA-coatedAA 2024-T3 is depicted in Figure 3.

A large magnitude of oxidation current (peak value of ca. 150μA/cm^2)was observed within the scribe after just 5 min immersion and the reduction current

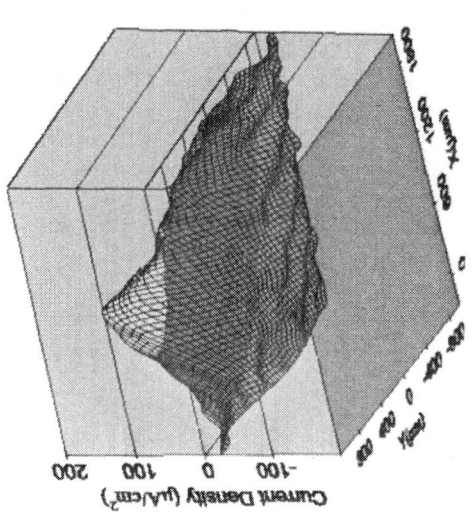

Figure 3. Current density map (left) and optical micrograph with superimposed current density vectors (right) for Ppy/DBA coated AA 2024-T3 (with scribe) after 5 minutes of immersion in DHS.

was distributed uniformly across the conducting polymer surface. This behavior is in contrast to that observed at a typical inert coating such as epoxy primer where both oxidation and reduction occur within the defect area (*15*). It is suggested that either an oxygen reduction reaction occurred atthe electrolyte/conductingpolymer interface and/or the Ppy film itself was reduced. On the other hand, it is interesting to note that gas bubbles were generated at the scribe area, where the oxidation current occurred. The amount of the gas generated and the corresponding oxidation current at the scribe both decreased as a function of time. Toward the end of the immersion period (typically 24 hours), the gas generation stopped and the oxidation current dropped to ca. 5 $\mu A/cm^2$, indicating that passivation was occurring. The scribed area visually looks shinythough somewhitecorrosion product was found at the scribe area after the immersion test.

This result is quite different from that observed with a poly(3-octyl-pyrrole) (POP) coated AA 2024-T3 where the coating inhibited the corrosion of exposed metal within the defect (*15*). This difference in behavior is most likely due to the significant differences in conductivity of the Ppy films. The conductivity of the POP film was several orders of magnitude lower than that of the Ppy/DBA film (2.5 S/cm) and oxygen reduction at the POP film was consequently more difficult. We conjecture that facile oxygen reduction at the Ppy/DBA film drove the oxidation reaction that was observed in the defect, which ultimately led to passivation of the exposed metal within the defect.

We are also curious about the gaseous product that was always generated at the exposed metal area in the scribe (rather than occurring at the polymer coated area) during the initial period of experiment. In our previous galvanic interaction study of Al/Ppy using the two substrate cell configuration (*12*), the gas bubbles were also repeatedly observed at the bare AA 2024-T3 substrate when it was electronically connected to a Ppy coated substrate (such as AA 2024, pure Al, or ITO) by a mechanical switch. Again, the gas bubble generation was associated with an oxidation current as measured with the SVET (*12*). The gaseous product appears to be H_2 (from SECM experiments; unpublished results) and may be due to the negative difference effect of the alloy (*16-18*), which is characterized by strong hydrogen evolution, the reaction rate of which increases with increase in the anodic polarization. It was proposed that Al oxidation may pass through a monovalent ion (Al^+) during the initial period of dissolution and then the Al^+ ion reacts with water to produce H_2:

$$2\ Al^+ + 8\ H_2O \rightleftharpoons Al_2O_3 \cdot 3H_2O + 2\ H_2 + 2\ H_3O^+ \qquad (1)$$

This would be a potential-independent chemical reaction and would occur simultaneously with the electrochemical corrosion reaction at the metal surface. Such a reaction would explain why the H₂ generation was always confined to the bare Al alloy surface and only observed when the polymer and the Al were electrically contacted. Eventual passivation of the Al surface, galvanically controlled by the Ppy/DBA polymer, would then lead to cessation of hydrogen generation as observed at the end of the immersion period.

The SIET was used to map the pH above the Ppy/DBA coated AA 2024-T3 so as to follow pH changes that occurred as a result of interaction between the Ppy/DBA and the Al alloy. As for the SVET experiment described above, a scribe extending through the coating to the alloy surface was introduced. A pH selective microelectrode based on a liquid ion exchange mechanism was positioned 100 µm above the sample surface. One typical SIET result for the Ppy/DBA-coated AA 2024-T3 is shown in Figure 4.

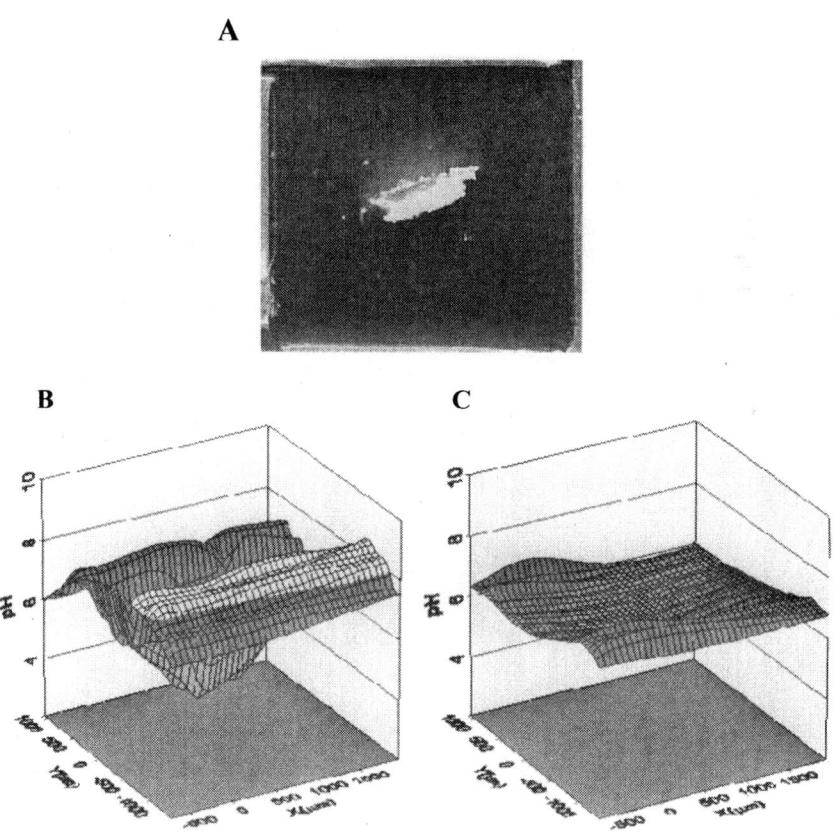

Figure 4. A typical SIET result for Ppy/DBA coated AA 2024-T3 (A: optical image of the sample with artificial defect. B, C: the pH distribution of the sample after 1.17 and 23 hours of immersion).

A decrease in pH was observed just above the scribe and its magnitude reached about 3.5 after 1.17 hours immersion. H_2 gas bubbles were also observed at the scribe area, as described above for the SVET experiments. The low pH observed at the exposed metal in the scribe is attributed to hydrolysis reactions of aquated Al^{3+} ion generated by the oxidation process occurring in the scribe area:

$$Al(H_2O)_6^{3+} + H_2O \rightarrow Al(OH)(H_2O)_5^{2+} + H_3O^+ \qquad (2a)$$

$$2\ Al(H_2O)_6^{3+} \rightarrow Al_2O_3 \cdot 3\ H_2O + 6\ H_3O^+ \qquad (2b)$$

Note that these reactions could also explain the observed H_2 generation if the pH right at the alloy surface were to reach ca. 0, in which case H_2 would be produced at the OCP of the substrate (ca. -0.2 V). Since our measurements were made at ca. 100 μm from the surface, the pH at the surface could be significantly lower. Experiments are underway to estimate the surface pH. We note that H_2 could also form in pits where extremely low pH may be encountered. However, no pits were observable in the defect region under optical magnification, consistent with the rather uniform current density observed in the scribe area by SVET.

A relatively higher pH (about 8) was distributed uniformly across the conducting polymer surface, suggesting that oxygen reduction occurred at the electrolyte/conducting polymer interface. Reduction of the polymer itself would lead to the expulsion of the DBA dopant ion, a weakly acidic species (pKa = 7.3) which would lead to a slight decrease in pH. Other experiments have suggested that polymer reduction occurs in the absence of oxygen (*12*). The higher pH above the conducting polymer surface decreased and simultaneously the lower pH above the scribe area increased with increased immersion time, reaching a more or less uniform pH of ca. 6 across the entire substrate after 23 hours immersion. This was consistent with the SVET results that also indicated diminishing anodic and cathodic currents toward the end of the immersion period (23 hours), i.e., passivation.

Mg-rich Coating

The analysis of a scribed/topcoated sample of Mg rich primer revealed (Figure 5) the presence of small anodic and cathodic currents at the defect area after 5 minutes of immersion. The anodic current disappeared with increasing immersion time and after 150 minutes the scribe area was completely cathodic (Figure 6). This behavior may be related to an induction period required for the electrolyte to penetrate the coating, after which the Mg particles started acting as a sacrificial anode. With increasing immersion time, more Mg was available for protection and the exposed area was completely cathodic.

Based on the above results, it was thought that the two-substrate configuration cell, with a Mg-rich primer coated AA 2024-T3 and a bare AA 2024-T3 substrate arranged as depicted in Figure 2, would be useful for studying the local electrochemistry and the related electrochemical/chemical interactions between primer and alloy. The experiment began with the electrodes disconnected and the bare Al corroded freely, showing anodic and cathodic areas characteristics of this type of alloy (Figure 7).

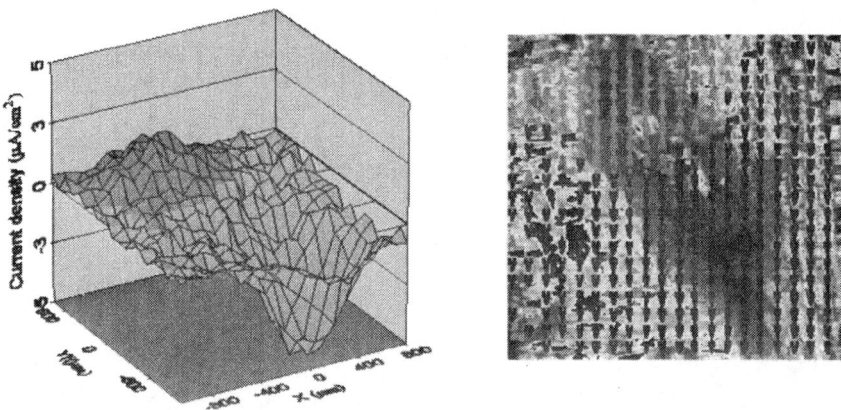

Figure 5. SVET scan over scribe of AA 2024-T3 coated with Mg rich primer and epoxy topcoat; 5 minutes after immersion in 3% NaCl.

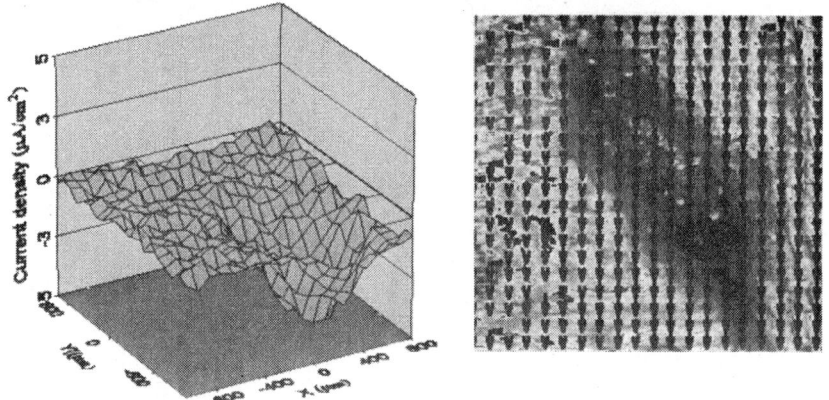

Figure 6. SVET scan over scribe of AA 2024-T3 coated with Mg rich primer and epoxy topcoat; 150 minutes after immersion in 3% NaCl.

After some hours the switch on the back of the electrodes was closed and the galvanic couple Mg-Al was established, at which point the Al surface quickly became cathodic (Figure 8). The surface remained cathodic for the entire period of the connection. The electrodes were then disconnected and a pit was immediately found to develop at the Al surface (Figure 9). The anodic current above the pit continued increasing as a function of time until a decision was made to reconnect the two substrates. At this point, the current over the pit started decreasing, reaching a minimum 2 h after the re-connection (Figure 10). It is clear that the Mg-rich primer cathodically controlled the localized corrosion at the AA 2024-T3 surface and can be used as a cathodic protection coating for the alloy.

Oxygen concentration profiles related to the galvanic interaction between the Al alloy and the Mg were obtained using the SECM and the two-substrate configuration cell. The tip potential was set -0.75 V, at which the oxygen

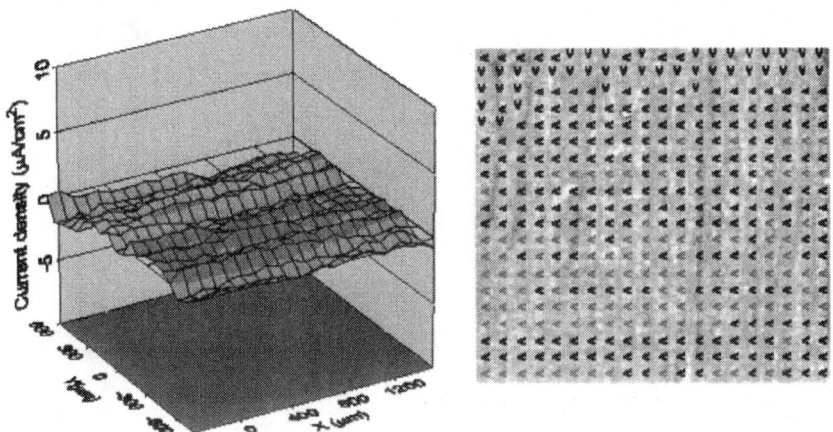

Figure 7. SVET scan over bare Al alloy substrate with the couple disconnected.

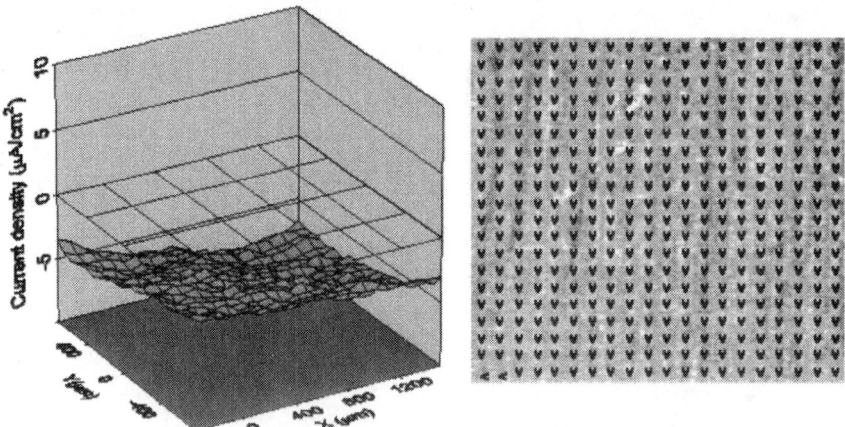

Figure 8. SVET scan over bare Al alloy substrate immediately after electrical connection with the Mg substrate; 18 h in immersion in DHS.

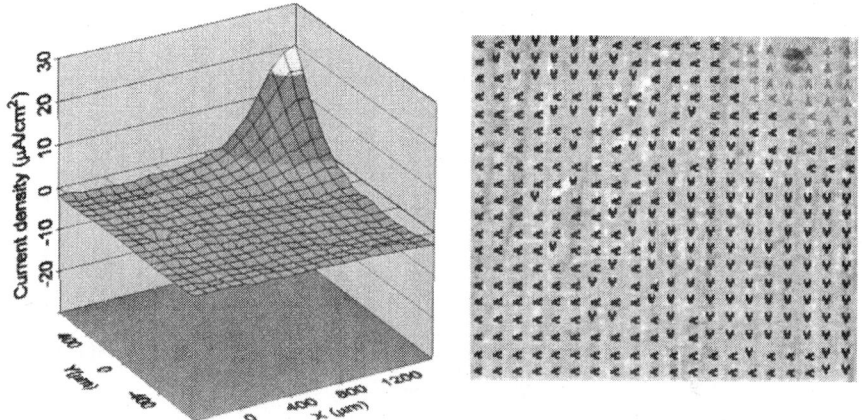

Figure 9. SVET scan over the Al alloy substrate 30 minutes after disconnecting the couple.

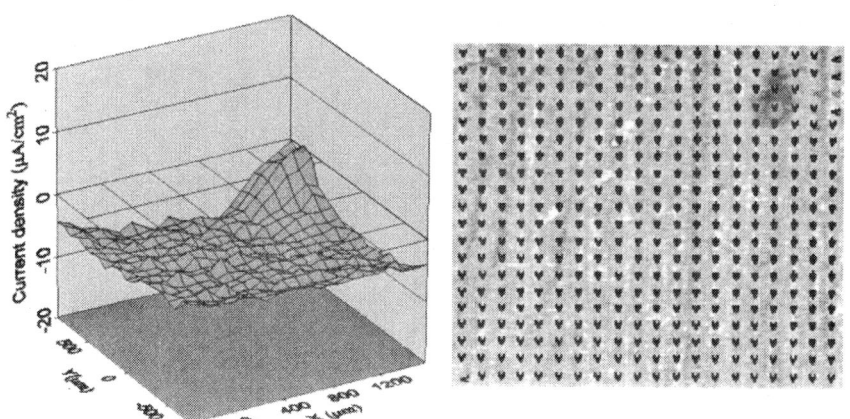

Figure 10. SVET scan over the Al alloy substrate 2 h after reconnection.

reduction reaction gave a steady-state current. Both metal substrates were at their OCP. The profile of oxygen reduction current (proportional to oxygen concentration) across the AA 2024-T3 electrode when it was electrically isolated from the Mg was practically constant, revealing only a slight depression above the aluminum compared to the neighboring epoxy resin (Figure 11, curve a). After connection to magnesium, the increased cathodic activity led to a rapid decrease of the oxygen concentration above the Al alloy (Figure 11, curve b). This reaction occurred in parallel with the reduction of protons, confirmed by some gas bubbling observed on the surface of the alloy electrode.

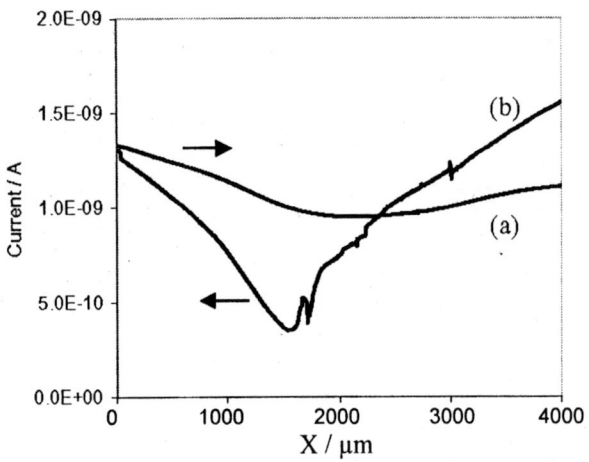

Figure 11. SECM scans for oxygen over the AA 2024- T3 electrode at a constant distance of 500 μm from the surface (scan direction as indicated). (a) isolated electrode; (b) electrode short-circuited with the Mg electrode.

On the magnesium electrode, some gas bubble formation on the surface was visually observed when the sample was electrically isolated. After electrical connection was made with the Al alloy, the bubbling at the Mg surface practically disappeared. Measurement of the dissolved oxygen profile evidenced a minimum in the cathodic tip current (i.e., oxygen concentration), as observed in Figure 12. This minimum, however, remained practically unaffected by the establishment of the electrical connection with the Al alloy electrode (compare curves a and b of Figure 12).

Aluminum, when passive, has very little electrochemical activity. The alumina passive film on the surface is insulating and, therefore, the rate at which reactions occur is slow and hard to detect. When it becomes polarized by the magnesium, it becomes cathodically active, and both the dissolved oxygen and

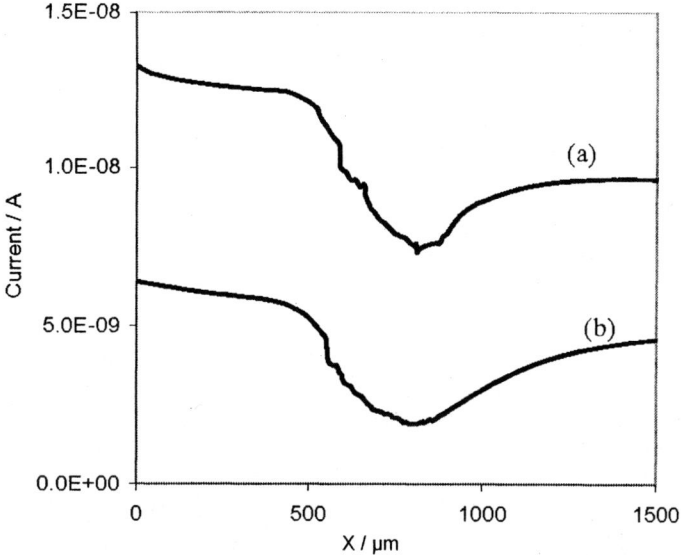

Figure 12. Scan over the magnesium electrode at a constant distance of 200 μm from the surface. (a) isolated electrode; (b) electrode short-circuited with the Al alloy electrode.

the protons in solution are reduced. With magnesium, however, the anodic polarization resulting from the coupling is not sufficient to stop completely the cathodic reactions occurring at its surface.

When comparing the two oxygen concentration profiles in Figure 12, a significant decrease was observed in the bulk oxygen concentration of curve (b) (i.e., the oxygen concentration measured away from the Mg electrode, for example at 0 μm or at 1500 μm). The SECM tip reduction current of curve (b) away from the Mg electrode was approximately half the value for curve (a). This derives from the fact that the total solution volume was small and, therefore, whenever the oxygen reduction rate was high, a steady-state oxygen concentration could not be maintained. The gradient of oxygen concentration spanned the entire electrolyte depth, and consequently the average concentration decreased with time, particularly if the experiments were of long duration.

Conclusions

Scanning probe techniques are very valuable tools for the study of local electrochemistry of active corrosion protection coatings at active metal surfaces.

These techniques provide more information on how the active coatings locally interact with the underlying metal, which traditional electrochemical techniques (EIS, ENM) can not offer.

From the SVET study of the electrodeposited polypyrrole/DBA-coated samples, it was observed that the oxidation current occurred entirely within the defect and the reduction current was distributed rather uniformly on the coating surface, consistent with the high electrical conductivity of these films (typically 2.5 S/cm). Such high conductivity facilitates transfer of electrons from the metal alloy to the coating-electrolyte interface where oxygen reduction can occur. In turn, this process drove the oxidation reaction observed in the defect, which ultimately led to passivation of the exposed metal within the defect. The SIET studies provided a pH mapping of the polypyrrole/DBA-coated sample. Low pH (3 to 3.5) was found at the coating defect area whereas higher pH (ca. 8) was observed at polymer coated areas during the initial period of immersion, consistent with the interpretation that Al was oxidized in defect and oxygen was reduced on polymer surface. On the other hand, the mechanism of H_2 generation at the exposed metal in the coating defect during the initial period of immersion is still not clear and is being explored further in our laboratory.

Using both the single substrate configuration cell and the two substrate configuration cell, SVET and SECM measurements revealed additional details about the interaction between AA 2024-T3 and the Mg-rich primer coating. It was demonstrated that this primer coating does inhibit the growth of localized corrosion and impedes the formation of new corrosion sites. Thus, we confidently conclude that the Mg-rich coating does provide protection through a cathodic protection mechanism.

References

1. Spinks, G. M.; Dominis, A. J.; Wallace, G. G.; Tallman, D. E. *J. Solid State Electrochem.* **2002**, *6*, 85.
2. Tallman, D. E.; Spinks, G.; Dominis, A.; Wallace, G. G. *J. Solid State Electrochem.* **2002**, *6*, 73.
3. Twite, R. L.; Bierwagen, G. P. *Prog. Org. Coat.* **1998**, *33*, 91.
4. Böhm, S.; Holness, R. J.; McMurray, H. N.; Worsley, D. A. In *Charge percolation and sacrificial protection in zinc-rich organic coatings*, Eurocorr 2000, Queen Mary and Westfield College, London, 10th-14th September, 2000.
5. Felix, S.; Barajas, R.; Bastidas, J. M.; Morcillo, M.; Feliu, S., In *Electrochemical Impedance Spectroscopy*, Scully, J.R; Silverman, D.C.; Kendig, M., Eds.; Amer. Soc. Testing and Materials (ASTM): Philadelphia, PA, 1993; pp 438.
6. Tallman, D. E.; Vang, C.; Wallace, G. G.; Bierwagen, G. P. *J. Electrochem. Soc.* **2002**, *149*, C173.

7. Nanna, M. E.; Bierwagen, G. P. *J. Coat. Technol. Res.* **2004**, *1*.
8. Costa, M. *Critical Reviews in Toxicology* **1997**, *27*, 431.
9. Mancuso, T. F. *American Journal of Industrial Medicine* **1997**, *31*, 140.
10. Bastos, A. C.; Simoes, A. M.; Gonzalez, S.; Gonzalez-Garcia, Y.; Souto, R. M. *Prog. Org. Coat.* **2005**, *53*, 177.
11. Jesse, C. S.; Daniel, A. B. *J. Electrochem. Soc.* **2003**, *150*, B413.
12. He, J.; Tallman, D. E.; Bierwagen, G. P. *J. Electrochem. Soc.* **2004**, *151*, B644.
13. He, J.; Gelling, V. J.; Tallman, D. E.; Bierwagen, G. P. *J. Electrochem. Soc.* **2000**, *147*, 3661.
14. Battocchi, D.; He, J.; Bierwagen, G. P.; Tallman, D. E. *Corros. Sci.* **2005**, *47*, 1165.
15. He, J.; Gelling, V. J.; Tallman, D. E.; Bierwagen, G. P.; Wallace, G. G. *J. Electrochem. Soc.* **2000**, *147*, 3667.
16. Drazic, D. M.; Popic, J. *J. Electroanal. Chem.* **1993**, *357*, 105.
17. Drazic, D. M.; Popic, J. P. *J. Appl. Electrochem.* **1999**, *29*, 43.
18. Drazic, D. M.; Popic, J. P. *J. Serb. Chem. Soc.* **2005**, *70*, 489.

Chapter 3

Optimization of One-Step, Low-VOC, Chromate-Free Novel Primer Coatings Using a Taguchi Method Approach

Anuj Seth and Wim J. van Ooij

Department of Chemical and Materials Engineering, University of Cincinnati, Cincinnati, OH 45221–0012

In this paper, we have used a Taguchi method approach for the optimization of 5 different novel room-temperature cured primers. These novel primers were developed for the replacement of the chromate-based Cr(VI) technology currently used for the corrosion protection of aerospace alloys such as AA 2024-T3. However, Cr(VI) has been identified by the Environmental Protection Agency (EPA) as toxic and carcinogenic and a replacement for it is being sought. The 5 systems differ in the type of organofunctional silane used for formulating the primer. The five different organofunctional silanes used were bis-(triethoxysilylpropyl) ethane (BTSE silane), bis-(triethylsilylpropyl) tetrasulfide (bis-sulfur silane), and combinations of the above silanes in the weight ratios 1:2, 2:1 and 1:1 respectively. The components of the room temperature primer were acrylate resin, an epoxy resin, one of the five organofunctional silane system and a crosslinker. An L9 orthogonal array was chosen with the parameters set as the components of the primer. The levels were varied through three different amounts for the resins and the silane or silane combinations. The crosslinker amount was fixed and was varied by using three different types of crosslinker. Various corrosion performance tests were conducted on the samples coated with the primers formulated. Electrochemical Impedance spectroscopy (EIS), 3.5 wt-% NaCl aqueous solution immersion test, DI water contact angle before and after exposure to 3.5 wt-% NaCl aqueous solution, tape adhesion test (ASTM D 3359), MEK double rub test (ASTM D 4752), chemical resistance test (ASTM D 3912) were used as performance evaluation tests. Using the results obtained from these experiments an optimized formulation was obtained for primers formulated with each of the five different primer systems.

Chromate conversion coating of aerospace alloys such as AA 2024-T3, followed by the application of a primer containing chromate corrosion inhibiting- pigments has been a technology used for preventing corrosion in the aerospace industry for over 40 years (*1*). Chromates outperform most of the available alternates in corrosion resistance properties however, they have been identified as carcinogenic and toxic and as such a replacement is needed (*2*).

Van Qoij et al. (*3-10*) have demonstrated that organofunctional silanes offer a promising alternate for meeting the challenge of chromate replacement. Simple rinses with dilute silane solutions result in a silane pretreatment which can be used for replacement of chromate conversion coatings (*3-7*). It has been successfully demonstrated that integration of the silane pretreatment with the primer applied on top of the silane pretreatment, results in a one-step, low-VQC, chromate-free novel primer system (*8-10*).

Genchi Taguchi's methods have become increasingly popular in the aerospace industry because of the ease of methodology and savings in time and resources (*11-16*). In this paper we have used an L9 orthogonal array Taguchi method approach for the optimization of five different room temperature cured novel primers (*8*). Various corrosion performance tests were conducted on the samples coated with the primers formulated. Electrochemical Impedance spectroscopy (EIS), 3.5 wt-% NaCl aqueous solution immersion test, DI water contact angle before and after exposure to 3.5 wt% NaCl aqueous solution, tape adhesion test (ASTM D 3359), MEK double rub test (ASTM D 4752), and a chemical resistance test (ASTM D 3912) were used as corrosion performance evaluation tests. Using the results obtained from these experiments an optimized formulation was obtained for primers formulated with each of the five different primer systems.

This paper is the second in a series of three papers. The first one has been submitted for publication and cited as reference (*8*). The third paper will discuss the characterization of the 5 optimized systems using FTIR-ATR and Si^{29} and C^{13} NMR in liquid and solid state. A model for the mechanisms will also be presented in the third paper and is therefore not included here.

Experimental Design using L9 Orthogonal Arrays for Optimization of the Novel Primer System

An L9 orthogonal array (*11*) was used for the optimization of the novel primer systems. The L9 orthogonal array allows optimization of four parameters by varying each parameter through three levels. A typical L9 orthogonal array is shown in Table 1.

Table 1. A typical L9 orthogonal array used in a Taguchi method approach

Sample Number	Parameter A	Parameter B	Parameter C	Parameter D
1	Level 1	Level 1	Level 1	Level 1
2	Level 1	Level 2	Level 2	Level 2
3	Level 1	Level 3	Level 3	Level 3
4	Level 2	Level 1	Level 2	Level 3
5	Level 2	Level 2	Level 3	Level 1
6	Level 2	Level 3	Level 1	Level 2
7	Level 3	Level 1	Level 3	Level 2
8	Level 3	Level 2	Level 1	Level 3
9	Level 3	Level 3	Level 2	Level 1

In this study the four parameters that were chosen for optimization were the four components that make up the primer system, the two resins, ECO-CRYL™ 9790 and EPI-REZ™ WD-510, and the organofunctional silane and the crosslinker.

The resin ECO-CRYL™ 9790, which is a 42 wt-% of an anionic dispersion of an acrylate copolymer in water, was varied through three levels by varying the weight used in the formulation. The three levels were fixed as 3g, 5g, and 7g. These levels were chosen because in our previous results we found that the coatings formulated with just the 7g ECO-CRYL™ 9790 and 3g of an organofunctional silane did not perform well and failed after a short duration of time due to excessive water-uptake when compared with the systems that had 7g of ECO-CRYL™ 9790, 2g of EPI-REZ™ WD-510, which is a bis-phenol A type epoxy, and 3g of organofunctional silane (8). As such it was evident that a large amount of ECO-CRYL™ 9790 present in the system results an inferior coating. The resin EPI-REZ™ WD-510 was varied through three levels of 1g, 2g, and 3g, respectively.

The organofunctional silane used was varied through 3 levels of 1.5g, 3g, and 4.5g. 5 systems of the novel primer were investigated which resulted in 5 L9 orthogonal arrays L9_1, L9_2, L9_3, L9_4 and L9_5, respectively. These 5 systems differed from each other in the type of silane used. The L9_1 orthogonal array was formulated using BTSE silane. The L9_2 orthogonal array was formulated using bis-sulfur silane. The L9_3, L9_4, and L9_5 orthogonal arrays were formulated using a combination of the BTSE and bis-sulfur silane in a ratio of 2:1, 1:2 and 1:1, respectively.

The final component of the novel primer system was the crosslinker. This parameter was varied in the type of crosslinker used for facilitating a room temperature cure. The level 1 was set as the A-Link 15 crosslinker, level 2 was set as the A-Link 25 crosslinker and level 3 was set as a combination of the above two crosslinkers in a ratio of 1:1. These cross linkers were chosen because in our previous study we had found these to give the best tape adhesion test results to a topcoat without compromising the pot life of the primer solution (8). The AA 2024-T3 samples coated with formulations of the 5 L9 orthogonal arrays L9_1, L9_2, L9_3, L9_4, and L9_5 were named according to the following nomenclature.

Each L9 orthogonal array had nine formulations in it. The 9 formulations belonging to the L9_1 coating system were named L1_1 through L1_9 with L1_1 having the formulation level 1 of parameter A ECOCRYL 9790, level 1 of parameter B EPI-REZ^{TM} WD-510, level 1 of the BTSE silane and A-Link 15 as the crosslinker. The L1_9 formulation corresponded to the formulation of sample 9 in Table I with parameters A, B, C, and D being ECOCRYL 9790, EPI-REZ^{TM} WD-510, BTSE silane and the crosslinker. Similarly the 9 formulations of the L9_2, L9_3, L9_4 and L9_5 systems were named L2_1 through L2_9, L3_1 through 9, L4_1 through L4_9 and L5_1 through L5_9, respectively.

Experimental

Materials

Metal Substrates

15 cm x 10 cm size, AA 2024-T3 panels were used for all testing in this study. Panels were obtained from Steel Metals and Supply Co., Stillwater, OK.

Resins for the formulation of the novel primer

An acrylate resin and an epoxy resin were used in this study. The acrylate resin was ECO-CRYL^{TM} 9790, which is a 42 wt-% anionic dispersion of an acrylate copolymer in water. It also contains 3 wt-% of triethylamine, 3 wt-% of xylene, and 7 wt-% of 2-propoxyethanol. The epoxy resin used was EPI REZ WD-510. This is a bis-phenol A type epoxy resin which is dispersed in water. Both the resins were procured from Resolution Performance Products, Inc. (17).

Organofunctional silanes and crosslinkers

The organofunctional silanes used in this study were bis(triethoxysilylpropyl) ethane (BTSE) and bis-(triethoxysilylpropyl) tetrasulfide (bis-sulfur silane). For facilitating a room temperature cure two crosslinkers were added to the formulation of the novel primer. These cross linkers were A-Link 15 and A-Link 25. The organofunctional silanes and crosslinkers were procured from GE Silicones Inc., Waterford, NY.

Primer formulations

The primers were formulated by weighing the suitable amounts of resins and organofunctional silanes in a beaker. To this DI water was added. The amount of DI water added to the resin was equal to 1/3 of the weight of the resins and silanes together. To this, 2 wt-% of the weight of the resins and silane mixture of a crosslinker was added. The amount of crosslinker was kept fixed and the type of crosslinker was varied through the various coating formulations. This was stirred until completely mixed using a high shear blender at 3000 rpm for 3 minutes.

Coating and curing of the coated samples

AA 2024-T3 samples were scrubbed using 3M Scotch Brite dipped in ethanol. The panels were then coated with the various novel primer recipes listed in Tables 2 through 6 using a R28 drawdown bar. The samples were left in the ambient conditions for 14 days for curing before any tests were conducted on them. The samples that had to be topcoated with a topcoat were topcoated after 24 hours of room temperature curing. A 24 hours room temperature curing results in a good cure of the system. The topcoat used was PRC DeSoto Desothane HS procured from the Wright Patterson Airforce Base, Dayton, OH.

Performance Evaluation

The response of varying the various parameters in the 5 systems being optimized was collected using the following standard corrosion performance evalution methods.

Electrochemical Impedance spectroscopy (EIS)

EIS was used to evaluate the corrosion behavior of the coating systems on AA 2024-T3 panels in a 3.5 wt-% NaCl solution. The EIS measurements were conducted using an SR 810 frequency response analyzer connected to a Gamry CMS 100 potentiostat. The measured range of frequency was from 10^5 to 10^{-2} Hz, with an alternating circuit (AC) voltage amplitude of ± 10 mV. A commercial Saturated Calomel Electrode (SCE) was used as the reference electrode coupled with a graphite counter electrode. The surface area exposed to the electrolyte was 5.16 cm^2 during the measurements. The value of modulus of impedance measured in Ohms, Ω, at 10^{-2} Hz on the day 30 was used for determining the efficacy with which a coating protects the metal substrate against corrosion. Ten times the logarithm of this value was used for the Taguchi analysis. This scaling was done to be able to represent the data on the same scale as the data obtained from other experiments described ahead. The higher the modulus the better is the resistance to corrosion (8). These results for the systems L9_1, L9_2, L9_3, L9_4 and L9_5 are shown in Tables 2 through 6 in Column A.

3.5 wt-% NaCl aqueous solution immersion test

Primer-coated and primer-coated with topcoat applied panels were scribed and immersed in a 3.5 wt-% NaCl aqueous solution for 30 days. The scribe simulates a damaged area in the coating. For a formulation, both topcoated and just primer-coated panels were visually examined and rated on a scale of 50. Evaluation was made on the basis of delamination, presence and extent of pit formation, extent of corrosion in the scribe, and extent of blistering. The values obtained on a scale of 50 for each of the topcoated and untopcoated sample were then added. A higher score meant a better capability of a coating to prevent corrosion of the substrate and scribe overall. These results for the systems L9_1, L9_2, L9_3, L9_4 and L9_5 are shown in Tables 2 through 6 in Column B.

Table 2. Results for various corrosion performance evaluation tests conducted on the formulations for the L9 orthogonal array for optimizing L9_1 system

Sample Number	Column A	Column B	Column C	Column D	Column E	Column F
L1_1	77.8	84	13.6	70	96	100
L1_2	76.2	89	14.3	100	25	100
L1_3	50.0	84	22.1	100	20	50
L1_4	69.5	84	10.6	50	10	25
L1_5	63.0	68	38.0	60	37	100
L1_6	77.0	83	37.6	97	34	100
L1_7	60.0	74	47.5	85	40	100
L1_8	83.0	86	19.3	90	67	100
L1_9	84.8	81	23.6	95	89	100

Column A = EIS, Column B = 3.5wt-% NaCl aqueous solution immersion test. Column C = DI water contact angle measurement, Column D = tape adhesion test, Column E = MEK double rub test, Column F = chemical resistance test

DI water contact angle

The static DI water contact angle was measured before and after exposure to 3.5 wt-% NaCl aqueous solution for 30 days. A drop of DI water was dropped on the coated samples and the contact angle was measured. Contact angle goniometer VCA2000 manufactured by AST Products Inc., Billeria, MA was used to perform these measurements. The contact angle is a measure of the

Table 3. Results for various corrosion performance evaluation tests conducted on the formulations for the L9 orthogonal array for optimizing L9_2 system

Sample Number	Column A	Column B	Column C	Column D	Column E	Column F
L2_1	68.5	89	9.3	100	13	75
L2_2	63.0	89	14.7	87	23	100
L2_3	45.4	81	-10.5	80	10	75
L2_4	74.8	92	8.6	60	63	50
L2_5	86.0	93	10.6	70	44	100
L2_6	89.0	93	11.8	100	197	100
L2_7	96.0	93	10.5	97	72	100
L2_8	83.0	86	10.2	98	119	100
L2_9	73.0	81	13.5	95	195	100

Column A = EIS, Column B = 3.5wt-%NaCl aqueous solution immersion test. Column C = DI water contact angle measurement, Column D = tape adhesion test, Column E = MEK double rub test, Column F = chemical resistance test

Table 4. Results for various corrosion performance evaluation tests conducted on the formulations for the L9 orthogonal array for optimizing L9_3 system

Sample Number	Column A	Column B	Column C	Column D	Column E	Column F
L3_1	73.6	78	12.8	55	34	100
L3_2	40.0	59	-1.4	60	17	0
L3_3	40.0	62	-1.5	100	7	0
L3_4	38.5	68	9.5	60	184	0
L3_5	39.5	56	17.6	70	23	0
L3_6	66.0	83	22.3	95	70	100
L3_7	27.0	49	56.4	80	57	0
L3_8	31.8	69	33.7	97	58	100
L3_9	93.0	83	14.9	98	43	100

Column A = EIS, Column B = 3.5wt-% NaCl aqueous solution immersion test, Column C = DI water contact angle measurement, Column D = tape adhesion test, Column E = MEK double rub test, Column F = chemical resistance test

hydrophobicity of the coating. A hydrophobic coating results in a higher contact angle, which implies that it can more efficiently keep the water and electrolyte from permeating to the metal-primer interface. This in turn results in a better corrosion resistance. As such the percentage change due to 30 days of exposure to electrolyte was recorded for each of the coatings. These results for the systems L9_1, L9_2, L9_3, L9_4 and L9_5 are shown in Tables 2 through 6 in Column C.

Tape adhesion test (ASTM D 3359)

The primer-coated and topcoated panels were scribed using a tungsten carbide scribing tool. These samples were then immersed in DI water for 24 hours and left to dry in ambient room temperature conditions for 4 hours. The tape adhesion test was carried out on these specimens in accordance with the ASTM D 3359 standards. The extent of delamination was graded on a scale of 100 and used a response to the variations of the parameters at the set 3 levels. These results for the systems L9_1, L9_2, L9_3, L9_4 and L9_5 are shown in Tables 2 through 6 in Column D.

Table 5. Results for various corrosion performance evaluation tests conducted on the formulations for the L9 orthogonal array for optimizing L9_4 system

Sample Number	Column A	Column B	Column C	Column D	Column E	Column F
L4_1	48.1	71	8.3	80	97	100
L4_2	57.8	82	5.7	60	19	75
L4_3	54.8	81	32.4	90	5	0
L4_4	73.0	86	3.0	50	181	25
L4_5	60.0	88	10.2	60	23	0
L4_6	56.5	89	18.7	100	69	100
L4_7	28.5	82	60.7	80	46	0
L4_8	54.0	81	11.5	80	52	100
L4_9	49.0	80	4.4	40	76	75

Column A = EIS, Column B = 3.5wt-%NaCl aqueous solution immersion test, Column C = DI water contact angle measurement, Column D = tape adhesion test, Column E = MEK double rub test, Column F = chemical resistance test

MEK double rub Test (ASTM D 4752)

The MEK double rub test was conducted by rubbing a primer-coated sample with cheesecloth dipped in methyl ethyl ketone in accordance with the

ASTM D 4572 standards. The MEK double rub number gives an indication of the extent of cure of a coating and is also an indication of the extent of cross link density in the coating. These results for the systems L9_1, L9_2, L9_3, L9_4 and L9_5 are shown in Tables 2 through 6 in Column E.

Chemical Resistance Test (ASTM D 3912)

The chemical resistance test was performed on all the primer-coated panels. The chemical resistance to 6N HCl and 6N NaOH was examined by putting a drop of each of the solutions on the panels and examining the area of the coating exposed to the chemical after 24 hours. The panels were rated on a scale of 50 with a high score for better resistance to each of the basic and acidic environments. The sum of the two was the overall score for that particular formulation/coating. Column F shows these results for the systems L9_1, L9_2, L9_3, L9_4 and L9_5 in Tables 2 through 6.

Table 6. Results for various corrosion performance evaluation tests conducted on the formulations for the L9 orthogonal array for optimizing L9_5 system

Sample Number	Column A	Column B	Column C	Column D	Column E	Column F
L5_1	84.8	76	11.1	50	73	100
L5_2	80.0	74	17.6	70	72	75
L5_3	44.5	79	8.8	100	9	75
L5_4	48.45	83	2.9	60	73	50
L5_5	44.0	79	6.9	40	14	100
L5_6	89.5	83	23.1	100	44	100
L5_7	31.8	56	55.4	80	53	0
L5_8	47.8	79	15.0	90	82	100
L5_9	36.0	82	14.1	90	159	100

Column A = EIS, Column B = 3.5 wt-% NaCl aqueous solution immersion test, Column C = DI water contact angle measurement, Column D = tape adhesion test, Column E = MEK double rub test, Column F = chemical resistance test

Optimization and Verification using the Taguchi Method Approach

The L9 orthogonal array is so designed that each parameter, when fixed at a given level, interacts with the other parameters at all the other 3 levels. For example the sample number 1, 2, and 3 in Table I have the parameter A fixed at Level 1. Inspection of the table reveals that by fixing the parameters in this way an interaction with all the other 3 parameters B, C, and D at Level 1, 2 and 3 is present. Hence for deriving the effect of a given parameter, at a given level, the average of the responses from the 3 occurrences of the parameter in the orthogonal array needs to be taken (*11-13*). For example, consider the L9_1 system. We calculate the effect of parameter ECO-CRYL™ 9790 at the level of 3g by taking the average of the values obtained in the tests for sample numbers L1_1, L1_2, and L1_3. When the parameter ECO-CRYL™ 9790 is set at the level of 5g the average value of the responses obtained from sample numbers L1_4, L1_5, and L1_6 are taken and similarly for the level of 7g for parameter ECO-CRYL™ 9790 the responses are the averages of L1_7, L1_8 and L1_9. Similar average values were calculated for each of the parameters at all the 3 levels at which they had been set. These responses were then plotted in a graph with the X-axis as the 3 levels of the parameters and the Y-axis as the value of the average response obtained at a given level for a given parameter. The result of the analysis for the 5 systems L9_1, L9_2, L9_3, L9_4, and L9_5 is shown in Figure 1.

Table 7. Optimization of the L9_1, L9_2, L9_3, L9_4, and L9_5 systems resulting in L1, L2, L3, L4, and L5 formulations, respectively

Parameter	L1	L2	L3	L4	L5
ECO-CRYL™ 9790	7.0g	7.0g	7.0g	5.0g	5.0g
EPI-REZ™ WD-510	3.0g	3.0g	3.0g	2.0g	3.0g
Silane	1.5g	1.5g	1.5g	1.5g	1.5g
Crosslinker	A-Link 15	A-Link 25	A-Link 15	A-Link 25	A-Link 15

Table 8. Results of the verification tests conducted on the optimized coatings L1, L2, L3, L4 and L5 respectively

Coating Name	Column A	Column B	Column C	Column D	Column E	Column F
L1	76.2	90	6.8	100	90	100
L2	85.7	92	3.1	100	89	100
L3	78.7	93	7.3	100	98	100
L4	86.2	97	3.2	100	94	100
L5	40.3	91	11.5	100	90	100

Column A = EIS, Column B = 3.5wt-% NaCl aqueous solution immersion test, Column C = DI water contact angle measurement, Column D = tape adhesion test, Column E = MEK double rub test, Column F = chemical resistance test

It is clear from the Figure 1 it can be seen that for any parameter there is no one level where all the 6 properties being optimized are the best. This tradeoff is due to functions each component plays in the coating systems. We believe that the silane crosslinks the resin component of the primer. This enhances the corrosion performance of the primer systems (*8*). However, when more silane is added the excess silane is present in the cured coating as tiny dispersed droplets.

This is detrimental for properties such as chemical resistance and MEK double rub test. Thus, trade-offs are resorted to and the optimized systems are chosen where most of the properties are at the best response. For the L9_1 system ECO-CRYL™ 9790 is chosen at level 3, EPI-REZ™ WD-510 is chosen at level 2, BTSE silane is chosen at level 1 and the crosslinker is chosen at level 1 for the optimized coating, which is referred to as L1 hereon. Similarly, for other systems the optimization is carried out and L9_2, L9_3, L9_4, and L9_5 result in the optimized formulations L2, L3, L4, and L5. The optimized formulations are listed in Table 7.

It is also noteworthy to note the trend of the systems formulated with BTSE silane or BTSE-rich silane combination systems performed better when formulated with the A-Link 15 crosslinker and the bis-sulfur silane-based system or bis-sulfur silane-rich combination system performed better when formulated with A-Link 25 crosslinker. It is speculated that the above phenomenon is due to faster rate of hydrolysis of the BTSE silane when compared with the bis-sulfur silane. This results in a generation of a large amount of ethanol (*18*). The A-Link 15 crosslinker is an amine-based crosslinker and is known to hydrolyze faster than the A-Link 25 crosslinker, which is an isocyante-based silane. The excess ethanol formed during the initial stages in the BTSE silane-based systems retards the hydrolysis of the A-Link 25 and consequently the resulting primer has a lower crosslink density. Experiments are under progress to identify the chemistry of these novel primer systems using techniques such as Reflectance Absorbance Infrared Spectroscopy (RAIR) and Nuclear Magnetic Resonance (NMR). The findings of these experiments will be published as a separate paper.

Verification of the optimized coatings

The optimized novel primer coatings were prepared using the formulations listed in Table 7. Table 8 shows the results of the verification tests conducted on the optimized coating systems using EIS (Column A), 3.5 wt-% NaCl aqueous solution immersion test (Column B), DI water contact angle measurement (Column C), tape adhesion test (Column D), MEK double rub test (Column E), and chemical resistance test (Column F).

Comparison of the data in Table 8 with the data in Table 2 through 6 clearly shows that all the properties of the coatings have been optimized. Very good values of the modulus of frequency at low frequencies are observed for coatings L1, L2, L3, and L4. The 3.5 wt-% NaCl aqueous solution immersion test also yields good result along with very low changes in the DI water contact angle measured before and after exposure to 3.5 wt-% NaCl aqueous solution. Figure 2 shows the scanned images of the 3.5 wt-% NaCl aqueous solution immersion test. It is seen from the images of the sample seem in the image the primer passes 30 days of salt immersion test successfully. The only areas of corrosion

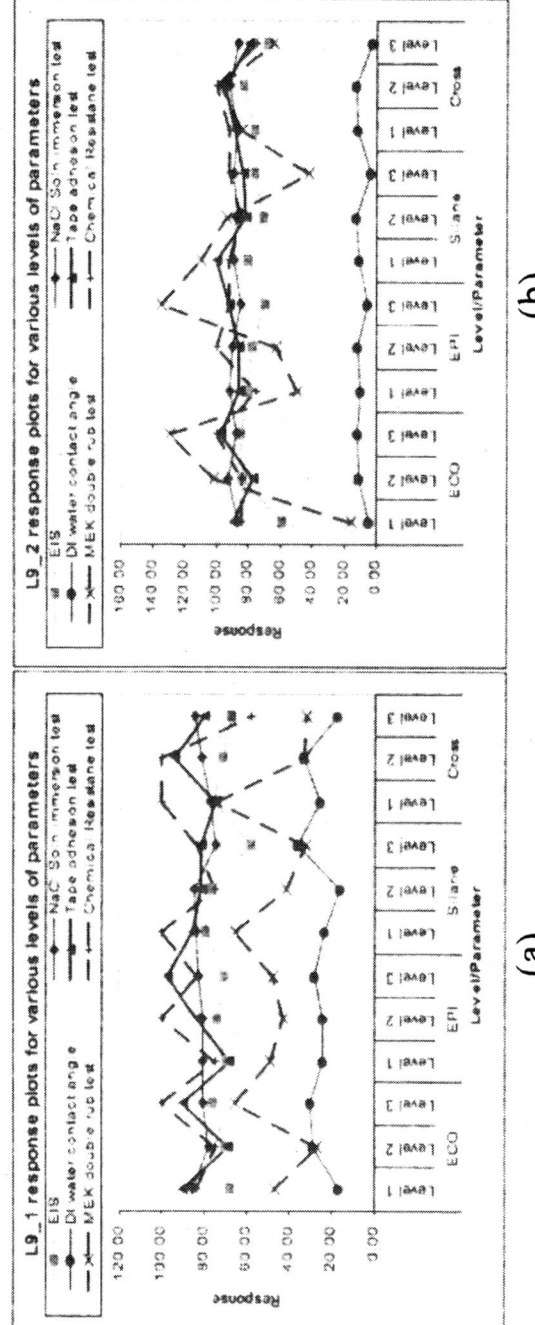

Figure 1. Analysis of the responses to varying the parameters through 3 levels for systems (a) L9_1, (b) L9_2, (c) L9_3, (d) L9_4, and (e) L9_5. Continued on next page.

36

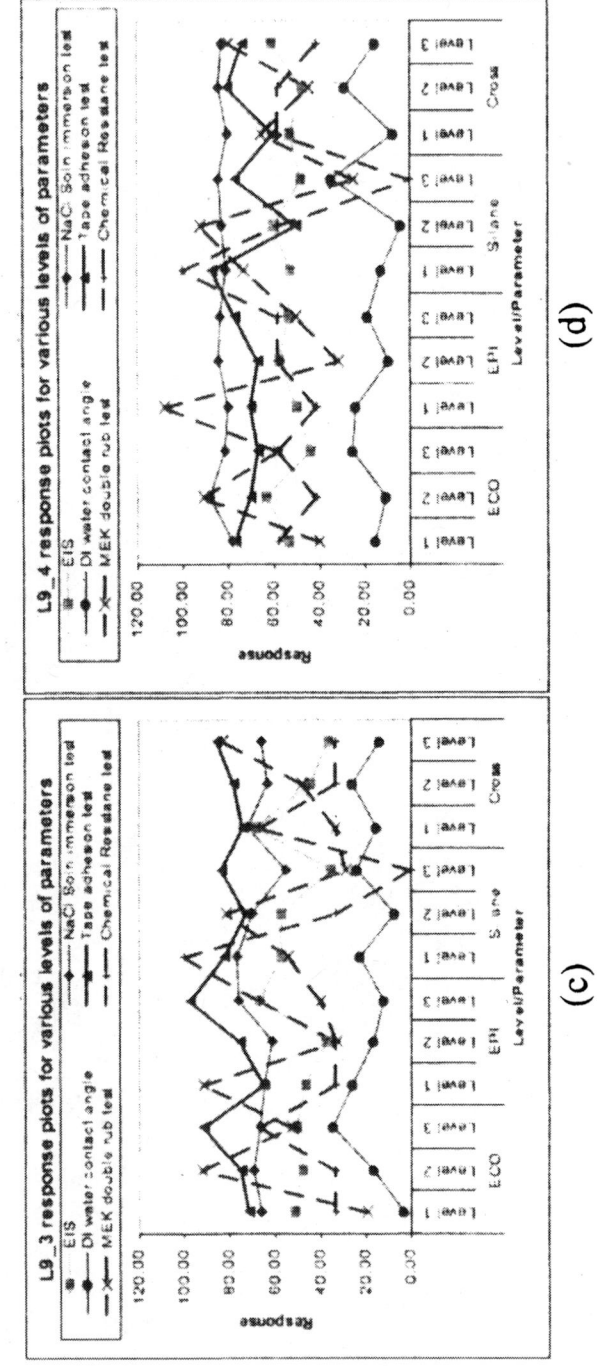

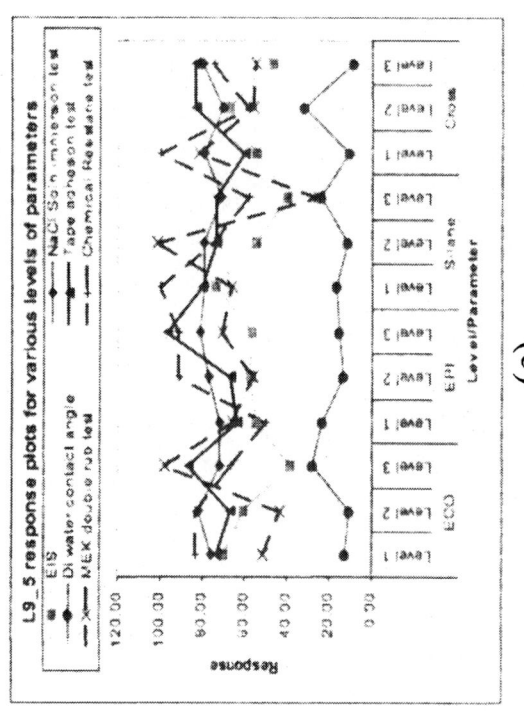

Figure 1. Continued.

observed are the scribes of the panel. This is due to the absence of any inhibitor in the primer. Typically primers contain corrosion inhibiting species which leach out to protect a damage in the coating which has been simulated here as a scribe. Also, it is noteworthy that no delamination is observed after the exposure in any panel away from the scribe and at the primer topcoat interface. Overall, the appearance of the optimized panels was found to be superior to the individual panels of the Taguchi analysis L9 orthogonal arrays tested with various formulations. This suggests that the identified optimized formulations result in a superior corrosion performance and the systems have been optimized.

Figure 2. Scanned images of samples exposed to 3.5-wt% NaCl aqueous solution after 30 days (a) without topcoat (b) with topcoat

Conclusions

It has been demonstrated in this paper that a L9 orthogonal array can be conveniently used for the optimization of an anticorrosion coating system. Five novel, one-step, low-VOC, chromate-free, primer systems were successfully optimized using a Taguchi model. In the testing of the various formulations prepared for the L9 orthogonal array of the various systems it was revealed that the primer is very sensitive to the ratio in which the components are mixed. This is evident from the wide variations seen in the results of the tests listed in Table 2 through 6. It was also seen that the primers formulated using BTSE silane were better in performance when A-Link 15 crosslinker was used and the bis-sulfur silane-based primers performed better when A-Link 25 was used.

Acknowledgments

The authors would like to acknowledge the Air Force Scientific Research Office for funding this project under the grant F49620-01-1-0352 (MURI). A special word of thanks goes to Jessica Lee-Ashley Martin for easing most of the toil of sample preparation, tapping and scribing.

References

1. Zhao, J.; Sehgal, A.; Lu, D.; McCreery, R. L. *Surf and Coatings Tech.* **2001**, *140*, 51.
2. LaPuma, P. T.; Fox, J. M.; Kimmel, E. C. *Reg Toxicology and Pharm.* **2001**, *33*, 343.
3. *Silane and Other Coupling Agents* Mittal, K. L., Eds.; VSP International Science publishers, Zeist, Netherlands, 2003; p 119.
4. Zhu, D.; van Ooij, W. J. *Prog in Organic Coatings* **2004**, *49*, 42.
5. Yuan, W.; van Ooij, W. J. *J. of Coll. and Int. Sci* **1997**, *185*, 197.
6. Zhu, D.; van Ooij, W. J. *Corr. Sci.* **2003**, *45*, 2163.
7. Nie, T.; van Ooij, W. J.; Gorecki, G. *Prog. in Org. Coatings* **1997**, *30*, 255.
8. Seth, A.; van Ooij, W. J. *Submitted for publication in the Silanes and Other Coupling Agents,* Volume 4., Mittal, K. L., Eds.
9. Seth, A.; van Ooij, W. J. *J. of Mat. Engg. and Perf* **2004**, *13*, 292.
10. Shivane, C.; van Ooij, W. J. *Submitted for publication in the Silanes and Other Coupling Agents,* Volume 4., Mittal, K. L., Eds.
11. *System of Experimental Design,* Clausing, D. Ed., UNIPUB/Kraus International Publications, Dearborn, MI 1987.
12. *Introduction to Quality Engineering,* Taguchi, G. Ed, American Supplier Institute, Dearborn, MI 1986.
13. *Quality Engineering in Production Systems,* Taguchi, G.; Elsayed, E., Hsiang, T. Eds., McGraw Hill, New York, 1989.
14. Bryne, D. M.; Taguchi, S. *ASQC Quality Congress Transactions,* Anaheim, **1986**, p168.
15. *A Primer on the Taguchi Method,* Ranjit, R. Eds; New York, Van Nostrand Reinhold, **1990**.
16. *Arrays and Linear Graphs,* Taguchi, G.; Konishi, S. Eds; American Supplier Institute, Dearborn, MI 1987.
17. Resolution Products Inc. www.resins.com
18. GE Silicones Inc. www.gesilicones.com

Chapter 4

Investigation of Electroactive Polymers and Other Pretreatments as Replacements for Chromate Conversion Coatings: A Neutral Salt Fog and Electrochemical Impedance Spectroscopy Study

P. Zarras[1], N. Prokopuk[2], N. Anderson[1], and J. D. Stenger-Smith[1]

[1]Polymer Science and Engineering Branch (Code 498200D), Naval Air Warfare Center Weapons Division, Department of the Navy, 1900 North Knox Road (Stop 6303), China Lake, CA 93555–6106
[2]Materials Chemistry Branch (Code 498210D), Naval Air Warfare Center Weapons Division, Department of the Navy, 1900 North Knox Road (Stop 6303), China Lake, CA 93555–6106

A series of coatings were examined as replacements for chromate conversion coating (CCC) pretreatments on aluminum alloy (Al 2024-T3). The pretreatment coatings that were studied include: Trivalent chromium pretreatment (TCP), poly(2,5-bis(N-methyl-N-hexylamino)phenylene vinylene), (BAM-PPV) and poly((2-(2-ethylhexyl)oxy-5-methoxy-*p*-phenylene)vinylene), (MEH-PPV). The pretreatments were coated onto Al 2024-T3 and subjected to neutral salt fog spray until failure. The BAM-PPV, MEH-PPV and TCP coated Al 2024-T3 panels each passed the minimum requirement (military pretreatment specification) of 336 hours exposure to neutral salt fog spray. Both the TCP coating and CCC showed no corrosion at 1000 hours in the neutral salt fog chamber. BAM-PPV coatings and CCC (as control) on Al 2024-T3 panels were also studied using EIS for a six month immersion study in 0.5 N NaCl. In both cases, the impedance values at low frequencies did not change over the exposure time. BAM-PPV shows both capacitive and resistive properties at high frequency, with the capacitive nature diminishing over time. The CCC showed purely resistive properties over time.

© 2007 American Chemical Society

Introduction

Aerospace and Department of Defense (DoD) currently use chromate conversion coatings (CCC) to inhibit corrosion of aluminum alloys (*1*). In addition these coatings provide excellent paint adhesion to the metal surface (*2*). The CCCs are applied via immersion or spraying onto both aluminum and steel substrates (*3*). Several recent studies have shown that residual hexavalent chromium (Cr(VI)) in chromate conversion coatings provide corrosion protection via a self-healing mechanism (*4-8*). However, Cr(VI) is a known carcinogen (*9-12*) and is highly regulated by the Environmental Protection Agency (EPA) and Occupational Safety and Health Agency (OSHA) (*13*).

Any viable alternative to Cr(VI) coatings must meet or exceed the performance of Cr(VI) (*1, 2, 14*). Ideally, these alternative coatings must be able to passivate the metal surface (*15*) and several alternatives to CCC have been investigated. One example is the trivalent chromium pretreatment (TCP). This coating was developed by the Naval Air Warfare Center Aircraft Division (NAWCAD) as an acceptable alternative to CCC on aluminum alloys (*16-18*). Other pretreatment coatings such as cerium films protect the metal alloy through a passivation mechanism (*19-21*).

During the past decade electroactive polymers (EAP) have received considerable interest as corrosion inhibiting coatings (22-25). Most of these studies have focused on polyaniline (PANI) applied as a primer onto steel substrates (*26-29*). More recent studies have focused on PANI and derivatives of PANI as replacements for chromated pretreatments (*30-32*). Additional EAP materials have also been prepared based on derivatives of poly(*p*-phenylene vinylene) (PPV). Poly(2,5-bis(N-methyl-N-hexylamino)phenylene vinylene, BAM-PPV and poly((2-(2-ethylhexyl)oxy-5-methoxy-*p*-phenylene)vinylene), (MEH-PPV) were prepared as described in the literature (*33, 34*). BAM-PPV coated onto aluminum alloys has shown corrosion inhibition in simulated seawater and exposure to neutral salt fog spray (*35-40*).

Experimental

BAM-PPV solutions were prepared with *p*-xylene as the solvent and were stirred for 2 days at 50°C. The solutions were filtered prior to use. The filtered solution was applied via spray gun onto Al 2024-T3 panels (.032 x 3 x 6"obtained from Q-Panel Lab Products Inc). After spraying, the BAM-PPV panels were placed in a vacuum oven and dried at 60°C for 2 hours. The average film thicknesses of the BAM-PPV coated panels were ~2 microns. MEH-PPV was obtained from Aldrich Chemical Co and used without further purification. MEH-PPV was also synthesized at the Naval Air Warfare Center Weapons

Division (NAWCWD) and solutions were prepared from cyclopentanone solvent. These solutions were stirred for 2 days at 50°C and filtered prior to use. The filtered solution was applied via spray gun onto Al 2024-T3 substrates and dried as described for the BAM-PPV panels. The average film thicknesses for the MEH-PPV coatings (Aldrich) were between 0.3-0.4 microns and ~4.0 microns (NAWCWD). TCP coated Al 2024-T3 panels were supplied by the NAWCAD. For each pretreatment system, 3 panels of each coating were placed in a neutral salt fog chamber in racks at a 6° angle. The panels were removed at selected time intervals for visual inspection. The coupons were examined for any delamination of the coating, blistering or corrosion. The pretreatments were tested with the current standard CCC as the control.

BAM-PPV and CCC pretreatments (CCC coupons obtained from Q-Panel Lab Products Inc.) on Al 2024-T3 were examined via electrochemical impedance spectroscopy (EIS). Each surface (12.56 cm^2) was exposed to 0.5 N NaCl (aq) solutions and impedance spectra were acquired over six months. Aluminum/liquid contacts were kept at room temperature and solutions were exposed to the ambient environment and nominal light. Impedance spectra were acquired with a Princeton Applied Research Model 2273 potentiostat/galvanostat. The frequency range extended from 2 MHz to 0.005 Hz with an rms amplitude of 20 mV. Two-electrode cells were employed with a platinum counter electrode. Data were fit using EQUIVCRT software.

Results and Discussion

Neutral Salt Fog Exposure of Alternative Pretreatment Coatings on Al 2024-T3

Several alternative pretreatment systems coated onto Al 2024-T3 panels were tested against CCC in a neutral salt fog chamber (see Table 1). The testing of these panels follows the guidelines found in ASTM B117 (*41*). These unpainted coatings were tested as an alternative to CCC (Type 1A) (*42*). The evaluation of these unpainted coatings follows a specific military specification (mil spec) for alternative coatings that contain hexavalent chromium. The mil spec requires that the unpainted coating pass 336 hours of neutral salt fog exposure. The accept/reject criteria from the mil spec is no corrosion of the underlying aluminum and that areas within ¼ inch from the edges are accepted. Any staining of the coating is not considered a failure and the coating must show no blistering, delamination or evidence of corrosion (*2, 42*).

During the neutral salt fog exposure tests for each pretreatment coating, CCC controls lasted well over 1000 hours showing no signs of corrosion. The TCP treated samples passed the minimum requirement set by the mil spec (Figures 1-4). At 1000 hours no delamination or visible corrosion was evident

Table 1. Pretreatment Alternatives to CCC

Pretreatment Method	Chemical Composition	Application
Alodine 1200S (CCC) *(2)*	Chromic acid, hexavalent and trivalent chromium complexes	Immersion, spray or wipe
NAVAIR-AD Trivalent chromium pretreatment (TCP) *(2)*	Chromium III sulfate basic, potassium hexafluorozirconate	Immersion, wipe or spray
BAM-PPV *(14, 15)*	Organic compound (C, H, N)	Immersion or spray
MEH-PPV	Organic compound (C, H, O)	Spray

on the TCP panels. At 1500 hours there were signs of corrosion along the edges of the panels and within the panels. In addition some discoloration of the TCP coating was also apparent.

Electroactive polymers, BAM-PPV and MEH-PPV were coated onto Al 2024-T3 panels and tested in the neutral salt fog chamber. BAM-PPV at various film thicknesses (0.5 -1.6 microns) have repeatedly passed 336 hours exposure to neutral salt fog *(14)* but failure before 500 hours was evident with these thinner coatings. In this study, the BAM-PPV coating thickness was increased to 2 microns to maximize its performance in the neutral salt fog chamber. At 336 and 840 hours of neutral salt fog exposure, the BAM-PPV coating shows similar corrosion resistance (Figures 5-8). The bulk of the panels at 840 hours did not show any blisters, corrosion or delamination of the coating. At 1304 hours, corrosion was evident along both the edges and in the bulk of the coating. The testing was stopped at this point due to failure of the BAM-PPV coating.

The next set of panels to be tested consisted of MEH-PPV coated onto Al 2024-T3. MEH-PPV has been extensively studied for applications in photovoltaics *(43-45)*, optical memory *(46)*, light-emitting devices *(47-49)* and lasers *(50)*. Since BAM-PPV shows promise as an alternative to CCC, a study into whether MEH-PPV could provide similar corrosion inhibition for Al 2024-T3 was investigated. Our initial studies focused on a thin film of MEH-PPV (film thickness ~0.3 micron). These MEH-PPV coated panels did pass the 336 hours for neutral salt fog exposure (Figure 9). In addition, MEH-PPV (film thickness 4.0 micron) also passed the 336 hours neutral salt fog exposure (Figure 10). In both cases, the MEH-PPV coatings showed no difference in corrosion performance at 336 hours. However, at this time, the optimal thickness for MEH-PPV has not been established for improved corrosion resistance in neutral salt fog chamber.

Figure 1. TCP on Al 2024-T3; Time = 0 hours
(See page 1 of color inserts.)

Figure 2. TCP on Al 2024-T3; Time = 336 hours
(See page 1 of color inserts.)

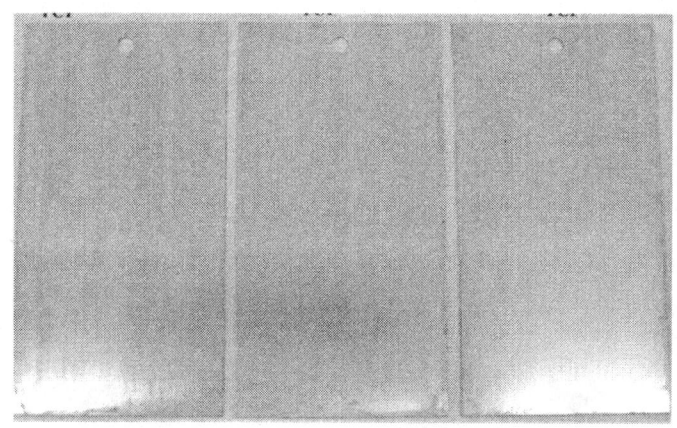

*Figure 3. TCP on Al 2024-T; Time = 1000 hours
(See page 2 of color inserts.)*

*Figure 4. TCP on Al 2024-T; Time = 1500 hours
(See page 2 of color inserts.)*

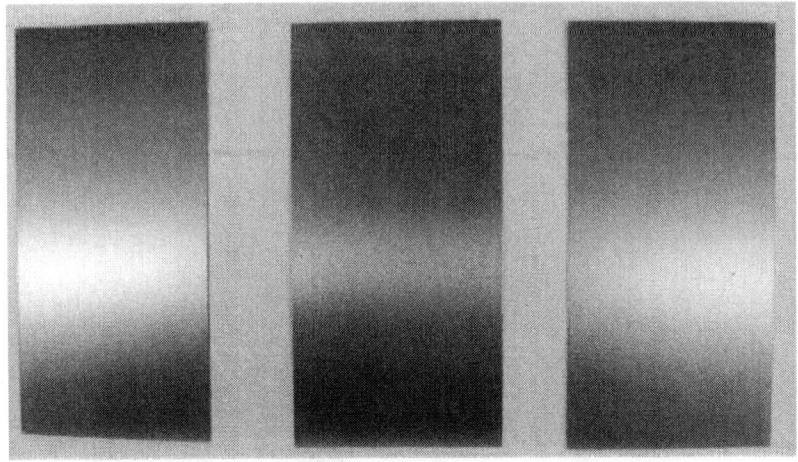

*Figure 5. BAM-PPV Coated Al2024-T3, Time = 0 hours
(See page 3 of color inserts.)*

*Figure 6. BAM-PPV Coated Al 2024-T3, T = 336 hours
(See page 3 of color inserts.)*

Figure 7. BAM-PPV Coated Al 2024-T3; T = 840 hours
(See page 4 of color inserts.)

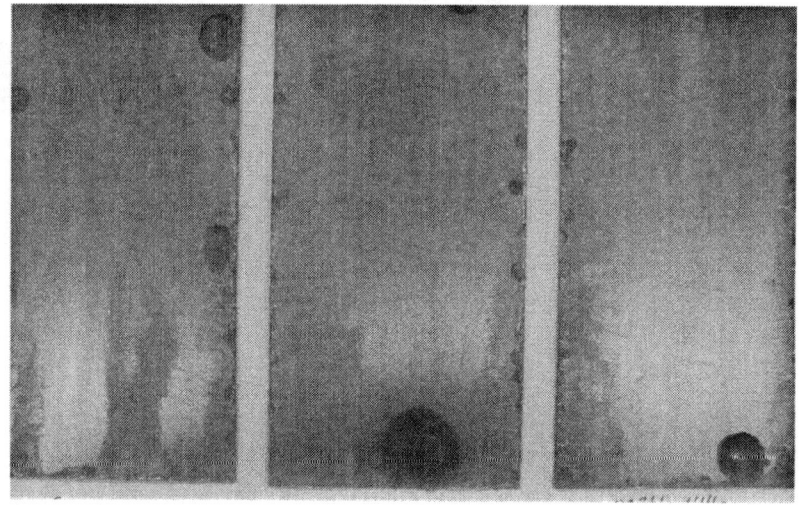

Figure 8. BAM-PPV Coated Al 2024-T3, Time = 1304 hours
(See page 4 of color inserts.)

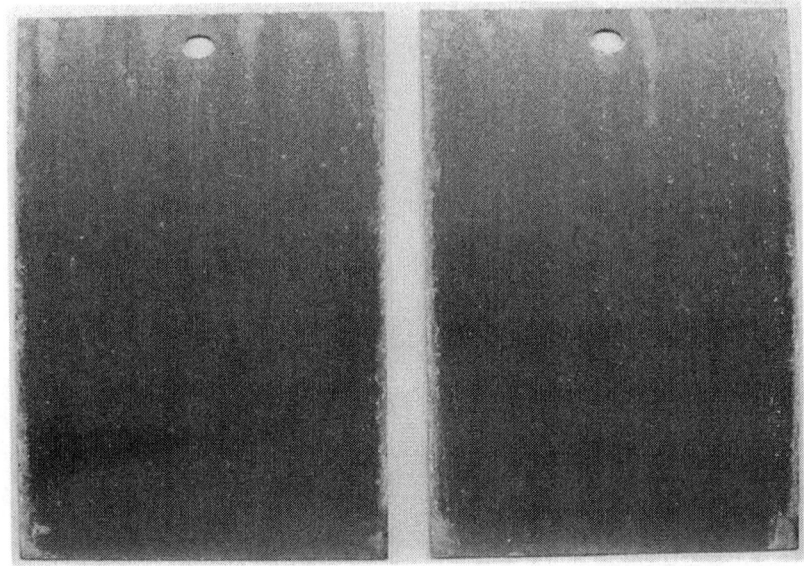

*Figure 9. MEH-PPV Coated Al 2024-T3; Time = 336 hours
(See page 5 of color inserts.)*

*Figure 10. MEH-PPV Coated onto Al 2024-T3; Time = 336 hours
(See page 5 of color inserts.)*

Electrochemical Impedance Studies (EIS) of CCC and BAM-PPV Coated Al 2024-T3 Coupons

CCC coated Al 2024-T3 were compared to BAM-PPV coated Al 2024-T3 (~1.5 microns) using EIS. The EIS measurements of the CCC shows low values for the impedance (Bode plot) which is consistent with published results (*51-53*). A similar result was obtained with the BAM-PPV coated panels. The Bode plots for Al treated with CCC and BAM-PPV exhibit significant changes over time (Figure 11). However, the impedances at low frequencies do not significantly change within the first six months of exposure to the salt solution. Specifically, the total resistance does not deviate significant from 10^4-10^5 ohms regardless of the coating (Figure 12). There are different frequency dependent processes occurring with the two surfaces. For example, at high frequencies (10^4-10^6 Hz) the total impedance of the BAM-PPV-aluminum surface initially has both resistive and capacitive elements. But over four months, the capacitive nature of this high-frequency process diminishes. By contrast the high-frequency impedance of the CCC treated aluminum is purely resistive at these frequencies. These results suggest that the two coatings are behaving differently yet still produce similar impedance and salt fog results.

Conclusions

All the pretreatment systems studied, TCP, BAM-PPV and MEH-PPV show promise as alternatives to CCC. The BAM-PPV has repeatedly passed 336 hours neutral salt fog exposure and this represents a potential non-chromium based alternative military pretreatment. An increase of the BAM-PPV coating thickness did extend the lifetime of the panels in neutral salt fog chamber. However, both the BAM-PPV and the MEH-PPV coatings did not match the performance of the CCC or TCP coating systems. These coatings (BAM-PPV and MEH-PPV) do offer a step in eliminating hexavalent chromium use in military pretreatment coatings. The optimal coating thickness for BAM-PPV and MEH-PPV has not been established at this time.

The EIS study on the BAM-PPV coating shows similar performance to the CCC in long term immersion studies. The BAM-PPV coating performed as well as the CCC in providing the minimum barrier protection in 0.5 N NaCl solutions. Both systems CCC and BAM-PPV as measured by EIS are inhibiting corrosion by two distinct mechanisms. Future work will include BAM-PPV and MEH-PPV pretreatments in full military coating systems with and without chromium (Cr (VI)). These systems will be studied using a combination of neutral salt fog and EIS.

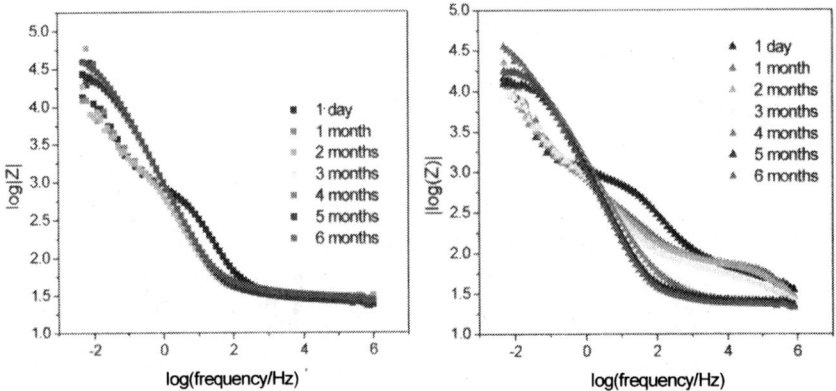

Figure 11. Bode plots of CCC on Al 2024-T3 panel (left) and BAM-PPV on Al 2024-T3 (right) in 0.5 N NaCl solution.
(See page 6 of color inserts.)

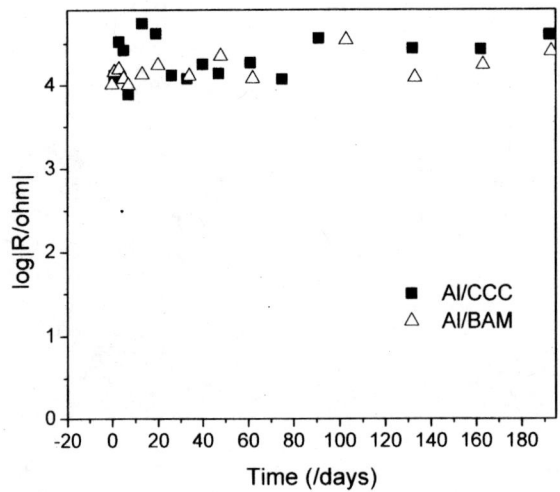

Figure 12. Total impedance obtained at low frequencies for CCC on Al 2024-T3 (black squares) and BAM-PPV coated Al 2024-T3 (white triangles) in 0.5 N NaCl solution

Acknowledgements

The authors would like to acknowledge the continuing support of the Office of Naval Research (ONR), Dr. A. Perez and the Strategic Environmental Research and Development Program (SERDP), Mr. C. Pellerin/Program Manager, Pollution Prevention. Mr. Craig Matzdorf (NAWCAD) for TCP coated aluminum panels, Mr. Bradley Douglas (NAWCWD) for the synthesis of the MEH-PPV compound and Ms. Cindy Webber (NAWCWD) for the coating of BAM-PPV and MEH-PPV onto Al 2024-T3 panels.

References

1. Anderson, N. ; Zarras, P. *Electroactive Polymers as Green Alternatives to Hexavalewnt Chrome*, Currents, Spring 2005, p. 60.
2. ESTCP , *Non-chromate Aluminum Pretreatments Phase I Report Project # PP0025*, August 2003, http//www.estcp.org.
3. Korinek, K.A. In *Chromate Conversion Coatings*, ASM Handbook, Volume 13 Corrosion, Eds. Korb, L. J. and Olson, D. L., ASM International, USA 1998 p. 389.
4. Katzman, H.A.; Malouf, G. M.; Bauer, R.; Stupian, G. W. *Appl. Surf. Sci.*, **1979**, 2, 416.
5. Kendig, M.; Jeanjaquet, S.; Addison, R.; Waldrop, J. *Surface and Coatings Technol.*, **2001**, 140, 58.
6. Kendig, M. W.; Davenport, A. J.; Isaacs, H. S. *Corrosion Science*, **1993**, 33(1), 41.
7. Xia, L.; Akiyama, E.; Frankel, G.; McCreery, R.; *J. Electrochem. Soc.*, **2000**, 147(7), 2556.
8. Illevbare, G. O.; Scully, J. R.; Yuan, J.; Kelly, R. G.; *Corrosion*, **2000**, 56(3), 227.
9. Blasiak, J.; Kowalik, J. *Mutat. Res.*, **2000**, 469, 135.
10. Wetterhahn, K. E.; Hamilton, J. W.; *Sci Total Envir.*, **1989**, 86, 113.
11. Leonard, A.; Lauwerys, R.R. *Mutat. Res.*, **1980**, 76, 227.
12. Tsapakos, M. J.; Hampton, T. H.; Wetterhahn, K. E.; *Cancer Res.*, **1983**, 43, 5662.
13. National Emissions Standards for Chromium Emissions from Hard and Decorative Chromium Electroplating and Chromium Anodizing Tanks. Environmental Protection Agency. *Federal Register.* RIN 2060-AC14, January 25, 1995.
14. Zarras, P.; He, J.; Tallman, D. E.; Anderson, N.; Guenthner, A.; Webber, C.; Stenger-Smith, J. D.; Pentony, J. M.; Hawkins, S.; Baldwin, L., *Electroactive Polymer Coatings as Replacements for Chromate Conversion*

Coatings, Ed., Provder, T., in Smart Coatings I, American Chemical Society, Washington DC, in press, 2006.
15. Zarras, P.; Stenger-Smith, J. D. In *Electroactive Polymers for Corrosion Control*, Zarras, P.; Stenger-Smith, J. D.; Wei, Y. Eds.; ACS Symposium Series 843, American Chemical Society, Washington DC; 2003, pp 2-17.
16. Agarwala, V. S.; Beckert, D. W.; Fabiszewski, A. S.; Pearlstein, F., Corrosion 94, Baltimore, Maryland, Paper No. 621, February 28-March 4, 1994.
17. Delaunois, F.; Poulain, V.; Petitjean, J-P., *Materials Science Forum*, **1997**, 213.
18. NDN-146-0025-4062-4 AIA CSA, *Adv. Mater. Process.*, 162(11), November 2004.
19. Decroly, A.; Petitijean, J-P., *Surface and Coatings Technology*, **2005**, 194(1), 1.
20. Mansfeld, F.; Lin, S.; Kim, S.,; Shih, H., *Corrosion*, August 1989, 615.
21. Wany, Y., *Corrosion Protection of Aluminum Alloys By Surface Modification Using Chrmate-Free Approaches*, NDN-135-0218-7230-8 DIS, 1994.
22. Wessling, B.; Posdorfer, J. *J. Electrochemica. Acta.* **1999**, 44, 2139.
23. Thompson, K. G.; Bryan, C. J.; Benicewicz, B. C.; Wrobleski, D. *Los Alamos National Laboratory Report,* LA-UR-92-360 , 1991.
24. Wrobleski, D. A.; Benicewicz, B. C.; Thompson, K. G.; Bryan, B. J. *ACS Polym. Prepr.***1994**, 35(1), 265.
25. McAndrew, T.P.; *TRIP*, **1997**, 5(1), 7.
26. Beard, B.C.; Spellane, P. *Chem . Mate.*, **1997**, 9, 1949.
27. Brumbaugh, D. *The AMPTIAC Newsletter*, **1999**, 3(1), 1.
28. Kinlen, P. J.; Ding, Y.; Silverman, D. C. *Corrosion*, **2002**, 58(6), 490.
29. Tallman, D. E.; Spinks, G. M.; Dominis, A. J.; Wallace, G.G. *J. Solid State Electrochem.*, **2002**, 6, 73.
30. Yang, S. C.; Brown, R.; Racicot, R.; Lin, Y.; McClarnon, F. In *Electroactive Polymers for Corrosion Control*, Zarras, P.; Stenger-Smith, J. D.; Wei, Y. Eds.; ACS Symposium Series 843 American Chemical Society, Washington DC; 2003, pp. 196-207.
31. Racicot, R. J.; Yang, S. C.; Brown, R., *Interfacial Engineering for Optimized Properties*, **1997**, 41,5.
32. Racicot, R. C.; Yang, S. C.; Brown, R., *Corrosion Protection Comparison of a Chromate Conversion Coating to a Novel Conductive Polymer Coating on Aluminum Alloys*, Corrosion 97, March 10-14, 1997, New Orleans, LA.
33. Stenger-Smith, J. D.; Zarras, P.; Merwin, L. H.; Shaheen, S. E.; Kippelen, B.; Peyghambarian, N. *Macromolecules*, **1998**, 31, 7566.
34. Heeger, A. J.; Parker, I. D.; Yang, Y. *Synth. Met.* **1994**, 67, 23.

35. Stenger-Smith, J. D.; Miles, M. H.; Norris, W. P.; Nelson, J.; Zarras, P.; Fischer, J. W.; Chafin, A. P. U. S. Patent 5,904,990, 1999.
36. Stenger-Smith, J. D.; Zarras, P.; Ostrom, P.; Miles, M. H.; In *Semiconducting Polymers*, Hsieh, B. R.; Wei, Y.; ACS Symposium Series 735, American Chemical Society, Washington DC, 1999, pp. 280-292.
37. Anderson, N.; Stenger-Smith, J. D.; Irvin, D. J.; Guenthner, A.; Zarras, P. Proceedings of the Navy Corrosion Information and Technology Exchange, Louisville, KY, July 16-20, 2001.
38. Anderson, N.; Irvin, D. J.; Webber, C.; Stenger-Smith, J. D.; Zarras, P. *ACS PMSE Prepr.*, **2002**, 86, 6.
39. Zarras, P.; Anderson, N.; Webber, C.; Guenthner, A.; Prokopuk, N.; Stenger-Smith, J . D.; Polymers in Aggressive and Corrosive Environments, Symposium Proceedings, PACE, September 8-9, 2004, p. 175-181.
40. Zarras, P.; Anderson, N.; Guenthner, A.; Prokopuk, N.; Quintana, R. L.; Hawkins, S. A.; Webber, C.; Baldwin, L.; Stenger-Smith, J. D.; He. J.; Tallman, D. E.; 2005 Tri-Service Corrosion Conference, Orlando, Florida, November 11-14, 2005.
41. ASTM B117, Annual Book of ASTM Standards, ASTM, West Conshohocken, PA, vol 03.02, 2001, p. 1.
42. Military Specification, MIL-DTL-81706B Chemical Conversion Materials for Coating Aluminum and Aluminum Alloys, October 25, 2004.
43. Halls, J. J. M.; Friend, R. H., *Synth. Met.*, **1997**, 85, 1307.
44. Dittmer, J. J.; Petritsch, K.; Marseglia, E. A.; Friend, R. H.; Rost, H.; Holmes, A. B., *Synth. Met.*, **1999**, 102, 879.
45. Gao, J.; Yu, Y.; Heeger, A. J., *Adv. Mater.*, **1998**, 10, 692.
46. Yoshino, K.; Kuwabara, T.; Iwasa, T.; Kawai, T.; Onoda, M., *Jpn. J. Appl. Phys.*, **1990**, 29, L1514.
47. Carter, S. A.; Angelopoulos, M.; Karg, S.; Brock, P. J.; Scott, J. C., *Appl. Phys. Letts.*, **1997**, 70, 2067.
48. Gustafsson, G.; Cao, Y.; Treacy, G. M.; Klavetter, F.; Colaneri, N.; Heeger, A. J., *Nature (London)*, **1992**, 357, 477.
49. Parker, I. D.; Kim, H. H., *Appl. Phys. Lett.*, **1994**, 64, 1774.
50. Hide, F.; Schwartz, B. J.; Diaz-Gracia, M. A.; Heeger, A. J., *Chem Phys. Lett.*, **1996**, 256, 424.
51. Campestrini, P.; Terryn, H.; Vereecken, J.; de Wit, J. H. W, *J. Electrochem. Soc.*, **2004**, 151(6), B370.
52. Meng, Q.; Frankel, G. S., *Corrosion,* **2004**, 60(10), 897.
53. Campestrini, P.; van Westing, E. P. M.; Hovestad, A.; de Wit, J. H. W., *Electrochim. Acta*, **2002**, 47, 1097.

Chapter 5

Reversible Addition–Fragmentation Chain Transfer Polymerization of 4-Anilinophenyl(meth)acrylates

Ru Chen, Chunzhao Li, and Brian C. Benicewicz

New York State Center for Polymer Synthesis, Department of Chemistry and Chemical Biology, Rensselaer Polytechnic Institute, Troy, NY 12180

The monomers 4-anilinophenyl methacrylate (**1**) and 4-anilinophenyl acrylate (**2**) were synthesized and polymerized by reversible addition – fragmentation chain transfer (RAFT) polymerization as models for oligoaniline side chain polymers. Two chain transfer agents (CTAs), 2-cyanopropyl dithiobenzoate (**3**) and α-cyanobenzyl dithiobenzoate (**4**), were investigated in the RAFT process. The methacrylate monomer exhibited a higher rate of polymerization than the acrylate monomer and produced polymers with a narrower polydispersity index. The chain transfer agents also had an effect on the polymerization and CTA **3** controlled the polymerization better than CTA **4**. An in-situ NMR study showed that there was an induction period of approximately two hours for the polymerization of the acrylate monomer in the presence of CTA **4**. With the proper selection of the chain transfer agent, the methacrylate monomer could be polymerized in a controlled manner to high conversion and produced polymers with low (<1.4) polydispersity index.

Introduction

Free radical polymerization is widely used in industry to produce polymers with limited control on the final polymer molecular weight and polydispersity. In the past decade, significant efforts have been put into the development of new radical polymerization techniques to prepare polymers with controlled molecular weight, narrow polydispersity and different molecular architectures (e.g., end group functionalization, block copolymers and star polymers). Reversible addition-fragmentation chain transfer (RAFT) polymerization was first reported by Rizzardo et al. in 1998 (*1*). RAFT polymerization is conducted in a manner similar to conventional free radical polymerization except that a small amount of chain transfer agent (CTA) is added. Many monomers, e.g., styrenes, methacrylates, acrylates, methacrylamides, acrylamides, and acrylic acid, can be polymerized by the technique. RAFT polymerization can be conducted in the presence of a wide variety of functionalities such as sulfonate, carboxylic acid and hydroxyl, and can be performed in bulk, solution, suspension and emulsion (*2,3*).

Aromatic amines have long been used as antioxidants in polymeric materials (*4*). In many applications, these low molecular weight antioxidants may be depleted by evaporation, extraction, migration, and irreversible absorption. Polymer-bound antioxidants are less susceptible to many of these depletion mechanisms and are retained in the polymer matrix for longer times. Polymer-bound antioxidants can be prepared by grafting or free radical copolymerization. The incorporation of aromatic amines by free radical polymerization was shown to improve the thermal stability of resulting polymers (*5*). (Meth)acrylamides and (meth)acrylates containing aromatic amines were synthesized and polymerized by free radical polymerization (*6,7*). Due to the hydrogen abstraction reaction of aromatic amine by radicals, the polymerization behaviors were quite different from ordinary monomers. In addition, copolymers of these special monomers, e.g., random copolymers and block copolymers, may help polymer-bound antioxidants disperse in the polymer matrix. Therefore, it is of interest from both academic and industrial points of view to investigate the possibility of extending RAFT polymerization to these special monomers and study the polymerization kinetics under RAFT conditions.

A series of poly(acrylate)s and poly(acrylamide)s containing oligoaniline side chains have been synthesized and characterized in our lab (*8,9*). Such polymers may be useful as anti-corrosion coatings, and may be more amenable to incorporate into acrylate-based paints and coatings. In this work, 4-anilinophenyl methacrylate (**1**) and acrylate (**2**) were used as model monomers to investigate the RAFT polymerization behavior of monomers containing oligoaniline side chains. Monomers **1** and **2** were polymerized in N,N-dimethylformamide (DMF) using 2,2'-azo-bisisobutyronitrile (AIBN) as an initiator with the presence of a chain transfer agent. The chain transfer agents studied include 2-cyanopropyl dithiobenzoate (**3**) and α-cyanobenzyl dithiobenzoate (**4**). The chemical structures of these monomers and chain transfer agents are shown in Figure 1. The result of this study showed that the polymerization of these models for oligoaniline side chain monomers could be conducted in a controlled manner by the careful selection of monomer and chain transfer agent. Further studies on longer side chain monomers will be presented in future works.

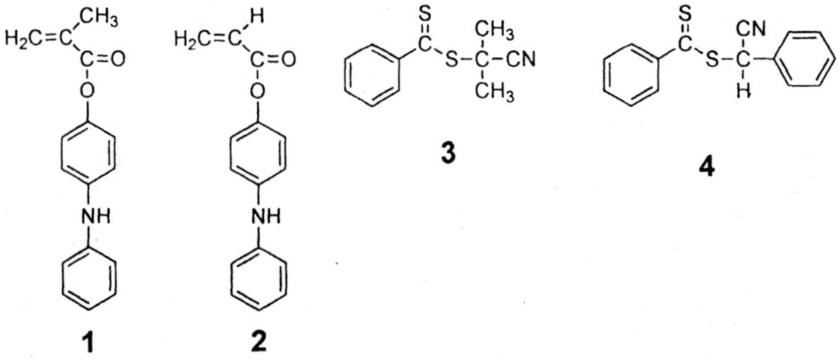

Figure 1. Chemical structures of monomers and chain transfer agents.

Experimental

Materials

All chemicals were purchased from ACROS Organics except 4-hydroxy diphenylamine (TCI America), methacryloyl chloride (Aldrich, 98+%), and acryloyl chloride (Aldrich), and used as received. 2,2'-Azobisisobutyronitrile (Aldrich) was recrystallized from ethanol. N,N-Dimethylformamide (DMF) was treated with 5 Å molecular sieves and distilled under reduced pressure.

Characterization

Melting points were measured by a capillary melting point apparatus without calibration. NMR spectra were recorded on a Varian 500 spectrometer using solvent residues as references. A Shimadzu GCMS-QP-5000 spectrometer was used to confirm the chemical structure and purity of all compounds prepared in this work. Fourier transform infrared (FTIR) spectrometry was performed using a Bio-Rad FTS3000 spectrometer. Elemental analysis was performed by Midwest Microlabs. The molecular weight of the polymers was measured on a Waters gel permeation chromatography (GPC) system equipped with a refractive index detector (Waters 2410) and 3 Styragel columns packed with 5 μm particles (HR1: effective molecular weight 100-5,000; HR3: effective molecular weight 500-30,000; and HR4: effective molecular weight 5,000-500,000) using tetrahydrofuran (THF) as the eluent at a flow rate of 0.3 mL/min. The column temperature was set at 30 °C. Molecular weights are reported relative to polystyrene standards.

Synthesis of monomers and chain transfer agents

The synthesis of monomers, 4-anilinophenyl methacrylate and acrylate were prepared according to the synthetic procedures reported previously (*4*). The synthesis of chain transfer agent 2-cyanopropyl dithiobenzoate (**3**) was carried out according to the methods in the literature (*10*). The preparation of α-cyanobenzyl dithiobenzoate (**4**) was described elsewhere (*11*). The synthetic procedures for the chain transfer agents used in this work are shown in Scheme 1.

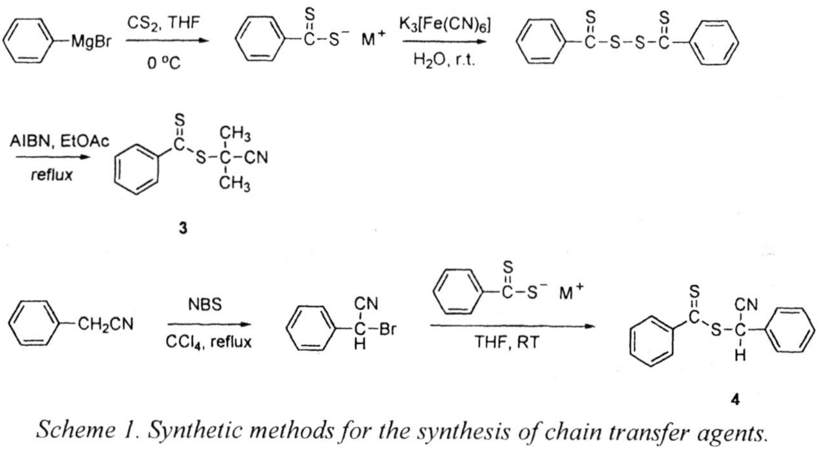

Scheme 1. *Synthetic methods for the synthesis of chain transfer agents.*

4-Anilinophenyl methacrylate (1)

4-Hydroxy diphenylamine (10.00 g, 53.99 mmol) was dissolved in THF (100 mL) with nitrogen protection. Triethylamine (10.92 g, 107.9 mmol) was added, and the reaction mixture was cooled in an ice bath. Methacryloyl chloride

(6.75 g, 64.57 mmol) solution in THF (50 mL) was added slowly into the solution with stirring. After the addition was completed, the reaction mixture was stirred at room temperature for 24 h. The precipitate was filtered off and washed with THF. The solutions were combined and the solvent was removed by rotary evaporation. The residue was dissolved in dichloromethane and washed with water. The solution was dried with anhydrous sodium sulfate. The solvent was removed and the residue was recrystallized from heptanes. The white needle-like crystals were dried in vacuo at r.t. for 24 h. The yield was 10.42 g (76%).

Melting point 72 – 73 °C (lit.(*4*) 68 – 70 °C). ^{1}H NMR (DMSO – d$_6$, δ): 8.19 (s, 1H, –NH), 7.23 (t, J = 7.9 Hz, 2H, –ArH), 7.14 – 6.96 (m, 6H, –ArH), 6.82 (t, J = 7.3 Hz, 1H, –ArH), 6.25 (s, 1H, cis =CH$_2$), 5.86 (s, 1H, trans =CH$_2$), 2.06 – 1.94 (m, 3H, –CH$_3$). GC/MS (EI): m/z 253 (M$^+$). IR (KBr): 3390, 3036, 2924, 1723, 1599, 1498, 1328 cm^{-1}. Anal. Calcd. for C$_{16}$H$_{15}$NO$_2$: C, 75.87; H, 5.97; N, 5.53. Found: C, 75.65; H, 6.00; N, 5.55.

4-Anilinophenyl acrylate (2)

4-Anilinophenyl acrylate (**2**) was synthesized using a procedure similar to **1**. The yield of recrystallized (heptanes) product was 53%.

Melting point 47 – 49 °C. ^{1}H NMR (DMSO – d$_6$, δ): 8.20 (s, 1H, –NH), 7.23 (t, J = 7.0 Hz, 2H, Ar), 7.14 – 6.96 (m, 6H, –ArH), 6.82 (t, J = 7.3 Hz, 1H, –ArH), 6.51 (dd, J = 17.3, 1.2 Hz, 1H, cis CH$_2$), 6.39 (dd, J = 17.3, 10.2 Hz, 1H, =CH), 6.12 (dd, J = 10.3, 1.2 Hz, 1H, trans =CH$_2$). GC/MS (EI): m/z 239(M$^+$). IR (KBr): 3385, 3034, 1728, 1595, 1507, 1404, 1295 cm^{-1}. Anal. Calcd. for C$_{15}$H$_{13}$NO$_2$: C, 75.30; H, 5.48; N, 5.85. Found: C, 74.91; H, 5.49; N, 5.82.

Reversible Addition-Fragmentation Chain Transfer (RAFT) Polymerization

Monomer (**1**: 6.325 g or **2**: 5.975 g, 1.0 M), AIBN (8.2 mg, 2.0 × 10^{-3} M) and CTA (**3**: 33.6 mg, or **4**: 40.4 mg, 6.0 × 10^{-3} M) were dissolved in DMF (25 mL). A small portion of the solution (1.0 mL) was transferred to separate Schlenk tubes and degassed by three freeze-pump-thaw cycles. The Schlenk tubes were sealed under nitrogen and heated to 70 °C with stirring for the indicated times and then stopped by placing in cold water. A sample volume of polymer solution (100 µL) was taken and evaporated in the fume hood and further dried in vacuum oven at room temperature for 24 h. The residue was dissolved in THF (1 mL) and filtered through 0.20 µm PTFE filter and analyzed by GPC to determine the polymer molecular weight, polydispersity and monomer conversion.

In-situ ^1H NMR Analysis of the Polymerization Reaction

A small portion of the solution described above (0.5 mL) was charged into a Young's tap NMR tube. Two drops of DMSO-d_6 were added as a NMR locking agent. The solution was degassed by three freeze-pump-thaw cycles and sealed under vacuum. The NMR tube was placed into a preheated NMR sample cavity at 70 °C. The sample was allowed to equilibrate for 6 minutes prior to recording spectra. The spectra were recorded every two minutes for approximately 12 h. For quantitative analysis, the pulse angle was set to 30 ° and the relaxation delay time was set to 27 seconds to insure complete relaxation of the nuclei between the individual pulses.

Results and Discussion

The mechanism of the RAFT polymerization is believed to involve a reversible addition-fragmentation equilibrium as shown in Scheme 2. The initiators decompose into free radicals, which add to monomers and grow into short chain propagating radicals $P_n\cdot$. The addition of $P_n\cdot$ to the chain transfer agent (in this case thiocarbonylthio compounds) gives adduct radicals, which fragment into polymeric chain transfer agents (polymeric thiocarbonylthio compounds) and new radicals R·. Radicals R· reinitiate and propagate to new short chain propagating radicals P_m before addition to the CTA. The subsequent addition-fragmentation steps establish an equilibrium between $P_m\cdot/S=C(Z)S-P_n$ and $P_n\cdot/S=C(Z)S-P_m$. The quick equilibration allows all polymer chains to grow simultaneously, resulting in polymers with narrow molecular weight distribution. In the end, most polymer chains are capped with the R- group on one terminus and a S=C(Z)S- group on the other terminus. The polymer chain can be further extended under appropriate conditions. Since the RAFT polymerization is conducted at a low radical concentration and a high ratio of CTA to initiator, the termination of propagating radicals and thus formation of dead chains is maintained at a very low level. The molecular weight of polymers is calculated according to the following equation: M_n = ([M]/[CTA]) × conversion x MW of monomer + MW of CTA.

The selection of the CTA is extremely important to perform a successful RAFT polymerization. The general requirement is that the rates of addition and fragmentation are relatively high compared to the rate of propagation, as described in the literature (*12-14*). Previous work has indicated that cumyl dithiobenzoate and cyanopropyl dithiobenzoate (**3**) were very effective in the polymerization of methacrylates and acrylates (*2,12-15*). Recently, our group discovered that compound **4** and its derivatives are also excellent agents for the RAFT polymerization of methacrylates and acrylates (*11*). In this study,

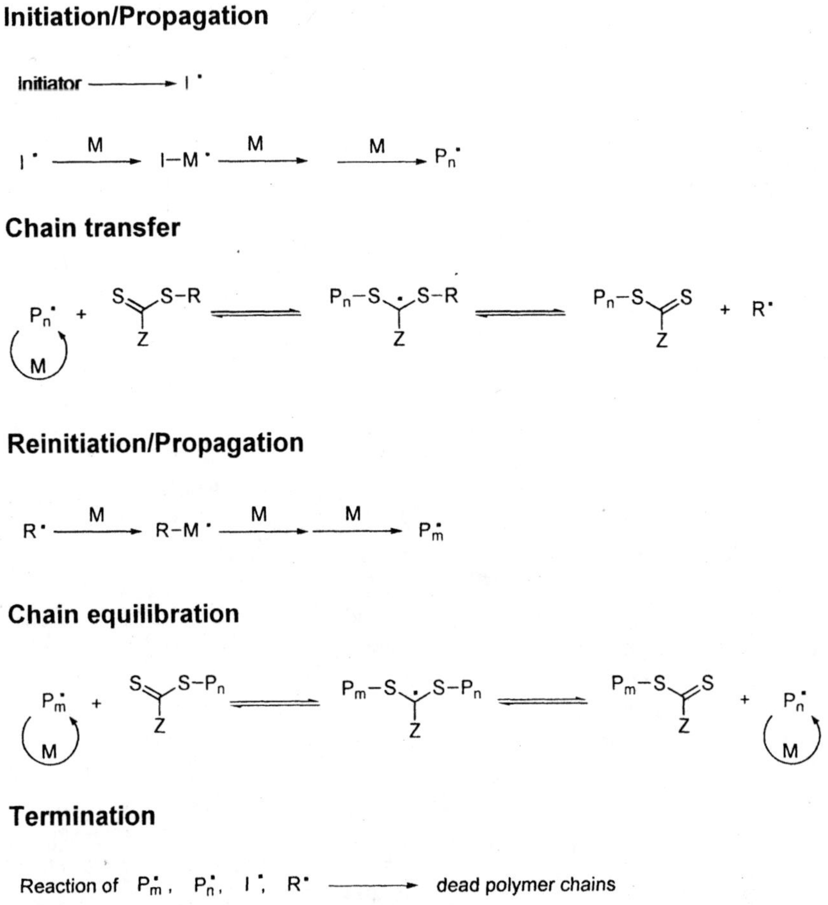

Initiation/Propagation

Initiator ⟶ I•

I• $\xrightarrow{M}$ I–M• $\xrightarrow{M}$ $\xrightarrow{M}$ $P_n^{\bullet}$

Chain transfer

$P_n^{\bullet}$ (M) + S=C(Z)–S–R ⇌ P_n–S–C(Z)(•)–S–R ⇌ P_n–S–C(Z)=S + R•

Reinitiation/Propagation

R• $\xrightarrow{M}$ R–M• $\xrightarrow{M}$ $\xrightarrow{M}$ $P_m^{\bullet}$

Chain equilibration

$P_m^{\bullet}$ (M) + S=C(Z)–S–P_n ⇌ P_m–S–C(Z)(•)–S–P_n ⇌ P_m–S–C(Z)=S + $P_n^{\bullet}$ (M)

Termination

Reaction of $P_m^{\bullet}$, $P_n^{\bullet}$, I•, R• ⟶ dead polymer chains

Scheme 2. General mechanism for RAFT polymerization.

compounds **3** and **4** were chosen as chain transfer agents for the RAFT polymerization of methacrylate and acrylate esters of aromatic amines, **1** and **2**.

Monomers **1** and **2** were synthesized by the reaction of 4-hydroxy diphenylamine with methacryloyl chloride or acryloyl chloride in tetrahydrofuran in the presence of triethylamine, and purified by recrystallization in heptanes. The synthetic procedures of CTAs are shown in Scheme 1. Compound **3** was prepared according to a literature method with slight modification (*10*). Phenylmagnesium bromide was reacted with carbon disulfide, followed by the oxidation with potassium ferricyanide (III). The resulting bis(thiocarbonyl)disulfide was reacted with AIBN in ethyl acetate under reflux conditions to afford the product. The synthesis of **4** was readily accomplished in high yield via nucleophilic substitution. Phenylmagnesium bromide was reacted with carbon disulfide, followed by the reaction with 2-bromo-2-phenyl acetonitrile, which was prepared by reaction of 2-phenyl acetonitrile and N-bromo-succinimide (NBS).

RAFT polymerizations were carried out in DMF at 70 °C using AIBN as an initiator and **3** and **4** as chain transfer agents. The polymerization conditions were [monomer] = 1.0 M, [AIBN] = 2.0×10^{-3} M and [CTA] = 6.0×10^{-3} M. For volatile monomers conversion is usually measured by gravimetry. After the polymerization is complete, the monomer and solvent in the reaction mixture are evaporated and the weight of polymer is used to calculate the conversion. The monomers used in this study are solids at room temperature and do not evaporate under vacuum at low temperatures. Precipitation is not often used to separate polymers because oligomers and low molecular weight polymers may be washed away, leading to inaccuracies of the molecular weight distribution. Thus, in this work the solvent was removed from the reaction mixture in vacuo at room temperature. The residue (a mixture of monomer and polymer) was dissolved and analyzed by GPC. The polymer peak and monomer peak were well resolved (Figure 2). The monomer peak area was used to calculate conversion via calibration with solutions of known monomer concentration. The molecular weight information (M_n, M_w and PDI) was obtained from the polymer peak with respect to polystyrene standards. The alternative was to use the

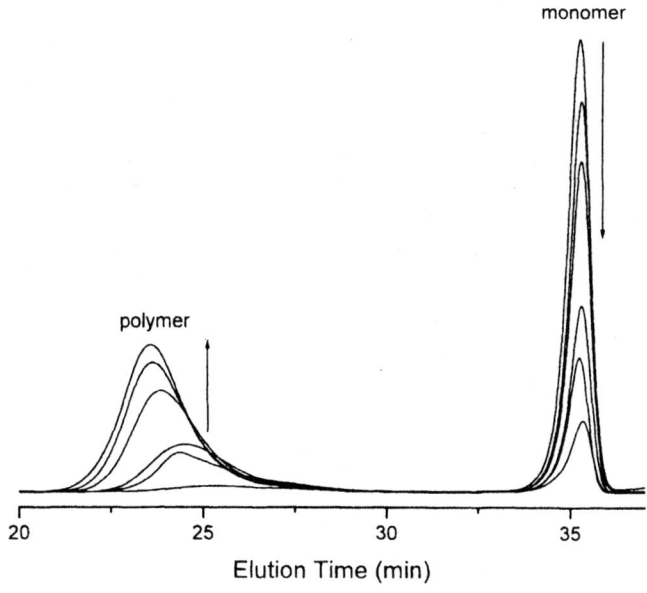

*Figure 2. GPC elution profiles for poly(4-anilinophenyl methacrylate) using CTA **3** at different polymerization times.*

relative percentage of polymer in the mixture to calculate the monomer conversion. It also gave the same values as the external calibration method.

In the RAFT polymerization of **1** and **2**, it was found that the polymerization rate of **1** was faster than that of **2**, which is in contrast to the polymerization rates of methyl methacrylate and methyl acrylate in conventional free radical polymerization. The monomer conversion increased gradually with

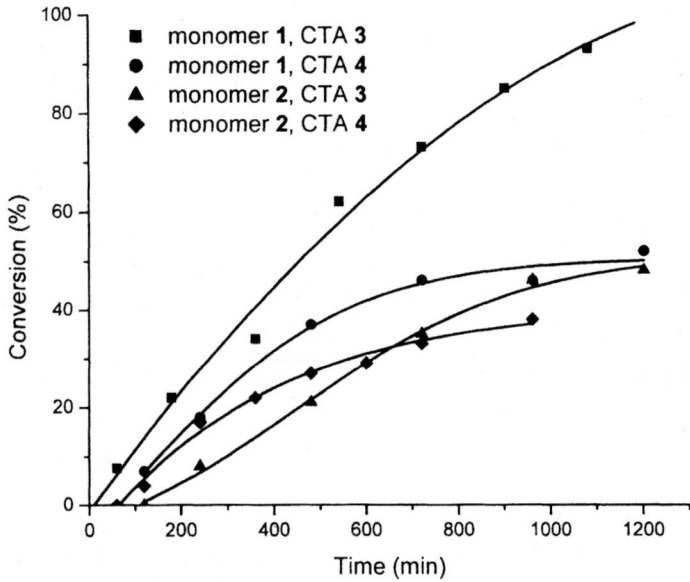

Figure 3. Time-conversion plot for the polymerization of monomers 1, 2 using CTAs 3, 4.

the reaction time up to 20 hours (Figure 3). When compound **3** was used as the CTA in RAFT polymerization, conversion of monomer **1** was up to 90% after 16 hours while the conversion of **2** was only 40%. The same trend was also observed when **4** was used as a CTA, i.e., the conversion of **1** was approximately 45% while monomer **2** was 35% after 16 hours. In both cases the molecular weights of the polymers increased with monomer conversions (Figure 4 and Figure 5). The polydispersity indices for the RAFT polymerization of **1** using CTA **3** were lower than 1.4 up to 85% conversion. However, the polydispersity indices of the polyacrylate from **2** using CTA **3** increased from 1.2 to 2.3 when conversions increased from 7% to 50%. Thus, CTA **3** was much more efficient in controlling the molecular weight of the methacrylate monomer **1**. This observation is different from the previous work on RAFT polymerization that described the limited number of CTA structures that are effective at

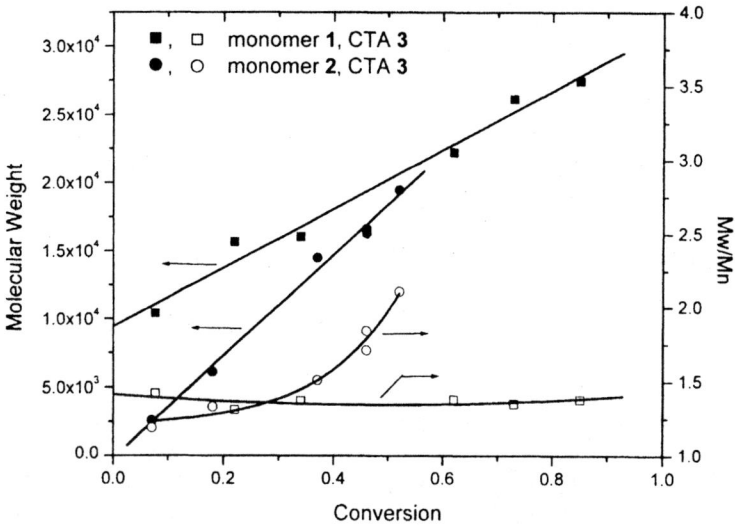

*Figure 4. Evolution of molecular weight and polydispersity for the polymerization of monomers **1**, **2** using CTA **3**.*

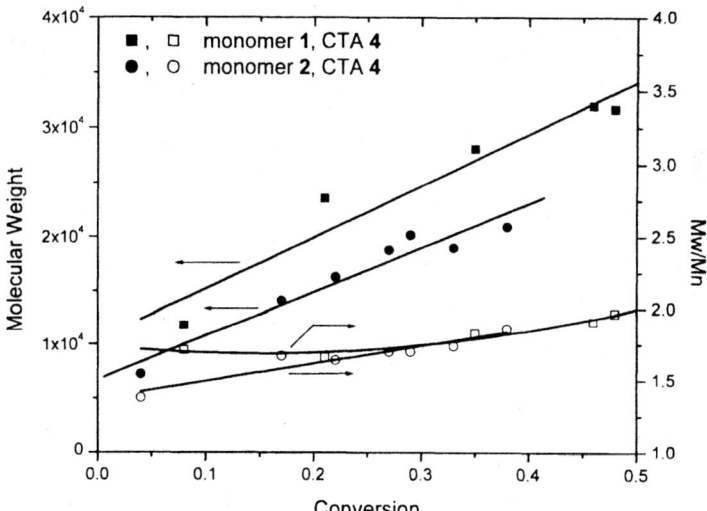

*Figure 5. Evolution of molecular weight and polydispersity for the polymerization of monomers **1**, **2** using CTA **4**.*

controlling the polymerization of methacrylates as compared to acrylates (12,13). This may be explained by the side reaction involving the monomers, or more accurately, the diphenylamine unit in the monomers. Diphenylamines are known to be free radical inhibitors and have been used as antioxidants in industry for many years (16). The aromatic amines undergo hydrogen abstraction upon reaction with free radicals and form non-radical species as well as stable aminyl radicals that can either combine with other radicals or self-couple with other aminyl radicals, resulting in non-radical products. These reactions have been shown to inhibit the free radical polymerization of vinyl monomers containing diphenylamine groups under certain polymerization conditions, e.g., polymerizations that use of dibenzoyl peroxide, cumene hydroperoxide, and tetraethylthiuram disulfide as initiators (7). It is well known that the propagating radicals of acrylates are relatively low in steric hindrance and high in reactivity compared to that of methacrylates, and thus more likely to abstract hydrogen from the diphenylamine units. This would lead to chain termination and an uncontrolled polymerization. For RAFT polymerization using CTA **4** (Figure 5) the molecular weight increased with monomer conversion, and PDIs for both polymers were in the range of 1.5 to 2.0, values that are typical for conventional free radical polymerization. Compound **4** showed only very limited control over the polymerization of both the methacrylate and acrylate monomers.

In-situ NMR spectrometry is an important technique to obtain real time structural and kinetic information on the RAFT polymerization process (17-20). It allows continuous determination of monomer conversion during the radical polymerization. Thus, the polymerizations of monomer **1** and **2** were carried out at 70 °C in a NMR tube fitted with a Young's tap. The change in NMR spectra with reaction time is shown in Figure 6 and Figure 7 for monomers **1** and **2**, respectively. The resonance signals of the vinyl groups *(5.8* and 6.3 ppm for monomer **1**; 6.0, 6.3 and 6.5 for monomer **2**) decreased gradually with polymerization time. A broad peak from the resulting polymer appeared at approximately 6.7 ppm and increased in intensity simultaneously with the decrease in the resonance at 6.8 ppm, which corresponded to the aromatic proton resonance in the monomer. The monomer conversion was measured by integration of the vinyl resonance (6.3 ppm for **1**, 6.5 ppm for **2**) relative to the combination of Ar-H resonances at 6.7 – 6.8 ppm from both monomer and polymer (Figure 6 and Figure 7).

The time-conversion plots from the NMR study (Figure 8) showed that the conversions of RAFT polymerizations using compounds **3** and **4** as chain transfer agents were lower than the conventional free radical polymerization without chain transfer agent under the same conditions. An induction time of 2 hours was observed in the polymerization of **2** using CTA **4**. In an ideal process (i.e., fast addition, fragmentation and reinitiation; no side reactions), the RAFT process should not have a significant influence on the rate of polymerization,

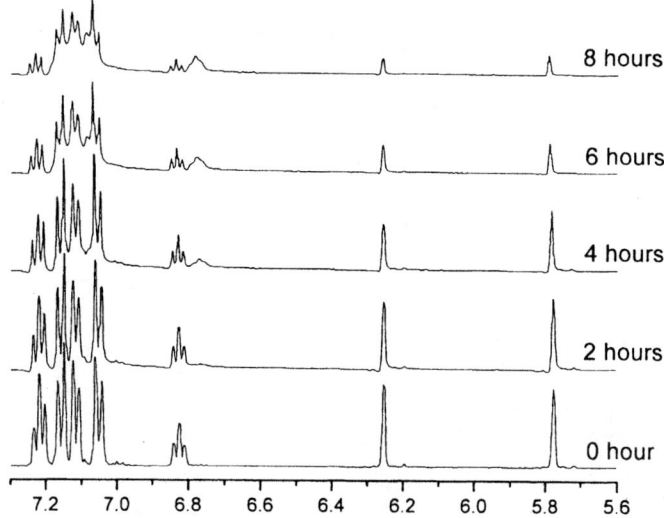

*Figure 6. Partial ^{1}H NMR spectra at different reaction times for the polymerization of monomer **1** in the absence of CTA.*

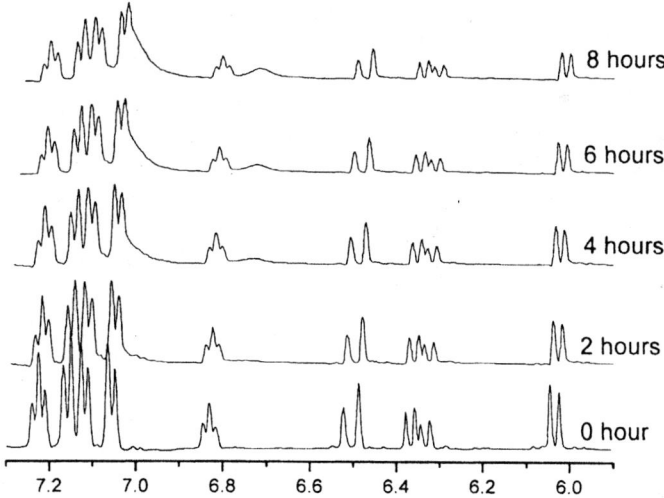

*Figure 7. Partial ^{1}H NMR spectra at different reaction times for the polymerization of monomer **2** in the absence of CTA.*

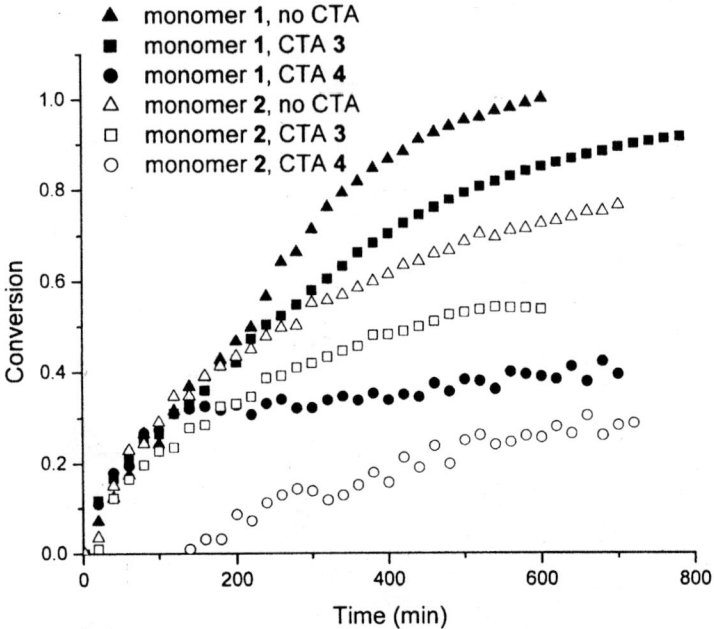

Figure 8. Time-conversion plots for the polymerization of monomers 1, 2 monitored by in-situ proton NMR.

and the stationary state kinetics should be observed shortly after the onset of the polymerization according to following equation $R_p = k_p(R_i/k_t)^{1/2}[M]$ (21). However significant retardation and inhibition have been observed in many cases (18,22,23). For example, polymerization was not observed for the first several hours in the RAFT polymerization of methyl acrylate in bulk using AIBN as an initiator and cumyl dithiobenzoate as a CTA (18). Cumyl dithiobenzoate-mediated bulk polymerization of styrene at 60 °C with AIBN was previously reported to have an induction period (24). The reasons for the retardation and inhibition are not well understood and under lively debate (2). The causes may be (1) slow fragmentation, (2) slow reinitiation, (3) cross-termination of propagating radicals and adduct radicals or (4) formation of adduct radicals of R· and S=C(Z)S-R. Additional studies outside the scope of the present work would be needed to define the cases of retardation and inhibition for these monomers.

The retardation for RAFT polymerization of the same monomer in the presence of CTA **4** is more significant than CTA **3**. The effects of Z and R in the chain transfer agent S=C(Z)S-R have been investigated recently (12,13). Since CTAs **3** and **4** have the same Z group (phenyl group), the addition and fragmentation in the chain equilibrium step is the same for both CTAs (Scheme

2). In the chain transfer step, the addition is mainly affected by the Z group, and CTAs **3** and **4** may have similar rate constants. The rate of fragmentation is mainly influenced by the steric and radical stability effects of the R group. Although the -C(CH$_3$)$_2$CN in CTA **3** may be more bulky and act as an excellent homolytic leaving group, the resonance stabilization also contributes to the facile fragmentation of the -CH(CN)Ph group in CTA **4**. From the above discussion, it is likely that the reinitiation capability of R· causes the difference in the retardation of polymerization mediated by CTAs **3** and **4**. The radical ·CH(CN)Ph may be more stable than the radical ·C(CH$_3$)$_2$CN, and thus slower in reinitiation.

Conclusions

In this article, 4-anilinophenyl methacrylate and 4-anilinophenyl acrylate were synthesized and polymerized by the RAFT polymerization for the first time. These monomers are models for oligoaniline side chain monomers which may have utility in anti-corrosion coatings and paints. The chain transfer agents 2-cyanopropyl dithiobenzoate and α-cyanobenzyl dithiobenzoate were investigated as controlling agents in the RAFT process. The rate of polymerization for the methacrylate monomer was higher than the acrylate monomer. Hydrogen atom transfer from the diphenylamine to propagating radicals may have been responsible for some loss of control over the polymerization. Overall, 2-cyanopropyl dithiobenzoate offered better control of the polymerization than α-cyanobenzyl dithiobenzoate and produced polymer from methacrylate monomer (**1**) at high conversion and low polydispersity. In-situ NMR showed that there was an induction period of two hours in the polymerization of the acrylate monomer in the presence of α-cyanobenzyl dithiobenzoate. Induction periods were not observed for other monomer/CTA combinations.

Acknowledgment

Financial support by the US NAVY, the Strategic Environmental Research and Development Program, and Rensselaer Polytechnic Institute is gratefully acknowledged.

References

1. Chiefari, J.; Chong, Y. K.; Ercole, F.; Krstina, J.; Jeffery, J.; Le, T. P. T.; Mayadunne, R. T. A.; Meijs, G. F.; Moad, C. L.; Moad, G.; Rizzardo, E.; Thang, S. H. *Macromolecules* **1998**, *31*, 5559.

2. Moad, G.; Chiefari, J.; Chong, Y. K.; Krstina, J.; Mayadunne, R. T. A.; Postma, A.; Rizzardo, E.; Thang, S. H. *Polym. Int.* **2000**, *49*, 993.
3. Kanagasabapathy, S.; Sudalai, A.; Benicewicz, B. C. *Macromol. Rapid Commun.* **2001**, *22*, 1076.
4. Kline, R. H.; Miller, J. P. *Rubber Chem. Technol.* **1973**, *46*, 96.
5. Parker, D. K.; Schulz, G. O. *Rubber Chem. Technol.* **1989**, *62*, 732.
6. Kato, M.; Takemoto, Y.; Nakano, Y.; Yamazaki, M. *J. Polym. Sci., Polym. Chem. Ed.* **1975**, *13*, 1901.
7. Tanaka, Y.; Noguchi, K.; Shibata, T.; Okada, A. *Macromolecules* **1978**, *11*, 1017.
8. Chen, R.; Benicewicz, B. C. *ACS Symp. Ser.* **2003**, *843*, 126.
9. Chen, R.; Benicewicz, B. C. *Macromolecules* **2003**, *36*, 6333.
10. Le, T. P.; Moad, G.; Rizzardo, E.; Thang, S. H. *PCT Int. Pat. Appl.* WO9801478, **1998**.
11. Li, C.; Benicewicz, B. C. *J. Polym. Sci., Part A: Polym. Chem.* **2005**, *43*, 1535.
12. Chiefari, J.; Mayadunne, R. T. A.; Moad, C. L.; Moad, G.; Rizzardo, E.; Postma, A.; Skidmore, M. A.; Thang, S. H. *Macromolecules* **2003**, *36*, 2273.
13. Chong, Y. K.; Krstina, J.; Le, T. P. T.; Moad, G.; Postma, A.; Rizzardo, E.; Thang, S. H. *Macromolecules* **2003**, *36*, 2256.
14. Benaglia, M.; Rizzardo, E.; Alberti, A.; Guerra, M. *Macromolecules* **2005**, *38*, 3129.
15. Rizzardo, E.; Chiefari, J.; Mayadunne, R. T. A.; Moad, G.; Thang, S. H. *ACS Symp. Ser.* **2000**, *768*, 278.
16. Tudos, F. In *Polymeric Materials Encyclopedia;* Salamone, J. C., Ed.; CRC Press: Boca Raton, FL, 1996; Vol. 5, p 3238.
17. Mitsukami, Y.; Donovan, M. S.; Lowe, A. B.; McCormick, C. L. *Macromolecules* **2001**, *34*, 2248.
18. Perrier, S.; Barner-Kowollik, C.; Quinn, J. F.; Vana, P.; Davis, T. P. *Macromolecules* **2002**, *35*, 8300.
19. Vosloo, J. J.; Tonge, M. P.; Fellows, C. M.; D'Agosto, F.; Sanderson, R. D.; Gilbert, R. G. *Macromolecules* **2004**, *37*, 2371.
20. Favier, A.; Barner-Kowollik, C.; Davis, T. P.; Stenzel, M. H. *Macromol. Chem. Phys.* **2004**, *205*, 925.
21. Goto, A.; Sato, K.; Tsujii, Y.; Fukuda, T.; Moad, G.; Rizzardo, E.; Thang, S. H. *Macromolecules* **2001**, *34*, 402.
22. Barner-Kowollik, C.; Vana, P.; Quinn, J. F.; Davis, T. P. *J. Polym. Sci., Part A: Polym. Chem.* **2002**, *40*, 1058.
23. Vana, P.; Davis, T. P.; Barner-Kowollik, C. *Macromol. Theory Simul.* **2002**, *11*, 823.
24. Barner-Kowollik, C.; Quinn, J. F.; Morsley, D. R.; Davis, T. P. *J. Polym. Sci., Part A: Polym. Chem.* **2001**, *39*, 1353.

Figure 4.1. TCP on Al 2024-T3; Time = 0 hours

Figure 4.2. TCP on Al 2024-T3; Time = 336 hours

Figure 4.3. TCP on Al 2024-T; Time = 1000 hours

Figure 4.4. TCP on Al 2024-T; Time = 1500 hours

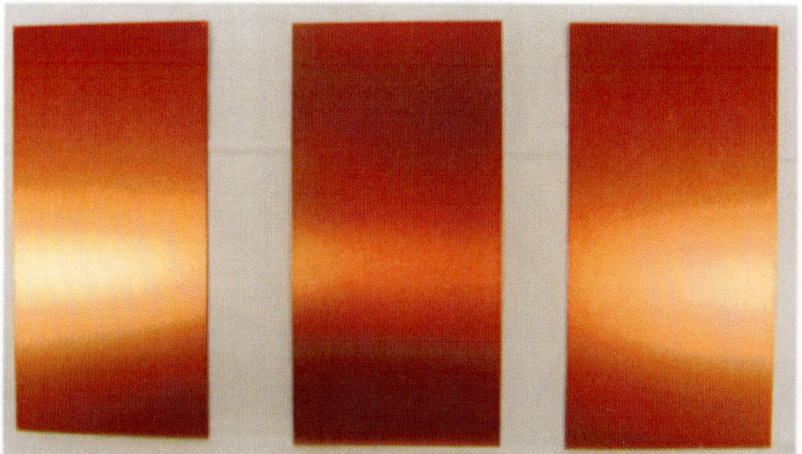

Figure 4.5. BAM-PPV Coated Al2024-T3, Time = 0 hours

Figure 4.6. BAM-PPV Coated Al 2024-T3, T = 336 hours

Figure 4.7. BAM-PPV Coated Al 2024-T3; T = 840 hours

Figure 4.8. BAM-PPV Coated Al 2024-T3, Time = 1304 hours

Figure 4.9. MEH-PPV Coated Al 2024-T3; Time = 336 hours

Figure 4.10. MEH-PPV Coated onto Al 2024-T3; Time = 336 hours

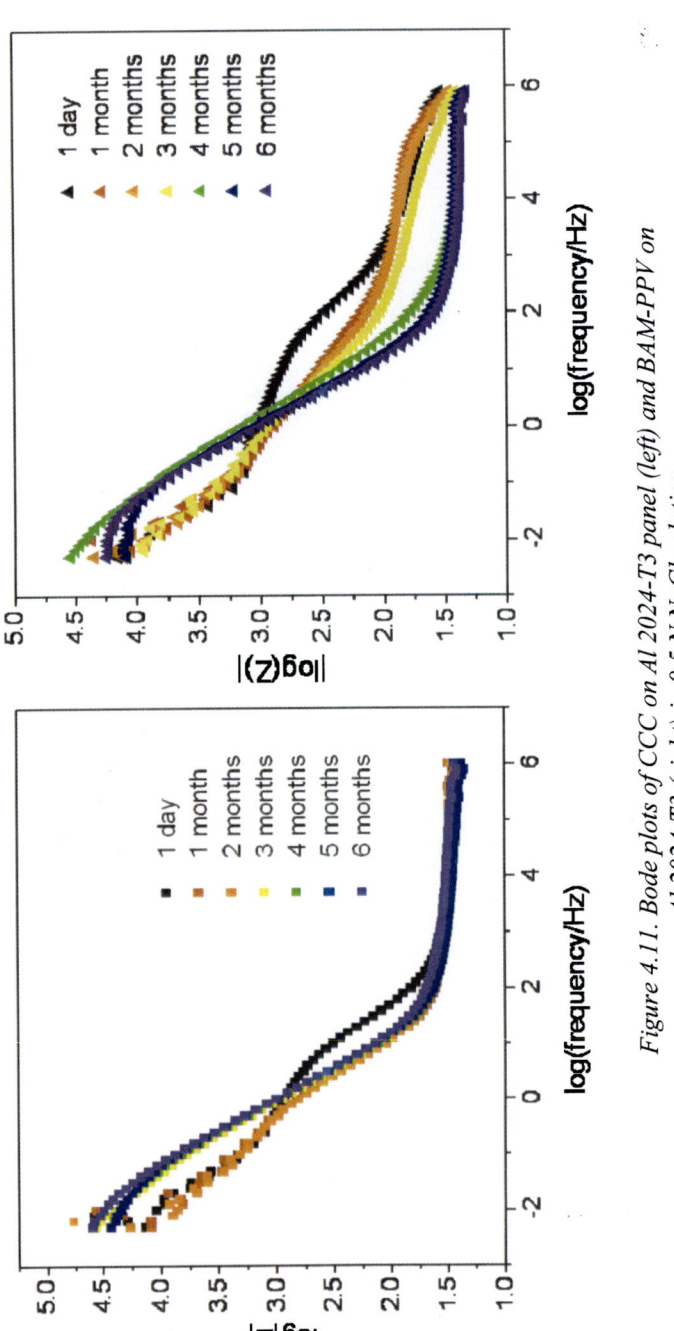

Figure 4.11. Bode plots of CCC on Al 2024-T3 panel (left) and BAM-PPV on Al 2024-T3 (right) in 0.5 N NaCl solution.

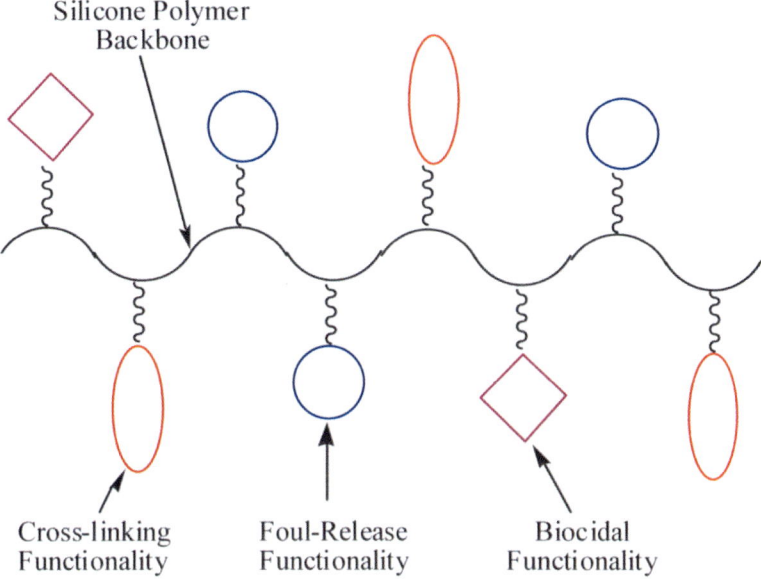

Figure 6.5. NDSU Multi-functional Silicone Resin

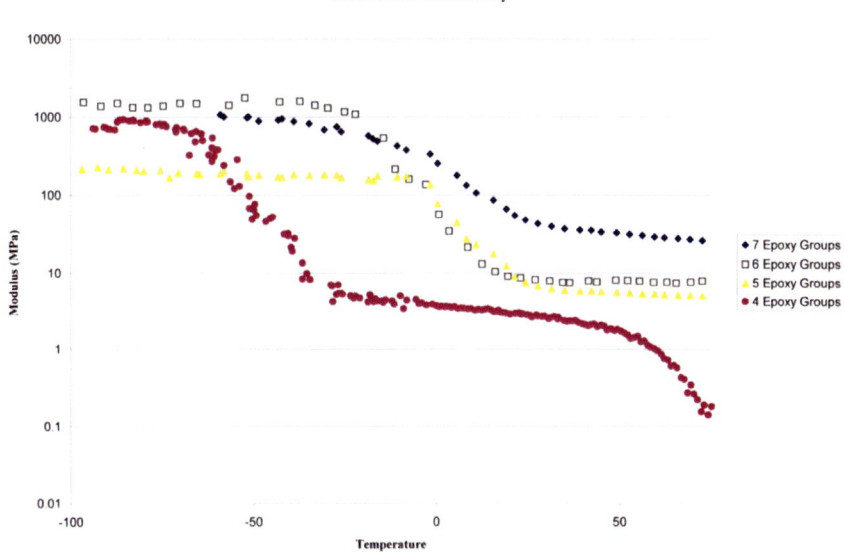

Figure 6.7. Variation of the Bulk Modulus of Elasticity with the Cross-Linking Density for NDSU Silicone Coatings

A.)

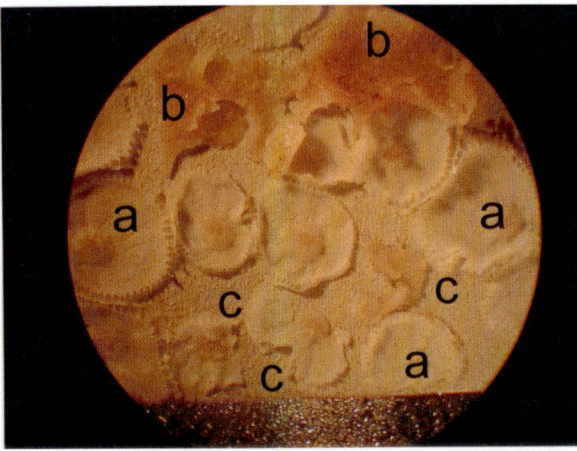

a) Barnacle bases
b) Experimental coating
c) Undercoating

B.)

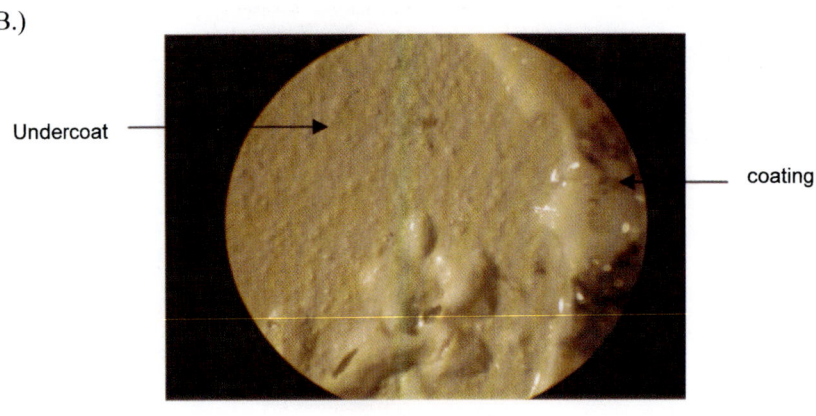

Figure 6.10. Barnacle Cutting and Corrosion in the Coatings.

Plate 7.1. See full caption on page 97 in chapter 7.

Figure 15.1. Crosslinked vs. non-crosslinked films

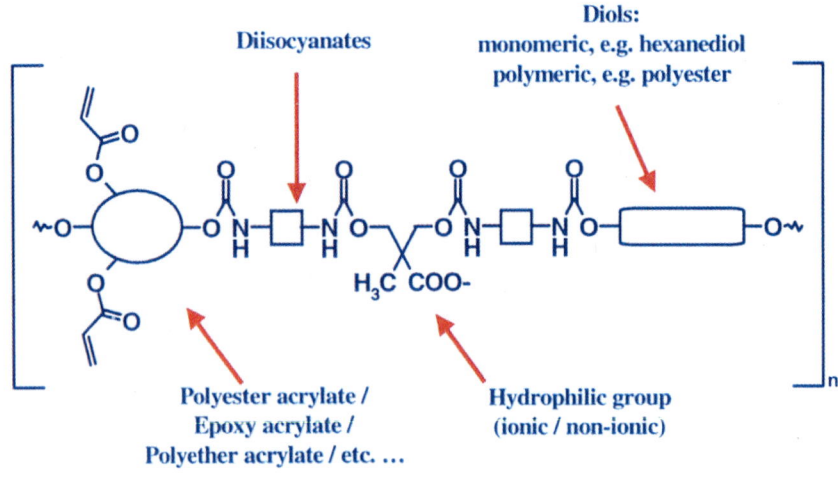

Figure 15.2. Generalized structure of UV-curable polyurethane dispersion

Chapter 6

New Technologies for the Analysis of Marine Coatings

Thomas E. Ready[1,*], Johnson Thomas[1], Seok-bong Choi[1], and Philip Boudjouk[2,*]

[1]Center for Nanoscale Science and Engineering, North Dakota State University, Fargo, ND 58102
[2]Department of Chemistry, North Dakota State University, Fargo, ND 58105

High-throughput methods for the fouling/fouling-release performance evaluation of marine coatings are desired to reduce the analysis cost, the amount of coating material needed, and the analysis time required. Combinatorial tools, methods, work-flows acquired and developed by North Dakota State Unversity to remedy these needs are described. Coatings based on silicones with grafted biocide (Triclosan) are described. The relationship between dynamic mechanical thermal analysis (DMTA) and settlement/release of barnacles on these coatings suggest that a bulk modulus of elasticity of approximately 0.1 – 10 MPa is optimum for coating resistance to fouling & fouling adhesion.

© 2007 American Chemical Society

Introduction

Biological fouling of ship hulls has serious consequences for U.S. Naval vessel performance and mission capabilities. Fouling increases ship fuel consumption by as much as 30 – 40% and necessitates extensive maintenance, which raises the overall costs of operation[1]. Fouling reduces ship speed, maneuverability, and range, which impede mission performance[2].

On another level, attachment of regionally specific aquatic nuisance species on ships that traverse the world can lead to the unwanted invasion and infestation of these entities to non-indigenous harbors[3]. In some instances, this can have severely adverse effects on local marine-based economies on the same order of magnitude as chemical pollution.

The fouling challenge to ship hulls is formidable as there are thousands of possible marine fouling entities which interact with the hull surface, and each other, in a complex and sometimes synergistic manner. There are multiple species within each genus of fouling entity. There is regional specificity to fouling entities. Each fouling entity may have completely different criteria for what it considers to be favorable conditions for settlement on a surface.

There are different levels of fouling which occur; microfouling (bacterial biofilms, slime, Ulva, bryazoans, diatoms, etc.) and macrofouling (e.g. barnacles, muscles, tubeworms, etc.). To a large extent, it is thought that fouling is sequentially hierachical, with microfouling occurring first, producing a surface suitable for subsequent macrofoulers. However, there is data to suggest that this might not always be the case. Although macro-fouling has been the primary area of concern due to the obvious increase in surface area & roughness it produces, there is also between a 5% - 80% increase in hull friction due to the slime layer alone[2b].

Traditionally, two parallel lines of coatings research and development aimed at reducing bio-fouling have predominated: biocide containing coatings and low surface energy, "non-stick", foul-release coatings. Each of these approaches has produced elements of success but continue to have serious problem areas, which demand solutions. Interestingly, while aimed at solving the same problem, these two lines of R & D have had a minimum of overlap.

Biocidal coatings contain metal or organic compounds toxic to fouling entities and hence, deter settlement on the hull surface. Since 1974, the most commonly used biocides in marine coatings have been organo-tin compounds and to a lesser degree, copper compounds. Currently, these agents are incorporated in hull coatings on the great preponderance of U.S. Navy ships and 70% of all maritime shipping[4]. Although highly effective at reducing hull fouling, these biocidal agents have been linked to environmental problems. While moored in harbors, paint chips and leaching have led to sediment accumulations of the toxins resulting in harm or destruction of non-targeted sea life (i.e. oysters)[3,5]. For this reason, most developed countries have already banned the use of organotin biocidal coatings on non-aluminum maritime craft smaller than 25 meters. The International Maritime Organization has adopted a ban on new applications of these paints to any ship starting in 2003, and their presence will be banned from all ships by the year 2008[3,6]. The average effective lifetime for tin containing ship coatings is 6 years while that for non-tin coatings is less than 3 years[3]. After 2.5 years, hull cleaning for tin-free painted ships has been required every 6 months at an annual cost of $15,000 to $20,000 per ship[3]. Hence, the development of an effective, non-toxic alternative to biocidal coatings in urgently needed.

Fouling-release coatings appear to be the leading non-toxic alternative to biocide containing coatings. Fouling-release coatings do not inherently protect against fouling settlement, but as the ship moves through the water the shear forces on the hull allow for some degree of fouling removal and self-cleaning. The most common formulations center on highly fluorinated hydrocarbon or silicone polymers with the silicone based coatings being slightly more effective at foul release[7]. Although these coatings do ease fouling release and removal, their inadequate adhesion to the substrate, durability, and high cost has limited their deployment in the field[1b].

An added technical challenge is that the U.S. Navy has called for coating lifetimes to be extended from the current 6 years to 12 years.

New Technologies for Coating Development

To address these technical challenges, The Center for Nanoscale Science and Engineering (CNSE) at North Dakota State University (NDSU), under the auspices of the U.S. Office of Naval Research (ONR), is adapting the evolving combinatorial technology developed by Symyx Technologies Inc. ® for rapid discovery, analysis, and evaluation of polymeric materials to meet the Navy's needs for advanced fouling and corrosion control coatings. These tools allow screening of coatings candidates on a scale that is hundreds of times faster than current technology. Our laboratory is using these tools to develop novel environmentally compliant, durable, effective coatings for the surface preservation of ships against fouling and corrosion.

There are different strategies which can be applied toward the utilization of combinatorial technology. Our current strategy is to explore the parameter variations of lead coating concepts, and down-select from this array of coatings to promising candidates (Figure 1).

The best, most effective testing environment for marine coatings is the ocean (or waterway of interest). However, ocean immersion testing of coatings can be lengthy and costly. The goal of the NDSU-CNSE's program is to establish a set of laboratory based screening protocols which measure structural & physical characteristics of coating variants and are tied to biological challenges which emulate the ocean environment. Establishment of such

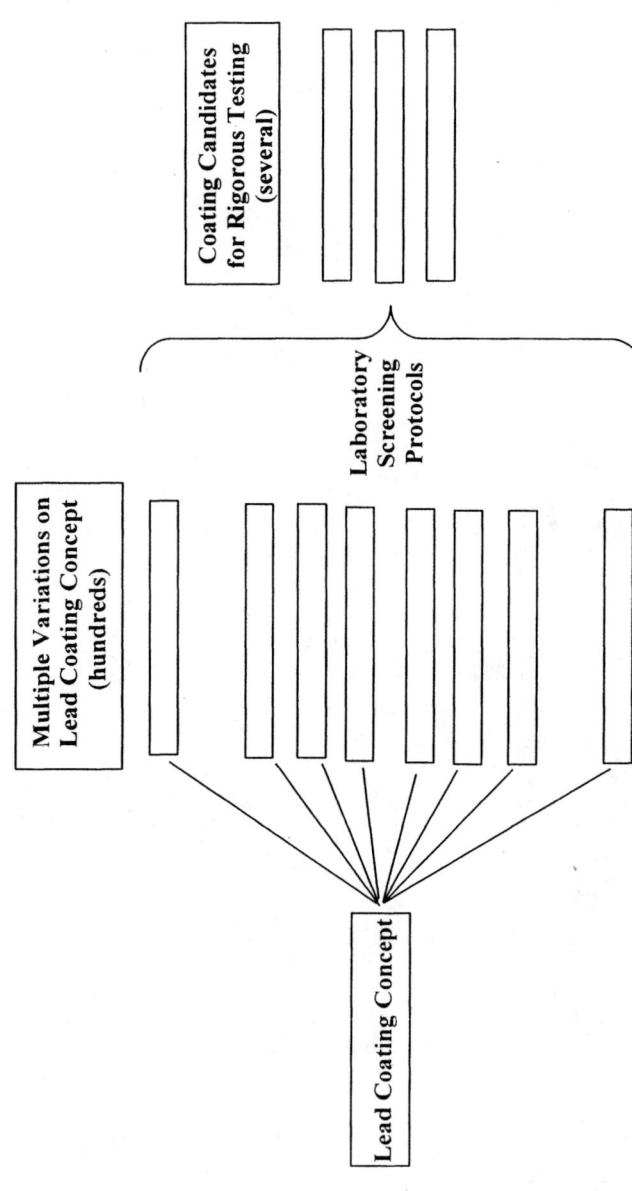

Figure 1. Strategy for Utilizing Combinatorial Tools and Screening Protocols to Down-select Coating Candidates

protocols could predict coating performance and rapidly screen-out poorly performing coatings (Figure 2), thus greatly reducing both the time and costs associated with coating development. The resulting promising coating candidates would then be sent to ocean sites for rigorous testing in a natural environment (Figure 2).

Four protocols have been selected as good candidates for screening marine coating candidates. For three of the protocols, we are using combinatorial tools designed and manufactured by Symyx® to assess surface energy, bulk modulus of elasticity, and "pseudo-barnacle" adhesion of coatings.

Surface Energy

Surface energy (usually measured via fluid contact angle with the surface) was first established as an important factor for fouling-release by Baier[8]. However, using this measurement in isolation to assess fouling performance yields results which must be interpreted in context, because some organisms prefer to settle on hydrophobic surfaces, while others prefer hydrophilic ones[9]. In addition, some organisms can alter the phenotype of their secreted adhesive based on the surface energy of the substrate. Importantly, contact-angle hysteresis measurements have been shown to be more indicative of fouling-release performance than static contact-angle measurements[10]. The Symyx® contact angle measurement tool performs automated static contact angle and contact angle hysteresis measurements on coating arrays.

The Bulk Modulus of Elasticity

It has been demonstrated that bio-adhesion to surfaces is directly proportional to the bulk modulus of elasticity[11a]. Coatings possessing a low elastic modulus facilitate marine adhesive failure via a peel mechanism which allows the adhesive to slip on the coating surface, lowering the force required for fouling-release[1,2a,11]. NDSU is using the Symyx® parallel dynamic mechanical thermal analyzer (DMTA) tool to perform automated measurements of Tg, and bulk modulus of elasticity of coatings. This unit can analyze 96 samples simultaneously.

Pseudo-Barnacle Testing

Swain et al. demonstrated that cylinders attached to coatings via a drop of epoxy glue, "pseudo-barnacles", seem to correlate well with barnacle adhesion to elastomeric coatings[12]. Several research groups[12] have utilized this test with success including Kohl & Singer[12b] who showed that the "pseudo-barnacle" test

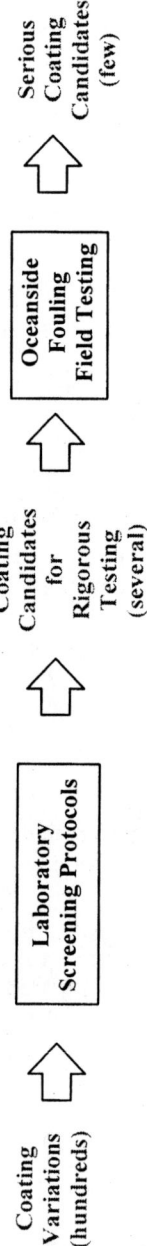

Figure 2. *A Prospective Scheme for Down-selecting Marine Coatings*

follows Kendall and Griffith fracture mechanics. NDSU had Symyx® custom-build an automated version of the "pseudo-barnacle" test which accommodates coating arrays.

Organism Bioassays

As a fourth testing protocol, NDSU, in conjunction with other laboratories, is engaged in the development of high-throughput laboratory bioassays to assess the settlement and removal propensities of several organisms on coating surfaces. The goal is to find correlations between these laboratory bioassays and ocean-side settlement and removal of macrofoulers, so that the laboratory assays can be used as initial biological screens for coating performance. Several types of organisms are being studied for this effort including *Hydroides elegans* (tubeworms)[13], (barnacle) cyprids[14], *Ulva* (*Enteromorpha*)[15], Diatoms[16], and bacteria[17] in the hopes that one or more of these assays will provide a first approximation of coating ant-fouling / fouling-release capability.

To be effective as a screen, an individual protocol or combination of protocols, should correlate to the fouling and/or fouling-release seen at ocean-side test sites. As of 2005, NDSU-CNSE is still in the process of method development and validation for these screening protocols (Figure 3).

Our laboratory is also developing new classes of polymers including silicones to be utilized as resins and surfactants in anti-fouling coating formulations. In contrast to historical R&D approaches, our initial approach is to design coatings that possess both non-leaching/non-metallic biocidal and foul-release elements. Leveraging available chemistries and our own experience in hydrosilation[18] & dehydrogenative coupling methods[19] for producing poly(dialkylsiloxane) materials (Fig. 4a) and poly(alkylalkoxysiloxanes) (Fig. 4b) respectively, multi-functional silicones are being synthesized which incorporate fouling-release, non-leaching biocidal, and crosslinking moieties (Fig. 5). This unified approach towards durable anti-fouling coatings is novel and untried. Both poly(dialkylsiloxane) and poly(alkylalkoxysiloxane) materials are being investigated in fouling-release coatings, although data suggest that alkoxy-modified silicones are less effective than their alkyl-modified counterparts for this application[20].

A typical example to demonstrate the types of materials being made and analysis being done is described below.

Triclosan (5-chloro-2-(2, 4-dichlorophenoxy) phenol), a broad spectrum antibacterial / antimicrobial agent, was modified and covalently attached to cyclic methylhydrosiloxane, methylhydrosiloxane-dimethylsiloxane copolymers and polyhydromethylsiloxane backbones through short and long alkyl chains to explore the possibility of using this as an antifouling agent in marine coatings without its release into the environment.

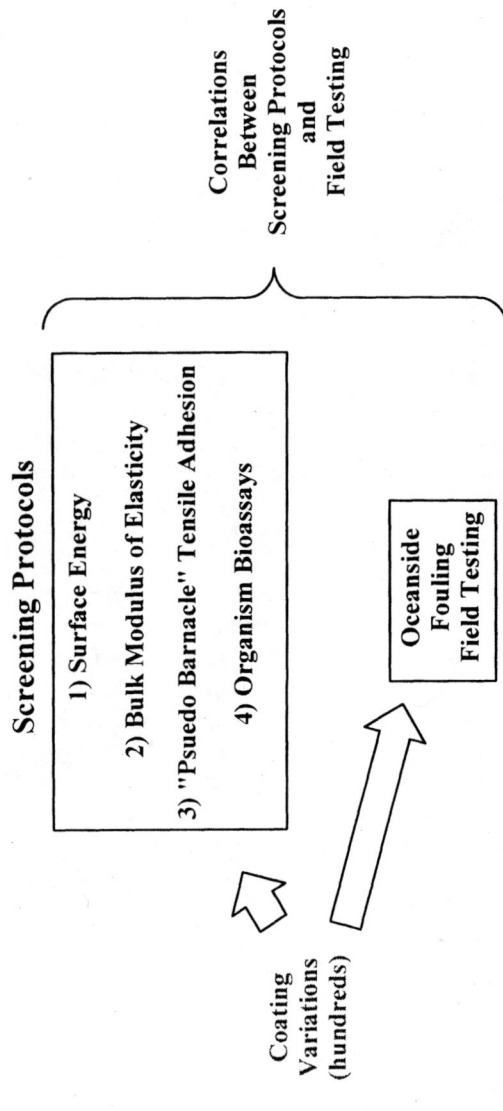

Figure 3. *Establishment of Correlations Between Screening Protocols and Oceanside Fouling*

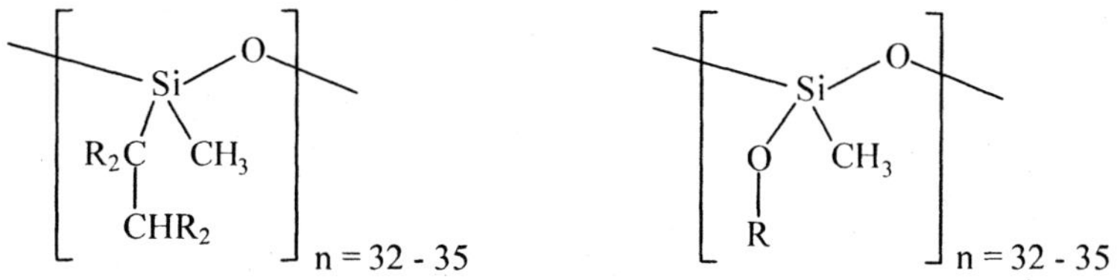

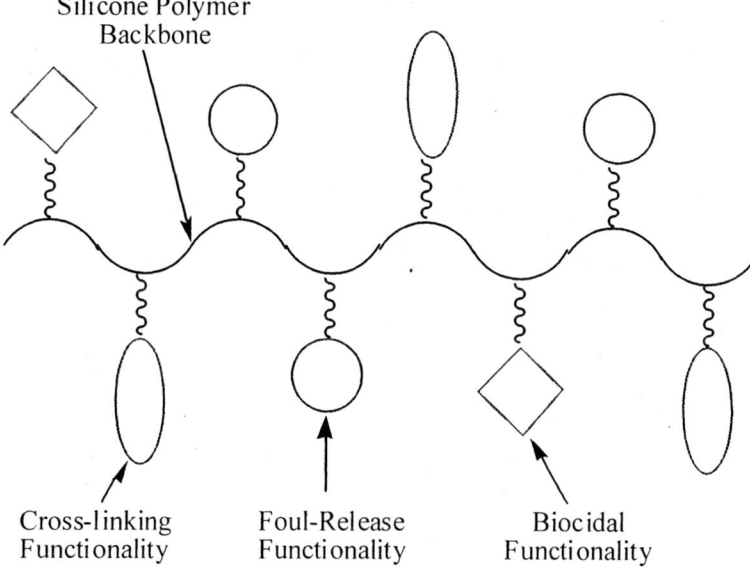

Poly(dialkylsiloxane) Poly(alkylalkoxysiloxane)

(a) (b)

Figure 4. Generic Types of Modified Silicones

*Figure 5. NDSU Multi-functional Silicone Resin
(See page 7 of color inserts.)*

First, Triclosan was modified by alkenyl bromide to facilitate hydrosilation. Allyl bromide (short chain) and 1-undecenyl bromide (long chain) were used to incorporate the biocide into siloxane. The long alkyl chain was selected to provide more flexibility so that the biocide will be more available on the top layer of the coating providing an antifouling surface.

Triclosan-alkenyl derivatives were incorporated onto the siloxane backbone by hydrosilation using Karstedt's catalyst[21]. The modification of the biocide and its incorporation into linear polydimethylsiloxane-co-methylhydrosiloxane and its further modification by allyl glycidyl ether are given in Figure 6.

Different cross linking moieties were used to tune the bulk properties of the coatings. In one set of coatings, the biocide incorporated resin (I) (Figure 6) with residual Si-H groups was used to prepare coatings using vinyl terminated PDMS or polybutadiene as the crosslinker. In another set of experiments, biocide incorporated siloxane resins were further modified by the incorporation of allyl glycidyl ether (II) (Figure 6) to produce another type crosslinking moiety. These glycidyl ether modified resins were crosslinked with bisamine.

Materials

Triclosan was purchased from Lancaster Chemicals (Windham, NH). 1, 3, 5, 7- Tetramethylcyclotetrasiloxane (D4), 50-55% methylhydrosiloxane-dimethylsiloxane copolymer (HMS-501), polyhydromethylsiloxane (HMS 992), methcrylated poly(dimethylsiloxane) (MCR-M11) and vinyldimethylsiloxy terminated poly(dimethylsiloxane) (V 05) were purchased from Gelest (Tullytown, PA). Karstedt's catalyst (platinum (0) -1, 3-divinyl-1,1,3,3-tetramethyl disiloxane complex), 1, 3-bis-aminomethyl-cyclohexane, allyl bromide, 1-undecenyl bromide, allyl glycidyl ether, and polybutadiene were obtained from Aldrich (Milwaukee, WI). All other reagents were obtained from Lancaster. These chemicals were used as received.

Modification of Triclosan by Bromoalkene

i) In a typical reaction, allyl bromide (5 g, 41 mmol) was added to a solution of Triclosan (10 g, 34 mmol) and potassium carbonate (6 g, 41 mmol) in 50 ml N,N-Dimethylformamide (DMF). The mixture was stirred at room temperature overnight. After the reaction, solvent was removed by evaporation and the residue dissolved in 50 ml hexane and washed with water four times (4 x 50 ml). The organic layer was separated, dried with anhydrous $MgSO_4$ and evaporated to yield allyl functionalized biocide as a white solid. (10.2 g, yield= 90%). ^1H NMR ($CDCl_3$): δ 4.58 (s, 2H), 5.2(q, 2H), 5.8(m, 1H), 6.67- 7.4 (m, 6H). ^{13}C NMR ($CDCl_3$): δ 70.09, 115.7, 117.48, 119.01, 120.42, 121.59, 122.00, 127.88, 128.43, 130.46, 132.31, 143.08, 151.34, 152.91.

ii) Triclosan was also modified by undecenyl bromide and methacryloyl chloride to obtain a long chain and ester linkage respectively.

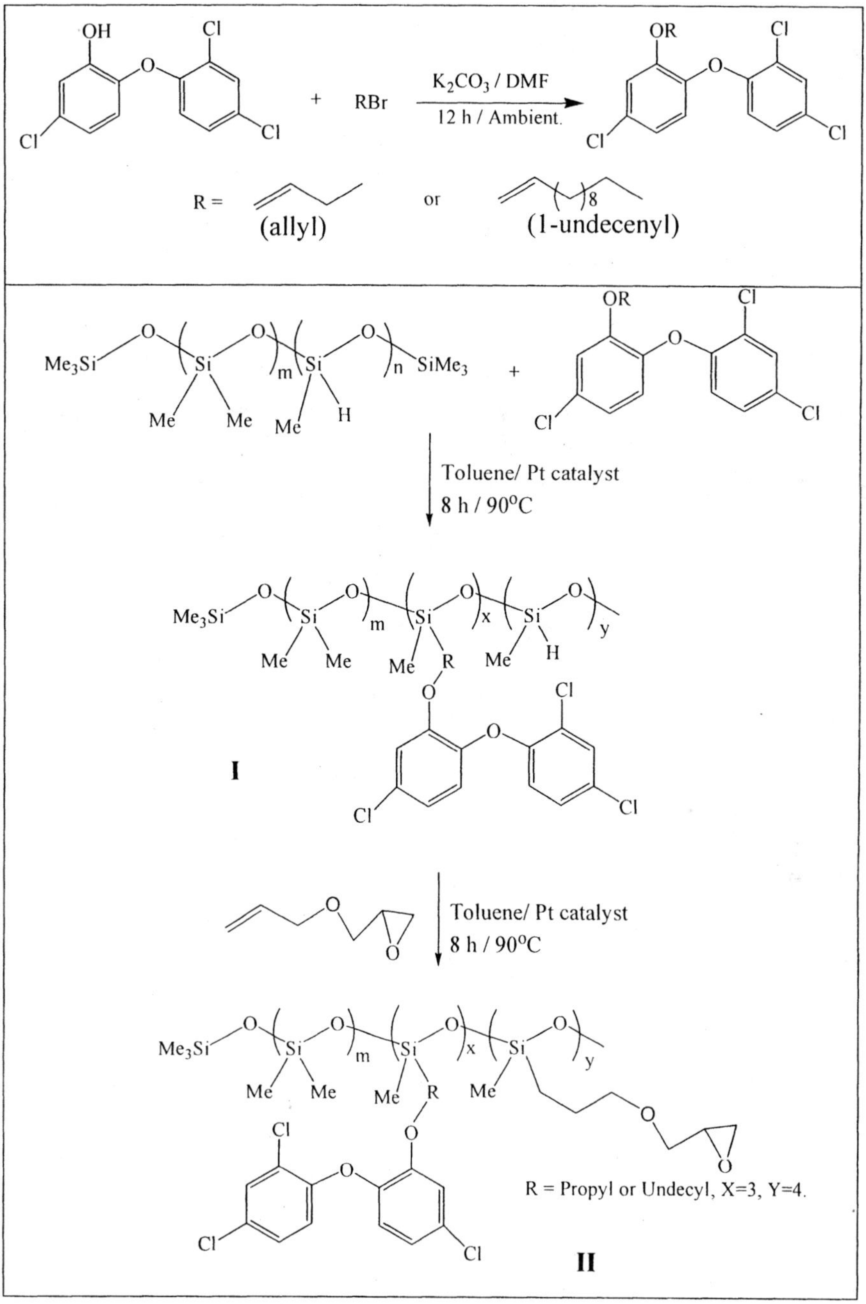

Figure 6. Overall Synthetic Scheme Showing the Modification of Biocide and Incorporation of Modified Biocide and Glycidyl Ether onto Siloxane.

Simultaneous Incorporation of Alkene Modified Triclosan and Allyl glycidyl ether onto Siloxane

i) Allyl functionalized Triclosan (4 g, 12.5 mmol) and allyl glycidyl ether (4.45 ml, 37.5 mmol) were added to a solution of D4 (3 g, 12.5 mmol) in 20ml of dry toluene and 2-3 drops of Karstedt's catalyst was added to the mixture and the reaction continued for 8 h at 90° C. After the reaction, the mixture was passed through neutral alumina column and solvent removed by evaporation to yield only pure product as a colourless viscous liquid (10.8 g, yield = 95%).
^{1}H NMR (CDCl$_3$): δ 0.09, 0.6, 1.53, 2.57, 2.75, 3.10, 3.36, 3.41, 3.60, 3.81, 6.7-7.45. ^{13}C NMR (CDCl$_3$): δ -0.46, 13.28, 23.35, 44.54, 51.06, 71.63, 74.20, 117.48, 119.01, 120.42, 121.59, 122.00, 127.88, 128.43, 130.46, 132.31, 143.08, 151.34, 152.91.

ii) Other functionalities were also grafted on to siloxane backbone by the same procedure.

Dynamic Mechanical Thermal Analysis

Bulk properties of the coatings were studied using a Symyx Technologies, Inc. parallel dynamic mechanical thermal analyzer (parallel DMTA^{TM}). This is a fully parallel instrument capable of making 96 simultaneous measurements. Modulus measurements were performed by measuring the force needed to deform a thin polyimide substrate by a given amount with and without a sample present. The method and theory of this instrument is documented by Kossuth *et al* [22]. Samples were deposited on standard DMTA plates supplied by Symyx Technologies (Santa Clara, CA). Coating samples were prepared by depositing 20-25 µl of the resin mixed with the appropriate crosslinker on to regions 5mm diameter on the DMTA plate. Four replicates were deposited on the plate for each sample. After the samples were deposited, the DMTA plate was kept in an oven and heated at 60° C for 48 h to completely cure all the coating samples. After curing, the thickness of each sample was determined using a laser profilometer. The height profile of each sample was recorded and fit to a square cross section and the height of the fitted profile was taken as the thickness of the film. The measured thicknesses range from 250 µm -300 µm. After the thickness measurements, the sample plate was introduced into the parallel DMTA and measurements were taken. These experiments were carried out in the range -125° C to 150° C with a ramp rate 5°C/min. at 10 Hz.

Preparation of Coatings

Table I shows a list of coatings used for this study. They were prepared by two methods. Glycidyl ether functionalized resins were mixed with 1,3-cyclohexane-bis (methylamine) (1epoxy equiv. /1amine equiv.) and applied on

Table I. List of Coatings and their Chemical Characteristics

Coating No.	Matrix	Biocide	Crosslinking group	MCR-M11
TJ01-A	HMS-501	1 Allyl	7 Epoxy	-
TJ01-B	HMS-501	2 Allyl	6 Epoxy	-
TJ01-C	HMS-501	3 Allyl	5 Epoxy	-
TJ01-D	HMS-501	4 Allyl	4 Epoxy	-
TJ02	HMS-992	8 Undecyl	6 Epoxy	-
TJ03	HMS-992	8 Allyl	Epoxy	2
TJ04	HMS-992	8 Undecyl	6 -Si-H	-
TJ05	HMS-992	8 Undecenyl	6 -Si-H	2
TJ07	HMS-992	8 Undecenyl	6 -Si-H	2
TJ08	HMS-501	4 Methacrylate	4 Epoxy	-
TJ09	HMS-992	8 Allyl	6 Epoxy	-
TJ10	HMS-992	8 Undecenyl	10 -Si-H	-
TJ11	HMS-992	8 Undecenyl	10 -Si-H	2

HMS-501- 50-55% methylhydrosiloxane-dimethylsiloxane copolymer (M_n = 900-1200), HMS-992- polyhydromethylsiloxane (M_n = 2000), MCR-M11-monomethacryloxypropyl terminated polydimethyl siloxane (M_n = 800-1000).

aluminum panels. The coatings were generally touch dry in 3 h and were further cured at 60° C for 24 h. Resins having residual Si-H groups were cured by hydrosilation using divinyl terminated poly(dimethylsiloxane) and polybutadiene (1.2 equiv. SiH/ 1 equiv. double bond). These coatings were generally touch dry in 6 h and were further cured at 60°C for 48 h. The dry film thicknesses of the coatings ranged from 250 μm to 300 μm.

Panel Preparation and Deployment Site

Coatings were applied on 4" x 8" marine grade aluminum panels. The panels were cleaned and roughed with sand paper (400 grits) followed by application of the anti corrosive epoxy primer coating, Macropoxy 646, from Sherwin Williams using airless spray equipment. All of the experimental coatings were applied on top of the epoxy primer layer without any tie-coat, using a draw down bar.

Two commercially available coatings were utilized as controls to compare the antifouling/fouling release properties of the experimental coatings in testing; Intersleek 425 (fouling-release coating from International Paints, U.K.) and Interspeed BRA 642 (copper ablative anti-fouling coating from International Paints, U.K.).

Static immersion tests were carried out by the Florida Institute of Technology in the Indian River Lagoon (Melbourne, FL). Before immersion, all the panels were subjected to 15 day pre-leaching in artificial sea water (Aquarium Systems Inc., Sarrebourg, France) under static conditions, to remove any unreacted materials including unreacted biocide from the coatings. The absence of unreacted biocide was confirmed by HPLC analysis. Water was changed after each five day period during pre-leaching.

All panels were held one meter below the water surface inside ½" galvanized mesh cages. This was done to protect the panels from fish and other aquatic organisms other than the foulants. Four replicates of each coating were used for the immersion studies and the fouling ratings of the coatings were calculated based on ASTM fouling rating. The static immersion studies of these coatings were carried out between April 26, 2004 and July 16, 2004. Two inspections were done on these coatings, one after 44 days and another one after 80 days of immersion. Fouling rating (FR) is defined by FR= 100- the sum of fouling % cover (not including slime), i.e., FR 100 is a surface free of macro fouling.

Results and Discussion

We have previously shown[23] that the bulk moduli of elasticity of silicone coatings are dependent on the polymer backbones and the types of cross links.

Coatings using cyclic siloxane resins show higher modulus and Tg values than coatings derived from linear siloxanes suggesting higher cross link density for these coatings. Amine cured coatings show higher values of Tg than coatings crosslinked via hydrosilation, indicating low chain flexibility. Modulii of elasticity and Tg were affected by the chain length of linking the biocide to siloxane; longer chains led to lower modulus and lower Tg. Marine fouling had a strong dependence on the bulk modulus of elasticity of the materials with fouling resistance being optimized for coatings with a modulus in the range of 0.1- 10 Mpa. It was postulated that for high modulus coatings, biocide may be trapped and unavailable to prevent macro- fouling in the highly cross linked polymer matrix, whereas for extremely low modulus coatings, barnacles appear to cut through the coatings and grow on the anti-corrosive primer coating.

Effect of Modulus and Biocide loading on Fouling Rating

As the fouling rating was found to be strongly dependent on the modulus of the coatings, we synthesized a set of coatings where the modulus as well as biocide loading in the coatings was systematically varied (Table I). As shown in Table I, the number of cross linking groups in the coatings from TJ01-A to TJ01-D decreases and this leads to a corresponding decrease in the modulus of these coatings (Figure 7). A few coatings were also prepared by grafting silicone fluid on to the siloxane backbone to further soften the coating. The fouling rating of these coatings after 44 days and 80 days static immersion studies at FIT are given in Figure 8 and Figure 9, respectively. It is interesting to note that at the first inspection, the fouling rating was strongly dependent on the modulus of the coatings. When the modulus decreased within the same set of coatings (TJ01-A to TJ01-D), the fouling rating showed a marked increase. However, increase in biocide loading or grafting of silicone fluid did not change the FR significantly. Statistically the coatings have similar FR.

Second inspection of the coatings after 80 days showed that all the coatings have lower FR except for TJ05 which contained silicone fluid grafted on the silicone backbone. The reason for the low fouling rating lies in the fact that all the coatings are soft silicone coatings and barnacles cut through the coatings and grow on the epoxy primer applied to the aluminum panels (Figure 10A). Added to this is the poor adhesion between the top coating and the primer which leads to blistering and delamination from the primer (Figure 10B). The combined biological corrosion and the poor adhesion between the primer and the top coating affect the durability of the coatings and they tend to foul after a period of eight weeks.

To summarize, the biocide Triclosan has been chemically tethered to silicone coatings for antifouling / fouling release applications. Coatings were characterized by dynamic mechanical thermal analysis and this study shows that the bulk properties of the coatings can be significantly varied by the nature of

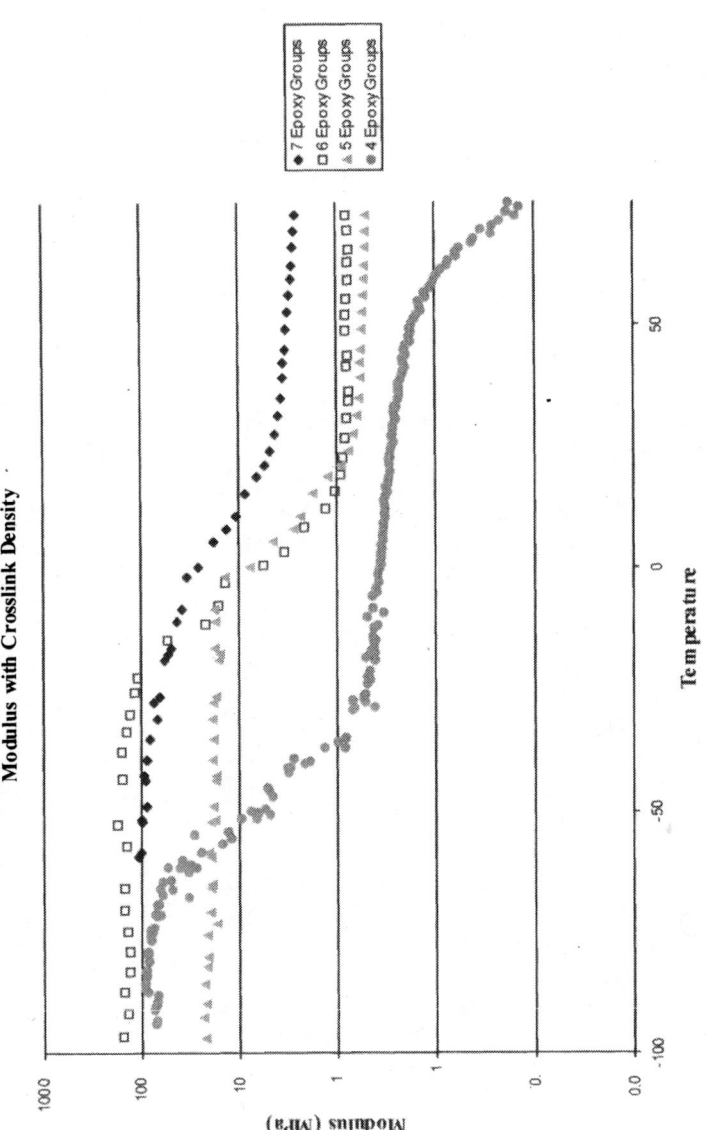

Figure 7. Variation of the Bulk Modulus of Elasticity with the Cross-Linking Density for NDSU Silicone Coatings (See page 7 of color inserts.)

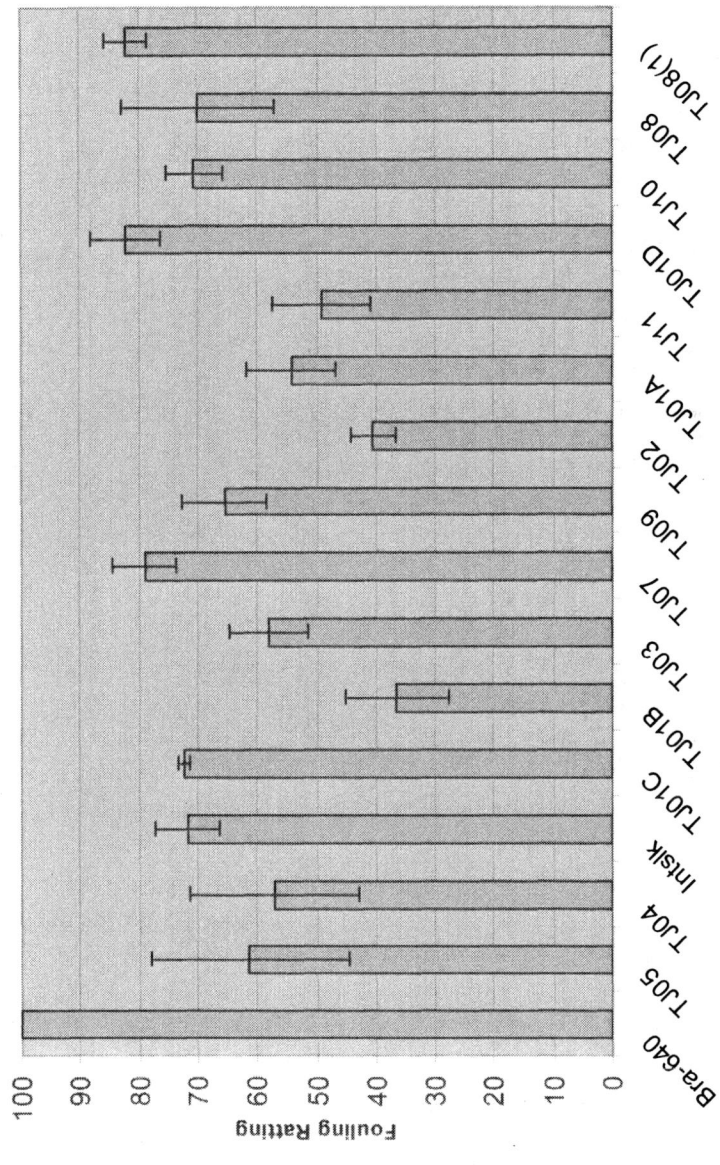

Figure 8. Average Fouling Rating of the Coatings after 44 Days Static Immersion in the Indian River Lagoon, Florida.

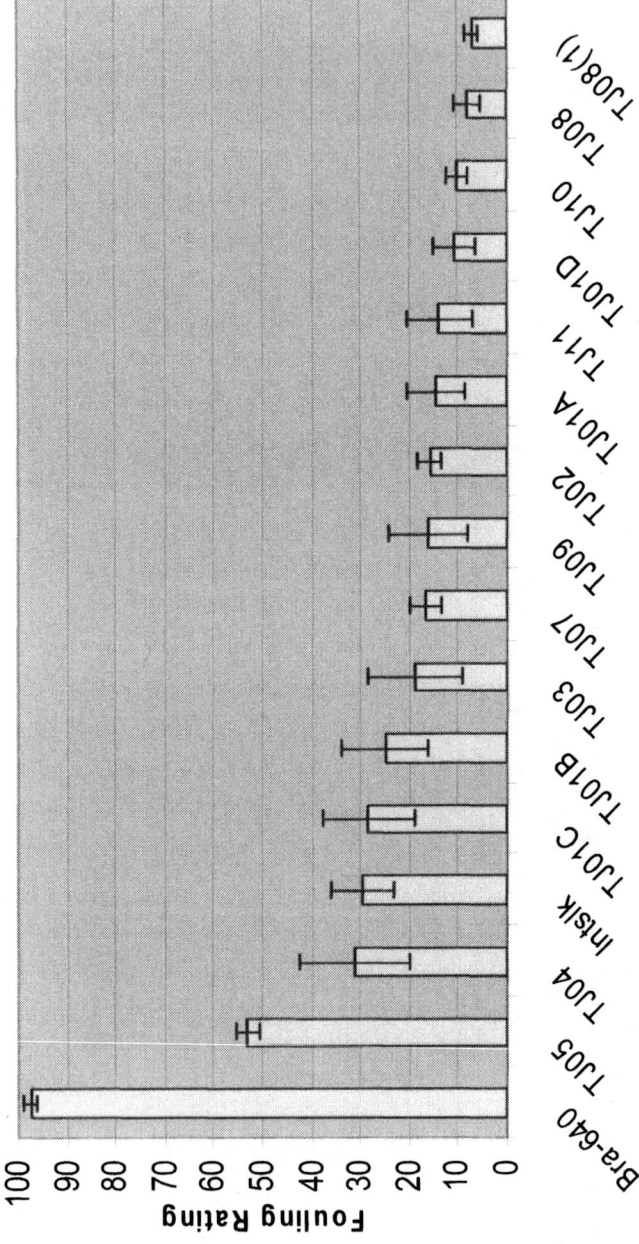

Figure 9. Average Fouling Rating of the Coatings after 80 Days Static Immersion in the Indian River Lagoon, Florida.

A.)

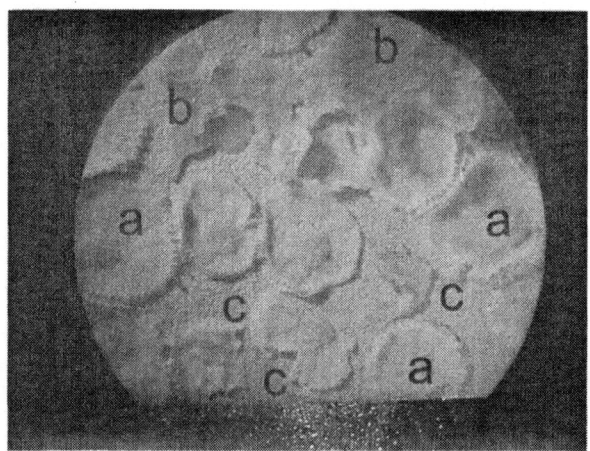

    a) Barnacle bases
    b) Experimental coating
    c) Undercoating

B.)

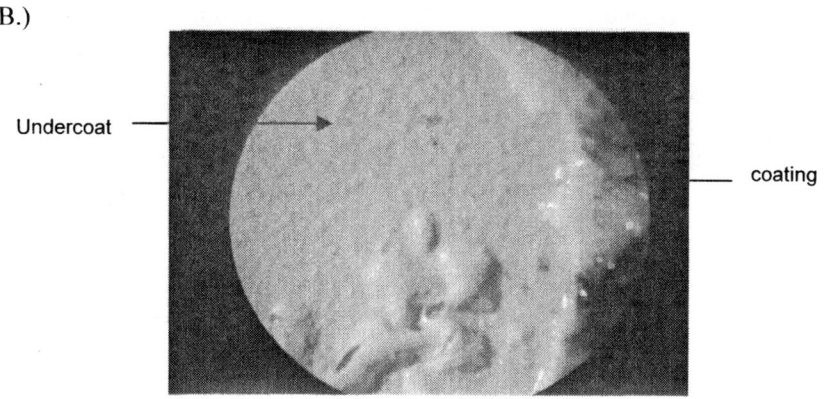

Undercoat corrosion

Figure 10. Barnacle Cutting and Corrosion in the Coatings.
(See page 8 of color inserts.)

the crosslink and the nature of side groups. Resistance to macrofouling was evaluated by static immersion tests in the Indian River Lagoon conducted by the Florida Institute of Technology. The static immersion study indicates that the coatings with covalently linked biocide and appropriate physical characteristics are effective in preventing macrofouling. Coating bulk properties, such as Tg and modulus of elasticity, strongly influence the fouling resistance. High modulus and very low modulus coatings demonstrate heavy macrofouling whereas coatings with modulus in the range 0.1 to 10 Mpa show significant reduction in barnacle recruitment. In the long term ocean immersion studies to date, our coatings have failed either as a result of poor adhesion to the primer, or to macrofouler induced damage to the soft coating.

Conclusion

The high-throughput work-flows for marine coating development are proving to be an effective and efficient means for the rapid development of marine coatings. Although much more work must be done to validate all of the methodologies, the example regarding the use of the bulk modulus of elasticity as a predictor of coating fouling-release bodes well.

Acknowledgment

Financial support from the Office of Naval Research through ONR grants # N00014-02-1-0794 and N00014-03-1-0702 are gratefully acknowledged. The authors are thankful to Prof. Geoffrey Swain and Mr. Kris Kavanagh at the Florida Institute of Technology (FIT), Melbourne, FL, for conducting the static immersion studies. The assistance of Jim Bahr and Christine Gallagher-Lein at Center for Nanoscale Science and Engineering (CNSE) for conducting the DMTA experiments is gratefully acknowledged.

References Cited

1. a) Brady, R. F. *Prog. Org. Coatings* **1999**, *35*, 31.
   b) Yebra, D. M.; Kiil, S.; Dam-Johansen, K. *Prog. Org. Coatings* **2004**, *50*, 75. (and references within)
2. a) Brady, R. F. *Prog. Org. Coatings* **2001**, *43*, 188.
   b) Townsin, R. L. *Biofouling*, **2003**, *19 (supplement)*, 9.
3. Champ, M. A. *The Science of the Total Environment* **2000**, *258*, 21 (and references within)

4. Bohlander, G. S.; Montmarano, J. A. *Naval Res. Rev.* **1997**, *49*, 9.
5. a) McClellan-Green, P.; Robbins, J. *Marine Environmental Research* **2000**, *50*, 243.
   b) Alzieu, C. *Ocean & Coastal Management* **1998**, *40*, 23.
   c) Nehring, S. *Journal of Sea Research* **2000**, *43*, 151.
   d) Marin, M. G.; Moschino, V.; Cina, F.; Celli, C. *Marine Environmental Research* **2000**, *50*, 231.
   e) Negri, A. P.; Smith, L. D.; Webster, N. S.; Heyward, A. J. *Marine Pollution Report* **2002**, *44*, 111.
6. a) Evans, S. M.; Leksono, T.; MeKinnel, P.D. *Marine Pollution Report* **1995**, *30*, 14.
   b) Abbot, A.; Abel, P. D.; Arnold, D. W.; Milne, A. *The Science of the Total Environment*, **2000**, *258*, 5.
7. a) Baier, R. E.; Meyer, A.E. *Biofouling* **1992**, *6*, 165.
   b) Brady, R. F. *J. Coatings Tech.* **2000**, *72 (no.900)*, 44.
8. Baier, R. E.; DePalma, V. A. Management of Occlusive Arterial Disease (Dale, W.A. ed.), Yearbook Medical Publishers 1971, 147-163.
9. a) Rittschof, D.; Costlow, J. D. *Sci. Mar.* **1989**, *53*, 411.
   b) Roberts, D.; Rittschof, D.; Holm, E.; Schmidt, A. R. *J. Exp. Mar. Bio. Ecol.* **1991**, *150*, 203.
   c) Cooksey K E, Wigglesworth-Cooksey B *Aquat. Microb. Ecol.* **1995**, *9*, 87.
10. a) Mera, A. E.; Goodwin, M.; Pike, J. K.; Wynne, K. J. *Polymer* **1999**, *40*, 419.
    b) Schmidt, D.L.; Brady, R. F. Jr.; Lam, K.; Schmidt, D. C.; Chaudhury, M. *Langmuir* **2004**, *20*, 2830.
11. a) Brady, R. F. *Prog. Org. Coatings* **1999**, *35*, 31.
    b) Swain, G. W.; Griffith, J.R.; Bultman, J. D. Vincent, H.L. *Biofouling* **1992**, *6*, 105.
12. a) Swain, G. W. J.; Shultz, M. P.; Griffith, J. R.; Snyder, S. in Brady, R.F. Jr., Park; Y. (eds) *Proc. Workshop on Emerging Non-Metallic Materials for the Marine Environment*, Office of Naval Research, Arlington, VA. March **1997**, pp. 1-60 to 1-69.
    b) Kohl, J. G.; Singer, I. L. *Prog. Org. Coatings* **1999**, *36*, 15.
    c) Stein, J.; Truby, K.; Wood, C. D.; Takemore, M.; Vallance, M.; Swain, G.; Kavanaugh, C.; Kovach, B.; Shultz, M.; Wiebe, D.; Holm, E.; Montemarano, J.; Wendt, D.; Smith, C.; Meyer, A. *Biofouling* **2003**, *19*, 87.
    d) Berglin, M.; Lonn, N.; Gatenholm, P. *Biofouling*, **2003**, *19 (supplement)*, 63.
13. a) Sundberg, D. C.; Vasishtha, N.; Zimmerman, R. C.; Smith, C. M. *Naval Res. Rev.* **1997**, *49*, 51.
14. a) Berglin, M.; Larsson, A.; Jonsson, P. R.; Gatenholm, P. *J. Adhesion Sci. Tech.* **2001**, *15*, 1485.
    b) Rasmussen, K.; Willemsen, P. R.; Ostgaard, K. *Biofouling* **2002**, *18*, 177.

c) Apsar, A.; De Nys, R.; Steinberg, P. *Biofouling,* **2003**, *19 (supplement),* 105.

d) Head, R. M.; Overbeke, K.; Klijnstra, J.; Biersteker, R.; Thomason, J. C. *Biofouling* **2003**, *19*, 269.

e) Head, R. M.; Berntsson, K. M.; Dahlstrom, M.; Overbeke, K.; Thomason, J. C. *Biofouling,* **2004**, *20*, 123.

15. Callow, M. E.; Callow, J. A.; Pickett-Heaps, J. D.; Wetherbee, R. *J. Phycol.* **1997**, *33*, 938.

16. Finlay, J.; Callow, M. E.; Ista, L. K.; Lopez, G. P.; Callow, J. A. *Integr. Comp. Biol.* **2002** *42*, 1116.

17. Shane J. Stafslien,* James A. Bahr, Jason M. Feser, Jonathan C. Weisz, Thomas E. Ready and Philip Boudjouk *J. Combinatorial Chem.* **2006**, *8*, 156.

18. a) Chauhan, M.; Hauck, B. J.; Keller, L. P.; Boudjouk, P. *J. Organomet. Chem.* **2002**; *645*, 1.

b) Chauhan, M; Chauhan, B. P. S.; Boudjouk, P. *Tet. Lett.* **1999**, *40*, 4127.

c) Boudjouk, P.; Han, B.- H.; Jacobson, J.R.; Hauck, B.J. *J. Chem. Soc. Chem. Commun.* **1991**, 1424.

d) Boudjouk, P.; Kloos, S.; Rajkumar, A. B. *J. Organomet. Chem.* **1993**, *443*, C41-C43.

e) Rajkumar, A. B.; Boudjouk, P. *Organometallics* **1989**, *8*, 549.

f) Boudjouk, P.; Choi, S.-B.; Hauck, B. J.; Rajkumar, A. B. *Tet. Letters* **1998**, *39*, 3951.

g) Chauhan M., Boudjouk, P. *Canadian J. Chem.* **2000**, *78*, 1396.

19. a) Ready, T. E.; Chauhan, B. P. S.; Boudjouk, P. *Macromol. Rapid Comm.* **2001**, *22*, 654.

b) Chauhan, B. P. S.; Ready, T. E.; Al-Badri Z.; Boudjouk, P. *Organometallics* **2001**, 20, 2725.

c) Chauhan, B. P. S.; Boudjouk, P. *Tet. Lett.* **2000**, *41*, 1127.

d) Chauhan, M.; Chauhan, B. P. S.; Boudjouk, P. *Org. Lett.* **2000**; *2*, 1027.

20. Mera, A. E.; Fox, R. B.; Bullock, S.; Swain, G. W.; Shultz, M. P. Gatenholm, P.; Wynne, K. J. *Naval Res. Rev.* **1997**, *49*, 4.

21. Chauhan M., Hauck B. J., Keller L. P., Boudjouk P., *Journal of Organometallic Chemistry* **2002**, *645(1-2)*, 1-13.

22. Kossuth M B, Hajduk D A, Freitag C, Varni J, *Macromolecular Rapid Communications* **2004**, *25*, 243.

23. Thomas, J.; Choi, S.-B.; Fjeldheim, R.; Boudjouk, P. *Biofouling,* **2004**, *20 (4/5)*, 227.

Chapter 7

Laboratory Methods to Access the Antifouling and Foul-Release Properties of Novel, Non-Biocidal Marine Coatings

M. E. Callow[1], J. A. Callow[1], and D. E. Wendt[2]

[1]School of Biosciences, University of Birmingham, Birmingham B15 2TT, United Kingdom
[2]Biological Sciences Department and Center for Coastal Marine Sciences, California Polytechnic State University, San Luis Obispo, CA 93407

In this article the advantages and disadvantages of laboratory-based methods to assess the performance of antifouling and foul-release coatings for application in the marine environment are discussed. Properties of the major groups of macrofouling organisms (algae and invertebrates) are briefly considered before the various assay methods are outlined. The nature of results obtained is illustrated by reference to sample data.

Introduction

Biofouling is the accumulation of microorganisms, algae and animals on man-made surfaces. The marine environment is unique because of the diversity of organisms that cause biofouling. Within minutes of immersing a clean surface in water it becomes 'conditioned' by adsorption of a macromolecular film. Bacteria colonise within hours, as may unicellular algae, cyanobacteria (blue-green algae) protozoa and fungi. These early small colonisers form a biofilm, which is an assemblage of attached cells often referred to as 'microfouling' or 'slime'. Macrofouling consists of multicellular algae (i.e. the 'seaweeds'), soft-bodied invertebrates such as sponges and tunicates ('soft' macrofouling), and calcified invertebrates such as barnacles and tubeworms ('hard' macrofouling). Although a microbial biofilm may moderate the settlement of spores and larvae of more complex organisms, it is not a pre-requisite (*1*).

On ships, most interest has focused on fouling by barnacles but other important invertebrate foulers are tubeworms, bryozoans, hydrozoans and sponges. *Ulva* (syn. *Enteromorpha)* is the most important macroalga because it is ubiquitously distributed and can withstand fluctuating environmental conditions. Microfouling shines are dominated by diatoms which are of interest because they are difficult to remove from non-biocidal, silicone-based foul-release coatings (*2*).

All fouling organisms secrete 'sticky' materials (adhesive proteins, glycoproteins and polysaccharides) to attach to substrata (*3*). Larvae of invertebrates and spores of algae need to quickly locate and attach to a surface in order to complete their life history. Non-biocidal coatings can be manipulated to deter settlement through for example, impressed microtopographic features (*4*). Controlling fouling without the use of biocides is essentially a problem of deterring settlement and/or controlling adhesion. To assess the potential of foul-release coatings, which release fouling under robust hydrodynamic conditions, necessitates the use of methods that measure the adhesion strength of attached organisms.

In this paper, we review the main laboratory-scale methods used to evaluate non-biocidal coatings for antifouling and fouling release properties. The key test organisms are the cypris larvae and young adult (juvenile) stages of barnacles, and the spores and young plants (sporelings) of the green alga *Ulva*.

Why is there a need for laboratory-scale evaluations of coatings?

This question is asked frequently. On the positive side, laboratory testing allows a rapid assessment of a coating or surface under controlled and reproducible conditions, thus the challenge is not site-specific or seasonal in character. Laboratory-scale evaluation is relatively low cost, both in terms of facilities, and the amount of test material required, an important factor where novel polymer systems are being investigated at an experimental scale. Furthermore, it allows the different stages of settlement, attachment and fouling release to be studied in detail. This is especially important, as the key to antifouling (cf. weak adhesion and fouling release) necessitates control of the settling stages (spores or larvae), which are not considered in field tests employing panels hung from rafts. Issues relating to film stability and bonding to the substrate also become apparent at an early stage the consequences of which may render expensive and time-consuming field trials worthless. In terms of coatings development, laboratory assays allow model systems to be evaluated that provide valuable data to synthetic chemists in terms of informing coatings development. Information from a failure can be just as valuable as from a success if it is possible to determine why a coating failed.

On the negative side, it can be argued that laboratory assays do not reflect the real world, where changing environmental conditions, season, competition and predation strongly influence the biofouling community structure as well as more subtle factors such as the adhesion strength of a particular organism. Assays with a single species may miss synergistic effects present in more complex communities

Key test organisms. 1: the green seaweed *Ulva*

Mature, reproductive plants release vast numbers of motile, unicellular, pear-shaped spores (technically 'zoospores'), 5-8 µm in length (Plate la). Swimming spores attach rapidly to a surface once the spore has 'detected" that it is suitable for settlement. Secretion of adhesive (Plate lb) results in firm attachment to the substratum. The adhesive cures rapidly by some form of cross-linking (5) which is manifest as increased adhesion strength with time after settlement (6). The spores germinate and grow into young plants (sporelings, Plate 1c), a few hundred microns in length, within a few days.

Key test organisms 2: Barnacles

Balanus amphitrite is the most common species of barnacle utilized in laboratory assays. It is easy to culture, fast-growing, and cosmopolitan in geographic distribution. Mature adults release thousands of free-swimming larvae ('nauplii'). The larvae actively feed and undergo a total of 6 instar stages. The final larval stage is the non-feeding cypris larva (Plate 1d). The cypris larva is approximately 500µm in length and is specialized for attachment and metamorphosis. It has a pair of sensory antennae that are used for determining physical and chemical properties of the substratum. The cypris larva explores the substratum by 'walking' on its antennae. When the cypris larva identifies a suitable substratum, it attaches by secreting an adhesive from cement glands located at the base of the antennae. After attachment the larva undergoes a complex metamorphosis into a juvenile barnacle and thence into an adult (Plate 1e).

Test sample format

The standard format for the assays described below is coated glass microscope slides. Acid-washed slides are usually included in assays as standards. Smooth coatings of uniform thickness, without sagging or 'picture framing' and well-bonded to the glass are essential if hydrodynamic studies in a flow channel apparatus are to be performed. Some coatings that contain toxic curing agents e.g. dibutyltin compounds, need to be leached before the assays are set up. This is usually accomplished in a tank of recirculating deionised water fitted with a carbon filter. Leachates are collected and monitored by bioassay (7) until no toxicity is measured. The slides are transferred to artificial seawater (ASW) for at least 1 hour before an assay is set up.

'Antifouling' Settlement Assays

Assays with Algae

Ulva plants, collected from the seashore, release zoospores on immersion in seawater. The spore concentration is adjusted to 1-2 $\times 10^6$ / ml (using absorbance at 660 nm). Six replicates of each coating are required if spore settlement and strength of attachment assays are to be performed; three for settlement and three for adhesion strength. For sporeling growth and strength of attachment assays, a further six replicates are needed. Slides are placed in individual compartments of Quadriperm dishes (Fischer) to which 10 ml zoospore suspension are added. The dishes are incubated for 1 h in darkness at room temperature (c. 20°C). Slides are gently washed in ASW to remove unattached spores and fixed in 2.5% glutaraldehyde in seawater *(v/v)*. Slides are air dried after washing in distilled water. Spore numbers are usually counted (30 counts per slide) at 1 mm intervals across the middle of the slide using a Zeiss Kontron 3000 image analysis system attached to a Zeiss epifluor-

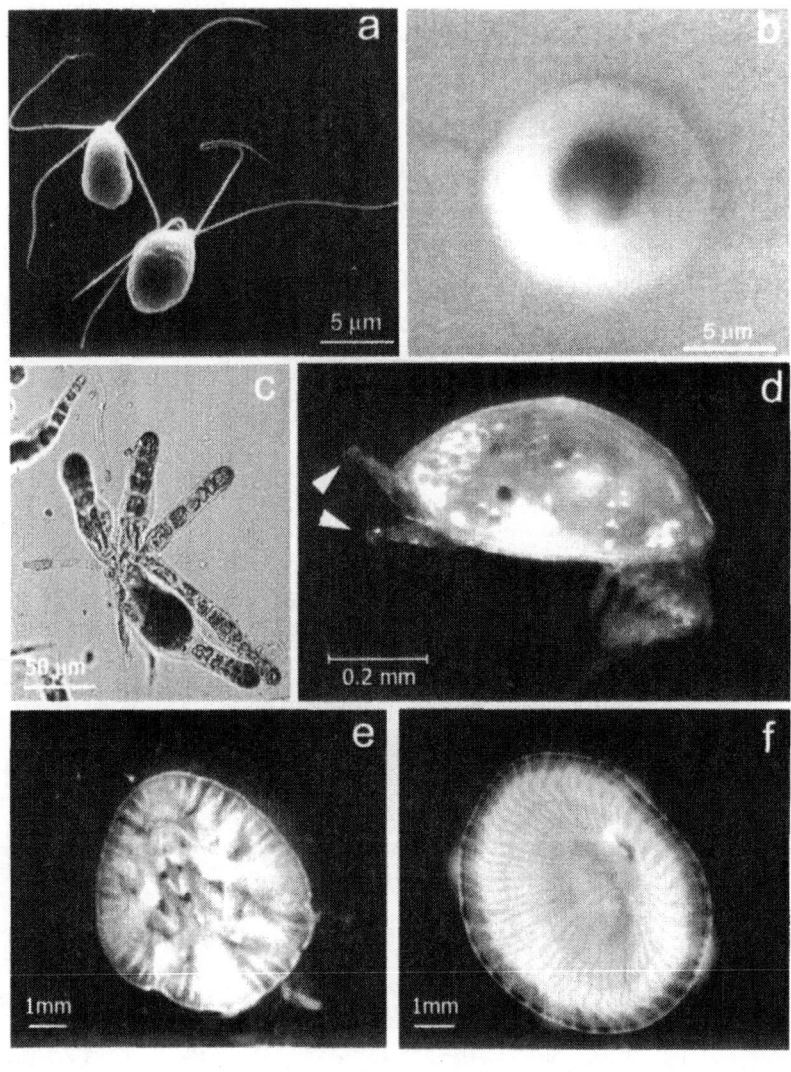

Plate 1. a) Zoospores of the green alga Ulva *showing the 4 locomotory flagella (false-colour scanning electron micrograph image). b) false-colour image of a settled spore of Ulva as seen by Environmental Scanning Electron Microscopy. The original spore body is now surrounded by an annulus of secreted, hydrogel-like adhesive. The flagella are absorbed during settlement. (Copyright 2005 from Walker et al., Journal of Adhesion **81**, 1101-1118; reproduced by permission of Taylor & Francis Group, LLC., http:llwww.taylorandfrancis.com). c) young plant ('sporeling') of* Ulva *which develops from germinated settled spores. d) The cypris larva of the barnacle* Balanus amphitrite *showing two anterior antennules (arrowed) and numerous posterior thoracic appendages used for swimming. Also apparent are multiple lipid droplets, the larval energy stores. e) top view of a young adult barnacle of B.* amphitrite, *which metamorphosed from the settled cypris larva. f) bottom view of an adult barnacle attached to a clear silicone coating showing the basal plate.*
(See page 9 of color inserts.)

escence microscope via a video camera. Spores are visualised by autofluorescence of chlorophyll.

Assays with barnacles

Antifouling settlement assays using barnacles (or larvae of the other invertebrates) are conducted using a "drop assay" or in certain situations a settlement preference experimental design. Which assay is utilized depends on the physical nature of the surface being tested. For example, a drop will not form on extremely hydrophobic or hydrophilic surfaces and thus the coatings need to be continuously submerged for testing. It is recommended that at least 5 replicate experimental surfaces and an additional 5 replicate control surfaces, uncoated glass slides, are used in antifouling assays.

For the drop assay, a 0.4ml aliquot of seawater containing 20-50 cyprids is placed on the surface of each replicate slide. The slides are placed in covered Petri dishes with moistened paper towels. The Petri dishes are placed in a constant temperature (25°C) and light (12 hour light-dark cycle) incubator. The assay lasts until 50% of the cyprids on the control surfaces have attached, or until 72 hours have elapsed. At the conclusion of the assay, the number of swimming larvae and the number of attached barnacles is enumerated for each replicate surface. Percent attachment is calculated by comparing the number of settled barnacles to total number of cyprids exposed to the surface.

The settlement preference assay differs from the drop assay in that larvae are presented with a choice: the experimental coating or the polystyrene surface of the Petri dish in which the coating is submerged. The basic protocol consists of adding 50-100 larvae in 40 ml of seawater to a Petri dish (100 x 15mm) containing an experimental or control slide. The duration of the experiment is determined as described above. At the conclusion of the assay, the number of swimming cypris larvae and the number of attached barnacles on the slide and Petri dish are enumerated. The percentage of individuals that settled on the slide and on the Petri dish are calculated. From these data a settlement preference ratio is calculated: the number of settled barnacles per available square millimeter of the experimental coating divided by the number of settled barnacles per available square millimeter of the Petri dish. A settlement preference ratio of 1 indicates no preference for either surface, whereas a settlement preference ratio less than one indicates a coating that may deter settlement.

Assays to evaluate foul-release potential

Assays with algae

Slides settled with *Ulva* zoospores are exposed to turbulent flow in a specially designed flow channel or exposed to the compressive forces of a water jet. The design and operating principles of the two apparatuses are described below. Cell density on three replicate slides before and after exposure is determined as described above. Percentage removal data are calculated from the mean number of spores remaining attached to the surface after exposure to turbulent flow (flow channel) or surface pressure (water jet) compared with the mean number before the slides were subjected to flow/water pressure (e.g. *8,9,10*).

Foul-release potential can also be assessed on *Ulva* sporelings (young plantlets formed from germinated settled zoospores after a few days growth). Zoospores are settled on 6 replicates of each test surface as described above. The washed slides are replaced in the Quadriperm dishes and 10 ml of growth medium added to each dish compartment (*8*). The dishes are incubated at 18 °C with a 16 h:8 h, light: dark cycle, the medium being refreshed every 2 days. After 7-8 days, bioinass on the surface of the slides is quantified before and exposure to flow in the flow channel. Biomass is recorded either directly in a TECAN plate reader using the autofluorescence of chlorophyll, or the sporelings are harvested by scraping with a razor blade. In the latter case, biomass is determined as extracted chlorophyll. A foul-release silicone elastomer, e.g. Silastic T2 (Dow Corning) is normally included as a positive standard whilst glass is included as a negative standard.

Assays with barnacles

Laboratory evaluation of foul-release properties of coatings using barnacles is done using an automated test stand that applies a shear force to animals attached to coatings (*11*). Barnacles with a basal plate diameter >3-5 mm are used as basal plate breakage occurs more frequently with smaller animals (*12*). Prior to dislodgement tests, the basal plates of barnacles (Plate 1f) are photographed with a digital camera mounted on a dissecting microscope. Digital analysis (NIH. ImageJ) is used to calculate the area of the basal plate. The coated slide is then clamped in the force-gauge testing apparatus. Data can be streamed continuously to a computer which allows creation of a force vs. time curve for analysis of fracture processes. Peak force is also recorded and is used to calculate critical removal stress (force per unit area) for each barnacle. In the event that the barnacle basal plate fractures during testing, the force at which it broke is recorded, and the percentage of area of the basal plate remaining attached is determined via digital image analysis. In addition to direct comparisons of critical removal stress, the linear relationship between basal plate area and percentage of basal plate remaining on a coating is a measure of the efficacy of foul-release surfaces (*12*).

Apparatus

The fully turbulent flow channel and water jet apparatus

The operating principle (*13*) is that of a high aspect ratio, turbulent channel flow since boundary layers around ships are turbulent in character. The fully-developed channel flow allows determination of accurate wall shear stress from a simple pressure gradient measurement. The flow channel holds 6 microscope slides with or without coatings. Turbulent flow is created in a 60 cm long low aspect ratio section of channel preceding the slides. Flows of ASW up to 4.9 m s^{-1} generate wall shear stresses up to 56 Pa. The relationship between three models and the results from laboratory scale tests to the self-cleaning of a ship coated with a foul-release surface are discussed in (14).

The water jet developed by Swain & Schultz (15), designed for assessment of fouled panels in the field, uses perpendicular rather than parallel flow and delivers a greater force/unit area than the flow channel. Modifications to the original field apparatus have produced an automated, standardized operation by computer-driven stepper motors that allow the jet nozzle to be raster-scanned across a batch of slides at a controlled rate, in a variety of reproducible patterns (6). The instrument is typically operated at a speed of 10 mm s^{-1} for 10 swathes, at the end of which an area of 500 mm^2 in the mid-region of each slide has been exposed to the jet of water. The water supply is housed in a pressure resistant tank and pressurised using a compressed air supply from a conventional SCUBA tank. The relationship between regulator setting and impact pressure exerted at the surface is described in (15).

Force Gauge Test Stand

Removal force testing with barnacles is done using an automated test stand (IMADA model number SV-5) and a digital force gauge (IMADA model number AXT-70ozR) (11). The capacity of the force gauge is up to 19.5 N with a sensitivity of 0.01 N. The force gauge is mounted on a motorized arm that allows for a constant rate of application at 60-75 microns/second. The reproducibility of angle and rate of application are critical to reducing variability between measurements. The test stand has a custom-made acrylic chamber in which test coatings are clamped.

The chamber ensures that the barnacles are continuously immersed in seawater during dislodgment tests. Data from the force gauge can be streamed using an RS-232 digital interface at 20 recordings/second, which allows determination of force vs. time curves as well as maximum force during dislodgement.

Examples of Lab-scale Evaluation

The application of these methods to the evaluation of actual test samples can be illustrated by reference to recent published data. Gudipati et al. (8) used *Ulva* to explore antifouling and foul-release properties of a series of cross-linked hyperbranched fluoropolymers (HBFP) and poly(ethylene glycol) (PEG) amphiphilic networks. Settlement of *Ulva* zoospores was consistently and markedly lower on all the cross-linked HBFP-PEG network formulations compared to glass controls (Figure 1). Removal of spores from the HBFP-PEG45 coating in the flow channel was significantly different from the rest (Figure 2). Greatest removal of sporelings was also obtained from the HBFP-PEG45 surface with a mean detachment of over 90% (Figure 3). This was approximately twice of that from the T2 Silastic PDMS standard which suggests that this nanostructured material has promising properties in the practical control of soft fouling.

The best illustration of the settlement preference assay using barnacles comes from unpublished data on proprietary coatings. The protocol can clearly resolve differences in settlement preference between coatings as is illustrated by vastly different ratios between surfaces (Figure 4). Tang et al. (9) used *B. amphitrite* to evaluate the foul-deterrent and foul-release properties of experimental hydrophobic and hydrophilic xerogel films. Settlement rates of cypris larvae on xerogels containing C3-TMOS, C8-TEOS, and enTMOS were lower than that of larvae on glass and polystyrene controls, but not significantly so (Figure 5). The 50/50 C8-TMOS/TMOS was significantly lower than glass controls. In the same study critical removal stress was evaluated for 50/50 C8-TEOS/TMOS xerogel coatings with different sol-

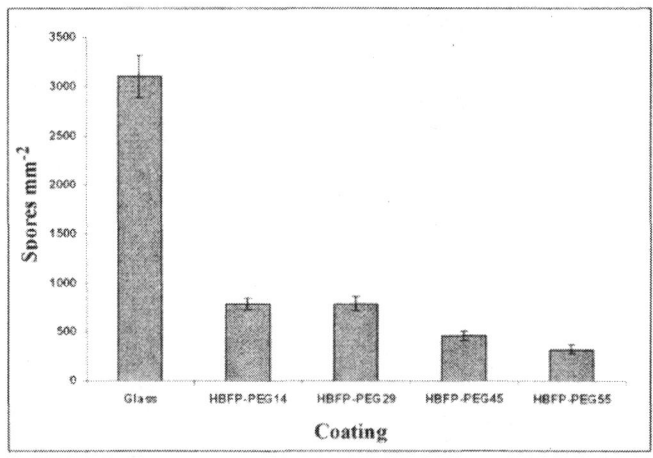

Figure 1. Settlement of Ulva spores upon cross-linked HBFP-PEG network coatings. (Reproduced from reference 8. Copyright 2005 American Chemical Society.)

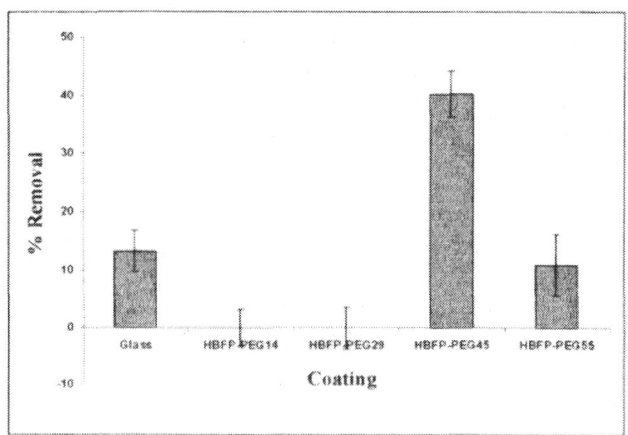

Figure 2. Removal of attached Ulva spores from HBFP-PEG networks in the flow apparatus (55 Pa wall shear stress). Removal was greatest from the coating with 45 wt % PEG, indicating its foul-release properties. (Reproduced from reference 8. Copyright 2005 American Chemical Society.)

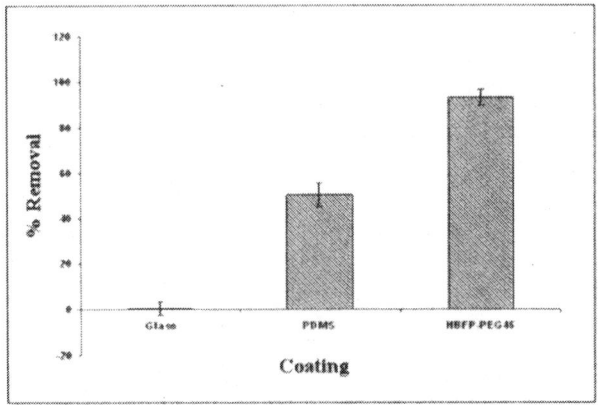

Figure 3. Percentage removal of Ulva sporelings in the flow apparatus. The crosslinked HBFP-PEG coating with 45 wt% PEG was more effective in fouling release than the standard PDMS coating. (Reproduced from reference 8. Copyright 2005 American Chemical Society.)

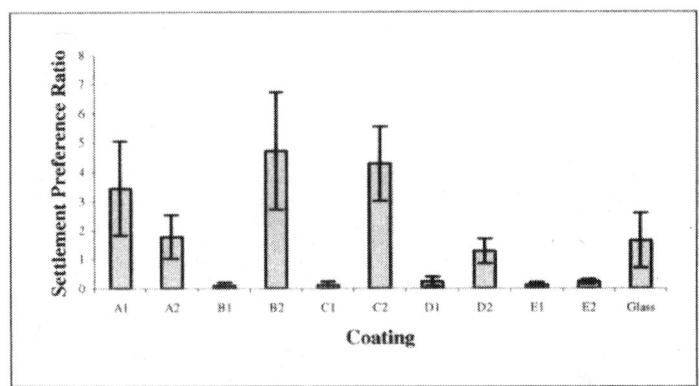

Figure 4. Settlement preference of cypris larvae of B. amphitrite exposed to proprietary coatings. A settlement preference ratio of indicates no preference for either surface, whereas a settlement preference ratio <1 indicates a coating that may deter settlement. Error bars are equal to one standard error. Unpublished data.

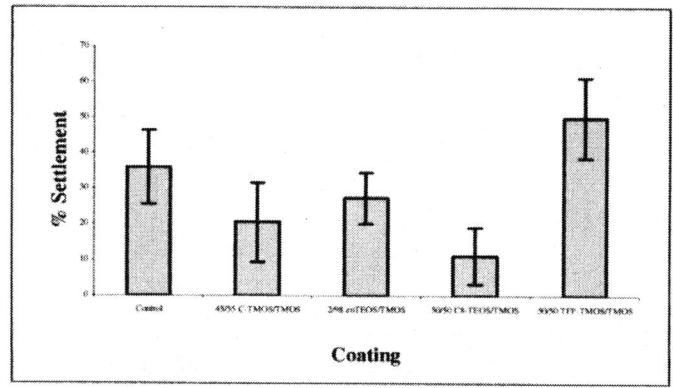

Figure 5. Percent settlement of cypris larvae of B. amphitrite *on xero gel coatings and glass and polystyrene controls. Error bars are equal to one standard error. (Reproduced with permission from reference 9. Copyright 2005 Taylor & Francis Ltd; http://www.tandf.co.uk/journals).*

processing times. All the surfaces performed well as foul-release surfaces compared to a standard Gelest silicone coating (9).

Conclusions

In this article we have provided examples of some of the commonly used laboratory-based methods to test the intrinsic antifouling and foul-release properties of novel surfaces and coatings for application in the marine environment. The examples chosen are not exhaustive and no one assay method should be used to assess performance. All laboratory assays should include suitable positive and negative calibration standards that give reproducible performance. Laboratory methods should be regarded as the first stage in a strategy to down-select surfaces for eventual testing in the field (e.g. through raft panels) since the extent to which they predict performance in the field has yet to be established.

Acknowledgements

The authors acknowledge support from the Office of Naval Research (awards N00014-99-1-0311 and N00014-02-1-0521 to JAC and MEC: award N00014-02-1-0935 to DEW).

References

1. Callow, J.A.; Callow, M.E.; 2006. *Biofilms.* In *Antifouling Compounds. Progress in Molecular and Subcellular Biology: Marine Molecular Biotechnology,* Editors, Fusetani, N.; Clare, A.S. Clare, Springer-Verlag, Heidelberg.
2. Holland, R.; Dugdale, T.M.; Wetherbee, R.; Brennan, A.B.; Finlay, J.A.; Callow, J.A.; Callow, M.E. *Biofouling,* **2004**, *20*, 323-329.
3. *Biological Adhesives;* Editors, Smith, A.; Callow, J.A.; Springer-Verlag, Heidelberg; 2006, (in press).
4. Berntsson, K. M.; Jonsson, P. R.; Lejhall, M.; Gatenholm, P. *J. Exp. Mar.Biol. Ecol.* **2000**, *251*, 59-83.
5. Humphrey, A.J.; Finlay, J.A.; Pettitt, M.E.; Stanley M.S.; Callow, J.A.; *J. Adhesion* 2005, *81*,791-803.
6. Finlay, J.A.; Callow, M.E.; Schultz, M.P.; Swain, G.W.; Callow, J.A. *Biofouling,* **2002**, *18*, 251-256.
7. Callow M. E.; Finlay J. A. *Biofouling,* **1995**, *9*, 153-165.
8. Gudipati, C.S.; Finlay, J.A.; Callow, M.E.; Callow, J.A.; Wooley, K.L. *Langmuir*, 2005, *21*, 3044-3053.
9. Tang, Y.; Finlay, J.A.; Kowalke, G.L.; Meyer, A.E.; Bright, F.V.; Callow, M.E.; Callow, J.A.; Wendt D.E.; Detty, M.R.. Hybrid xerogel films as novel coatings for antifouling and fouling release. *Biofouling,* **2005**, *21*, 41-48.
10. Youngblood, J.P.; Andruzzi, L.; Ober, C.K.; Hexemer, A.; Kramer, E.J.; Callow, J.A.; Finlay, J.A.; Callow, M.E. *Biofouling,* 2002, *19* (supplement), 91-98.
11. Wendt, D.E.; Kowalke, G.L.; Singer, I.L. *Biofouling,* **2006**, *22*, doi: 10.1080/08927010500499563

12. Berglin, M.; Larsson, A.; Jonsson, P.R.; Gatenholm, P. *J. Adhesion. Sci. Technol.* **2001**, *15*, 1485-1502.
13. Schultz, M.P.; Finlay, J.A.; Callow, M.E.; Callow, J.A. *Biofouling,* **2000**, *15*, 243-251.
14. Schultz, M.P.; Finlay, J.A.; Callow, M.E.; Callow, J.A. *Biofouling,* **2003**, *19* (supplement), 17-26.
15. Swain, G.W.; Schultz, M.P. *Biofouling,* **1996**, *10*, 187-197.

Chapter 8

Ion Exchange Compounds for Corrosion Inhibiting Pigments in Organic Coatings

R. G. Buchheit and S. P. V. Mahajanam

Fontana Corrosion Center, Department of Materials Science and Engineering, Ohio State University, 477 Watt Hall, 2041 College Road, Columbus, OH 43210

This paper summarizes emerging work on the use of synthetic inorganic ion exchange compounds as pigments in organic coatings for storage and release of non-chromate corrosion inhibitors. The basic principles of ion exchange and new synthetic ion exchange compounds being explored are described. The two main research thrusts that currently comprise this research area are presented. The first is based on the use of cation exchanging clays such as bentonite, which have been explored for the delivery of the Ce^{3+} cation. The second is based on the use of anion exchanging hydrotalcites, which have been explored for delivery of a range of anionic inhibitors with a focus on delivery of vanadate oxoanions. How the use of these compounds leads to the ability to sense water uptake by coatings and exhaustion of the inhibitor reservoir is also presented.

Introduction

An important component of corrosion protection in many types of corrosion resistant organic coatings is derived from the release of soluble inhibitors into an attacking environment. Inhibitor release is normally from chemical dissolution of sparingly soluble inorganic compounds dispersed as finely divided pigment particles in the coating once moisture from the environment has been taken up. A well known example is the use of $SrCrO_4$ pigment, which is widely used in aerospace primers. $SrCrO_4$-epoxy primers are not necessarily intended to be good moisture barriers. When sufficient moisture is taken up by the coating, inhibitor ions are released by dissolution of the pigment. The inhibitor ion concentration that can be developed in the coating electrolyte can increase up to the solubility limit of the inhibiting compound. Once released, the inhibitors can diffuse and migrate to the coating-metal interface to reinforce passivity or to heal defects in the coating system such as scratches that expose unprotected bare metal substrate. The maximum solubility of the $SrCrO_4$ is low, which keeps to concentration of the coating electrolyte low and reduces the tendency for osmotic blistering. Fortunately, Sr^{2+} and CrO_4^{2-} are both inhibitors and the mixture of the two is powerful enough to provide good corrosion protection at low concentrations.

The case of $SrCrO_4$-pigmented primers highlights the significant constraints imposed on inhibiting compounds in organic coatings. First and foremost, inhibitor pigment solubility must be low enough to avoid the risk of blistering. In $SrCrO_4$-pigmented coatings, the resin phase must be permeable enough to enable access to and release of the corrosion inhibitor. However it is essential to avoid conditions where the coating functions like a semi-permeable membrane, where large solute ions are trapped but smaller water molecules may diffuse. Should this occur, water will diffuse from the external environment into the base of the coating to relieve the concentration difference thus leading to blistering. This process is a manifestation of osmosis, and the resulting form of coating degradation is osmotic blistering. To minimize the risk of osmotic blistering as much as possible, it is necessary to limit the choice of inhibiting pigment compounds to those that are typically considered to be "sparingly soluble". In practical terms, the upper limit on inhibitor pigment solubility has been estimated to be 200 mM [1] The constraint on inhibitor compound solubility leads to another important constraint—the need for the pigment compound to release inhibitor ions that provides high levels of corrosion protection at low concentrations.

These two constraints–sparing solubility and inhibitor potency are severe and eliminate a large number of inorganic inhibiting pigment compounds. Sinko has gone as far as to say that these constraints render all inorganic inhibiting compounds inferior to sparingly soluble chromates [1] One strategy not considered by Sinko at the time his analysis was published was the use of substances that release inhibitor ions by means other than chemical dissolution. One such class of substances are inorganic ion exchange compounds (IECs). IECs are molecular sieves that are comprised of a fixed host cage structure and an exchangeable gallery of ions or molecules loosely contained within the cage. Species in the galleries can be released from the compound in low doses into a contacting electrolyte while the host cage structure remains intact. The thermodynamics and kinetics of release are subject to their own sets of constraints that must be considered for application to corrosion protection. However, a key advantage associated with the use of IECs is the great latitude in selection and combination of host cages and the exchangeable species and the possibility for achieving low doses of potent inhibitor combinations that cannot be delivered using approaches based on chemical solubility alone.

Strictly speaking, the use of IECs as corrosion inhibiting pigments in paints is not new. Commercial ion exchange pigments such as Shieldex^{TM}, Dowex^{TM} and Activox^{TM}, deliver corrosion inhibiting cations in coating systems. Additionally, it is likely that some complex oxide inhibitors posses some ion exchange capacity. However, the recently reported delivery approaches based on the use of synthetic ion exchanging silicate clays and hydrotalcites constitutes a new direction in coating science. While it is too early to determine the significance this new direction will have on commercial formulations, it is clear that these approaches represent an opportunity to address core issues such as replacement of toxic inhibitors like chromates and addition of new functionalities to coatings like sensing of water uptake in coatings and inhibitor exhaustion.

Basics of Ion Exchange

IECs are insoluble molecular frameworks that contain anionic or cationic species that can be exchanged with a contacting electrolyte without loss of the essential framework structure [2]. IECs can be cation exchangers, typically aluminosilicate compounds, or anion exchangers, such as hydrotalcite, which are sometimes referred to as layered double hydroxide compounds. Certain IECs are capable of exchanging both cations and anions. These are referred to as amphoteric ion exchangers. A few hydrotalcite compounds fall in this latter category [2,3].

Aluminosilicate-based clays such as montmorillonite and bentonite typify layered cation exchangers. As shown in Figure 1, the clay host structure is layered and consists of a mixed tetrahedrally coordinated silicate layer and an octahedrally coordinated Mg^{2+}/Al^{3+} layer. These layers are positive charge-deficient due to cation substitution in the layer lattice (e.g., Mg^{2+} on an Al^{3+} site). The net negative charge on the layers is offset by absorption of hydrated cations in the interlayer gallery of the clay host. It is these interlayer cations that are exchangeable with cations in a contacting electrolyte.

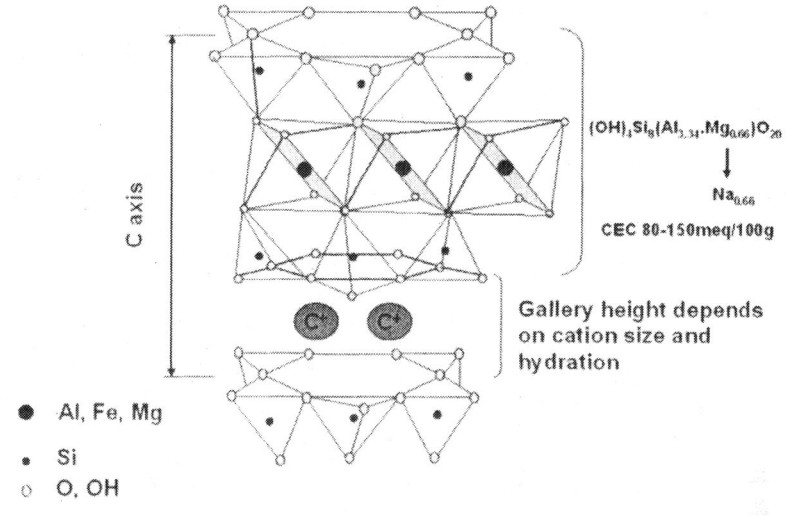

Figure 1. Schematic structure of bentonite. Adapted from reference [4].

Hydrotalcites are the inverse of cation exchanging clays in some important respects. As shown in Figure 2, the host layers in these compounds consist of a mixed M^{2+}/M^{3+} metal hydroxide layer. This layer possesses net positive charge due to substitution of the trivalent cation on a divalent cation lattice site. This charge is offset by absorbed, exchangeable anions in the interlayer gallery. A naturally occurring compound, hydromagnesite, $Mg_6Al_2(OH)_{16}CO_3 \cdot 0.5H_2O$, has a mixed Mg^{2+}/Al^{3+} hydroxide layer and an interlayer comprised of carbonate, CO_3^{2-}, and water. Many artificial hydrotalcite compounds with mixed M^{2+}/M^{3+} host layers have been synthesized and are generally of the form $[M^{2+}_{1-x} M^{3+}_x(OH)_2]^{n+} \cdot A^{n-}_{x/2} \cdot yH_2O$ (x = 0.1 – 0.33) [5]. M^{2+}/M^{4+} and M^+/M^{3+} (e.g., Al-Li) host structures have also been synthesized. In the Al-Li M^+/M^{3+} host structure, Li^+ occupies octahedral voids in $Al(OH)_3$ [6]. Like the more common M^{2+}/M^{3+} hosts, a range of anions can occupy the interlayer of these frameworks.

The ion exchange process is a redistribution of exchangeable ions between the host and the contacting electrolyte. Generally, the exchange process (in a sodium chloride electrolyte) is characterized by the following reaction:

$$Host\text{-}X + NaCl(aq) \Leftrightarrow Host\text{-}Cl + NaX(aq) \quad (1)$$

where X represents an exchangeable anion, *Host* refers to the insoluble IEC framework, and *Na* refers to the sodium ion from a sodium salt. A comparable reaction can be written for cation exchangers. The equilibrium in eq. 1 can be characterized by an equilibrium or exchange constant whose value depends on environment chemistry and experimental conditions. Adsorption or exchange isotherms are also used to characterize exchange behavior. Other important characteristics of the reaction are exchange kinetics, which are often diffusion-controlled (see results for Ce-exchanged bentonite below), and the exchange capacity, which ranges from about 1 to 5 meq ions/g and greater for some cation exchangers [7-10]. The exchange process is subject to constraints imposed by electroneutrality in the IEC crystal. While the ion exchange process is amenable to characterization using familiar chemical equilibrium principles, exchange involving ions of different valency, or exchange in amphoteric compounds [3] adds significant complexity to analytical characterization of the process [11].

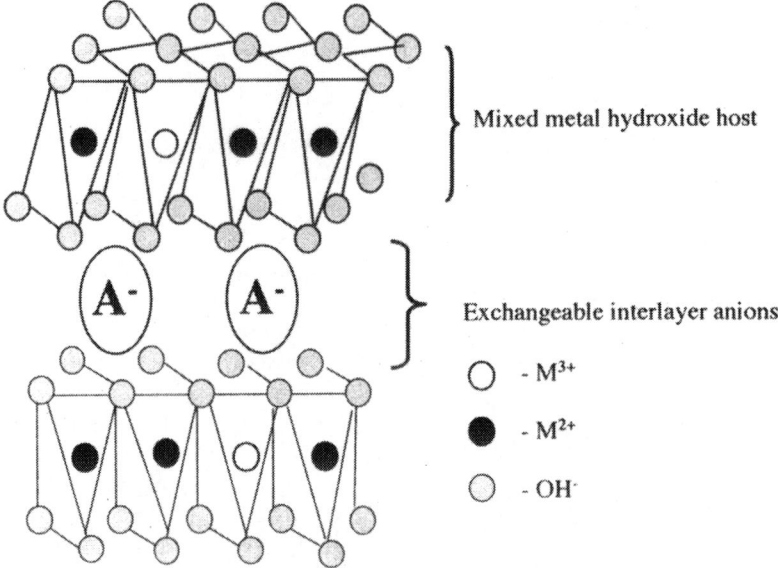

Figure 2. Schematic structure of hydrotalcite.

Synthesis often Exchange Compounds

There are three main synthesis routes for artificial IECs. These are direct synthesis by co-precipitation, and indirect synthesis by ion exchange of an existing IEC in a solution containing an ion of interest, or rehydration of a calcined IEC in a solution containing the intercalant ion of interest. Different

IECs are more or less amenable to synthesis by one of the different strategies. The literature is replete with detailed recipes for IEC synthesis. Recent reviews on IECs are good starting points into the body of literature on synthesis (6, 12)

Ion Exchange Compounds and Corrosion Inhibition

Commercial ion exchange pigments

The model for corrosion inhibiting pigments in organic coatings on galvanized steel substrates is sparingly soluble chromate compounds. Chromates are anodic inhibitors for zinc dissolution, and cathodic inhibitors that suppress oxygen reduction on steel. These two modes of inhibition work in concert in the case of cut edge corrosion, which is a potential problem in components manufactured from galvanized steel. In the galvanic cell at a cut edge, steel supports oxygen reduction and the sacrificial zinc layer is subject to anodic dissolution. Inhibition of oxygen reduction reduces the current demand placed on the sacrificial zinc layer and anodic inhibition of the zinc reduces any tendency for self-corrosion. This makes chromate well-suited for corrosion protection of galvanized steel structures.

A commercially available ion exchange compound, Shieldex™ (W.R. Grace Co.), is an example of a compound that has been the subject of recent scientific study (13-17) Shieldex™ pigments are silica gel compounds that exchange calcium ions. This pigment was developed in part as a replacement for toxic anti-corrosive pigments such as zinc chromate, strontium chromate and phosphate-based pigments. Corrosion inhibition is derived from calcium ion precipitation under alkaline conditions due to formation of protective $Ca(OH)_2$-type deposits at sites of cathodic activity. As such, it only mimics a part of the function of chromate inhibitors and is therefore not usually as effective as a chromate pigment for corrosion protection.

Corrosion protection from Shieldex™ pigments is improved significantly when it is used in conjunction with other inhibitors. A Ca^{2+}-Zn^{2+}-PO_4^{3-} inhibitor combination, in which the Ca^{2+} ion is delivered by the Shieldex™ ion exchange pigment, is an effective synergistic mix for steel and galvanized steel surfaces, perhaps due to additional anodic inhibition from the supplemental ingredients [17]. In electrochemical testing, evidence for both anodic and cathodic inhibition is found, although inhibition is more profound and longer lasting on zinc than on steel, and cathodic inhibition tends to be the more significant component of overall inhibition mechanism. This inhibitor mixture therefore is more potent for galvanized surfaces and Zn-Fe galvanic couples at cut-edges than on steel alone. Inhibition is ultimately due to the formation of a thin surface film comprised of a mixture of Ca and Zn phosphates. The film is sparingly soluble and forms a barrier that suppresses electron transfer reactions at cathodic sites thereby slowing the corrosion cell process.

In this application, the use of Ca^{2+}-exhanged silica as an inhibiting pigment appears to be incidental compared to the synergy of the Ca^{2+}-Zn^{2+}-PO_4^{3-} inhibitor mix. Nonetheless, mediated delivery of Ca^{2+} over a range of pH may be very important in this type of an inhibition scheme. Ca^{2+} solubility is high in acidic conditions and decreases sharply as pH increases. Cation exchange is not as pH-sensitive as chemical precipitation making it possible to deliver low and consistent dosages of inhibitor across a corrosion cell where pH can vary widely. The ability to keep inhibitor dosage low under acidic and neutral conditions is necessary to reduce the tendency for osmotic blistering. The ability to release inhibitor at all under alkaline conditions may distinguish ion exchange pigments in a critical way from more traditional metal salt or oxide pigments that rely on solubility and chemical dissolution to release inhibitor species.

An additional, though perhaps minor component corrosion inhibition may be present in the case of silica-based ion exchanging hosts. Chemical surface analysis of protective deposits formed in the presence of Shieldex™ pigmented coatings indicates the presence of silicon, which is a cathodic inhibitor for steels (as silicate) and may promote film formation in the presence of Ca^{2+} and PO_4^{3-} ions [15]. Although dissolution of the host structure was not ruled out, the presence of Si in protective films in these experiments suggests the possibility that this exchange compound exhibits some amphoteric character.

Ce, and Ce-exchanged Clays and Related Rare-earth Inhibitors and Pigments

A range of studies carried out over the past 20 years has established and characterized the corrosion inhibiting behavior of lanthanide rare earth element cations for engineering alloys [18-25] Good characterizations of the inhibition phenomenon exist, and there is general agreement on the inhibition mechanism.

Soluble salts of Cerium [26] Yttrium (Y), Lanthanum (La), Neodymium (Nd), and Praesodymium (Pr), among others when added to aqueous solutions inhibit corrosion of steel, zinc (galvanized steel) and aluminum alloys to varying degrees depending on inhibitor type and concentration, environment chemistry, and substrate composition. Trivalent Ce, La and Y added in concentrations in the range of 100 to 1000 ppm provide levels of corrosion protection that approach protection levels provided by chromate inhibitors in electrochemical testing [19, 22, 27] The lanthanide cations are primarily cathodic inhibitors. Trivalent lanthanide cations are characterized by decreasing solubility with increasing pH. On freely corroding metal interfaces, these cations precipitate as hydroxides or carbonates at locations where the pH is sufficiently high due to oxygen reduction:

$$O_2 + 2H_2O + 4e^- \rightarrow 4OH^- \qquad (2)$$

which is the main cathodic reaction in a corrosion cell process in aerated solutions. Under mass transport-limited conditions, the interfacial pH has been estimated to range from 10.35 to 10.65 [28] This is more than sufficient to induce precipitation of Ce^{3+} hydroxide from a 2.5mM Ce^{3+} solution [29] Precipitated layers form preferentially at cathodic sites associated with Cu-rich particles in high strength aluminum alloys [30], because oxygen reduction and its associated alkalinization is supported preferentially at these locations [23, 31] Precipitation is more uniform on zinc and steel substrates because the cathodic reaction is supported more uniformly across the surface.

Among the lanthanides, Ce and Pr may deserve special attention because they can be oxidized to a tetravalent state. The solubility of the tetravalent cations is exceedingly low (e.g., $K_{sp} = 10^{-50}$ for Ce^{4+}) [26], although it should be noted that Ce^{4+} can be complexed by a number of simple ligands, such as nitrates and sulfates, and is can remain supersaturated with respect to oxide precipitation [32]. In the absence of complexing agents, it is speculated that Ce^{3+} in solution is oxidized to Ce^{4+} by peroxide, which can be an intermediate in the oxygen reduction reaction. This leads to the formation of an exceedingly insoluble compound that contributes to that stability and protectiveness of the precipitated film.

Corrosion protection due to precipitated deposits that form during long-term immersion in lanthanide-bearing solutions results in latent corrosion protection [19] This observation led to the development of inorganic conversion coatings or pretreatments processes. Most notable among these are cerium conversion coatings [33, 34]. Peroxide additions and pH control have been used to accelerate the film formation process from tens of hours to minutes. These coatings can provide to high levels of corrosion protection, which approach that of chromate conversion coatings.

The use of Ce inhibitors and pigments in corrosion resistant organic coatings is a logical extension of the work on soluble inhibitors and conversion coatings. The notion is that Ce released from coating pigments would lead to corrosion inhibition as it is understood for Ce inhibitors and conversion coatings. Naturally occurring inorganic Ce compounds with suitable solubility characteristics are few, hence the use of cation exchanging compounds for storage and release of cerium inhibitors is potentially quite enabling. Early evidence is that Ce-releasing pigments in organic coatings enhance corrosion protection to some extent, but mechanistic studies suggest that the high corrosion resistance of Ce inhibitors and conversion coatings does not extrapolate directly to Ce pigments in coatings.

In one example, McMurray et al. have reported that cut edge corrosion is more effectively reduced by a Ce-exchanged bentonite pigment than by a $SrCrO_4$ pigment, or commercial Ca-exchanging inhibiting pigment (Shieldex^{TM})[29]. In these experiments, a Ce-bearing bentonite pigment was prepared by cation exchange in a Ce salt solution producing a compound containing over 30,000 ppm Ce^{3+}. The Ce-exchanged bentonite was then added to a polyester resin to form a coating with a 38% pigment volume concentration. This coating was applied to galvanized steel substrates with cut edges to test for corrosion protection. Performance of this coating was compared to coatings made with a Shieldex™ ion exchange pigment, a $SrCrO_4$ pigment and a Ca^{2+}-exchanged bentonite. As the solution permeated the coating, it was expected to trigger a Ce^{3+}-releasing exchange reaction with Na^+ ions in solution. Once released, Ce^{3+} was expected to act at cathodic sites to slow the corrosion cell process and the resulting delamination. Delamination of a polyester coating applied to galvanized steel surfaces from cut edges that exposed underlying bare metal appeared to be controlled by a 38% PVC Ce-exchanged bentonite pigment addition. The delamination rates were comparable to those observed when the coating was pigmented with $SrCrO_4$ and was decreased by about 50% compared to a commercial Ca^{2+}-exchanged silica pigment.

In another example, Chrisanti and Buchheit, prepared epoxy coatings pigmented with 25 wt.% Ce-exchanged bentonite and applied them to deoxidized and degreased Al 2024-T3 substrates and evaluated them as corrosion resistant coatings using electrochemical impedance spectroscopy (EIS) [35]. EIS was carried out in aerated 0.5M NaCl solutions using a three electrode measurement. A 10 mV sinusoidal voltage perturbation modulated over frequencies ranging from 10^5 to 10^{-2} Hz was used to interrogate electrochemical behavior of electrodes that presented a 14.6 cm^2 area to the electrolyte. In these experiments, low frequency impedances were high suggesting good corrosion protection. However, these coatings showed evidence of the onset of substrate corrosion as indicated by an increase in the coating capacitance associated with water uptake, the emergence of a pore resistance and the development of a well articulated resistance at the lowest measured frequencies (10^{-2} Hz) (Figure 3). By comparison, $SrCrO_4$-pigmented epoxy coatings showed very low pore resistances indicating rapid water uptake, but fully capacitive behavior at intermediate and low frequencies indicating perfect interfacial passivity in the presence of what is likely to be a chromate-saturated solution within the epoxy coating. Samples coated with epoxy containing a Ce-exchanged bentonite pigment were also subject to neutral salt spray exposure testing for 1200 hours. At the conclusion of the test, the scribed areas were corroded, with a slight amount of paint creep back and under-coat corrosion that had initiated as scribe marks. No blistering or corrosion under the coating was evident by visual inspection (Figure 4). Overall, these results suggest enhanced protection by Ce^{3+}-exchanged bentonite pigments, however the protection appears to be at levels considerably lower than that provided by $SrCrO_4$-pigmented coatings.

Several experiments were carried out to verify Ce release and to characterize any protective capacity of the Ce inhibitor released from the

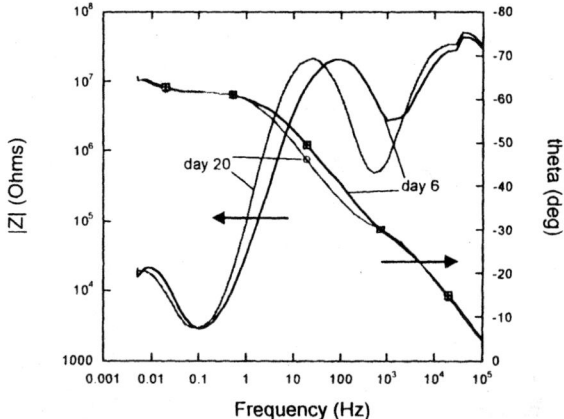

Figure 3. Bode magnitude and phase angle plots for an epoxy coating loaded with 25 wt.% Ce^{3+}-exchanged bentonite on 2024-T3 exposed to 0.5 M NaCl solution.

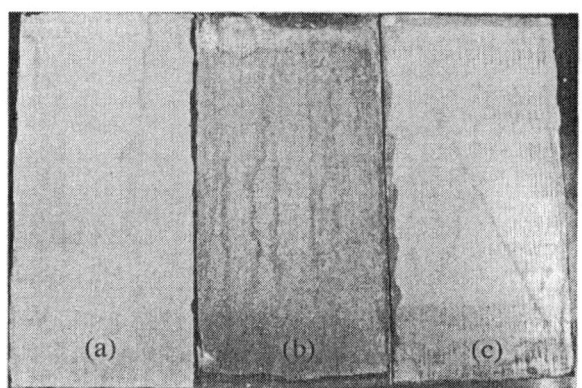

Figure 4. Photographs of epoxy coatings with 25 wt. % pigment loading levels after 1200 hours of salt spray exposure: (a) $SrCrO_4$, (b) Al-Zn hydrotalcite with exchangeable $V_{10}O_{28}^{6-}$, (c) Ce^{3+}-exchanged bentonite.

pigment, the Ce-exchanged bentonite pigment was soaked in 0.05 or 0.5M NaCl solution for 24 hours during which time Ce^{3+}/Na^+ exchange occurred releasing inhibitor into solution. The amount of inhibitor released was not quantified, but addition of hydrogen peroxide to an aliquot of the supernatant solution eventually produced precipitation of an orange solid consistent with the presence of Ce that was oxidized to the tetravalent oxidation state. The remainder of the supernatant and was used as an electrolyte for measurement of the anodic and cathodic polarization curves of Al 2024-T3. The pH of the bulk solution was estimated to be about 6.0. In aerated solutions, there was essentially no inhibition of anodic or cathodic reactions (Figure 5). This result likely reflects the low level of Ce^{3+} in solution and the unfavorable (low) bulk pH for precipitation of insoluble Ce hydroxides.

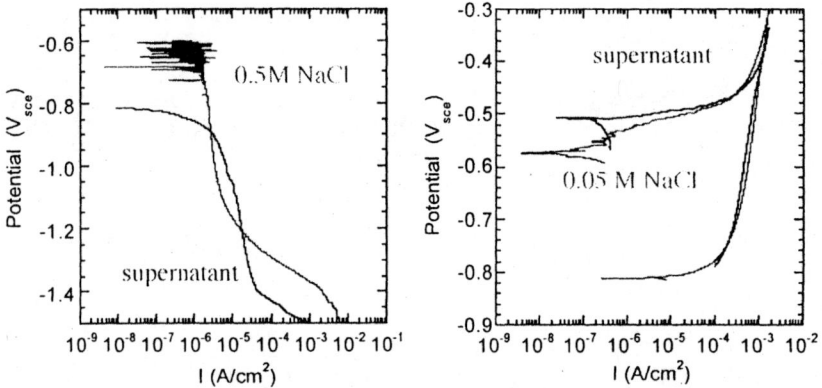

Figure 5. Cathodic (left) and anodic (right) polarization curves for 2024-T3 in Ce^{3+}-exchanged bentonite soaked in 0.05 or 0.5 M NaCl solutions for 24 hours.

On the basis of kinetic studies and characterization of delamination of polyvinyl butyral coatings with Ce-exchanged pigment from galvanized steel substrates, Williams, McMurray et al. concluded that the Ce released from the pigment in the coating was not capable of inhibiting corrosion of exposed surfaces in large defects such as scribes or cut edges [25]. However, coating delamination was significantly reduced. The observations suggested that Ce^{3+} concentrations at the exposed defect were too low and pH was too acidic for appreciable Ce^{3+} inhibition; consistent with the polarization results of Chrisanti and Buchheit [35]. While Ce^{3+} released into an established under-film electrolyte in the delaminated region did not migrate to and did not interact with the acidic environment at the delamination front to slow corrosion, it was observed to precipitate as $Ce(OH)_3$ under alkaline conditions and accumulated near the interface of the delaminated region and the bulk electrolyte. This deposit increased the length and tortuosity of the migrational pathway between the delamination front and the bulk electrolyte, which slowed the transport of Na^+ ions away from the front. As a result, delamination kinetics were slowed though they remained subject to mass-transport control.

These results suggest several important characteristics of Ce inhibition within an organic coating system. First, Ce^{3+} released from the pigment is able to act only by precipitating in the high pH electrolyte that develops at sites of local cathodic activity. This means that Ce^{3+} can inhibit only after a corrosion cell has been established. As such, Ce^{3+} does not act to inhibit corrosion initiation, but rather slows corrosion propagation. The implications of these mechanistic findings is that Ce-bearing pigments (and lanthanide-bearing pigments by extension) may be expected to work best in conjunction with systems with good inherent adhesion to the underlying substrate where coating delamination kinetics are slower. From a practical perspective, this would place a premium on good surface cleaning and the used of adhesion-promoting pretreatments. Second, it appears that Ce^{3+}, by itself does not possess the inhibiting power of chromates. It is primarily a cathodic inhibitor, which only acts on the cathodic half of the corrosion cell process. For these reasons, Ce^{3+} inhibitors in organic coatings will be best used in conjunction with supplemental anodic inhibitors.

Hydrotalcite Pigments and Corrosion Protection

The use of hydrotalcite compounds for corrosion protection originated in observations of unusual passivity of aluminum alloys in alkaline lithium salt solutions and passivity of Al-Li alloys in alkaline solutions [36-40]. This unexpected passivity was attributed to the formation of an M^+/M^{3+} host aluminum-lithium-carbonate hydrotalcite, $Li_2[Al(OH)_6]_2 \cdot CO_3 \cdot nH_2O$, that formed spontaneously on Al alloy substrates in the presence of a source of Li^+ and alkaline conditions [41]. This phenomenon was later exploited as the basis for an alkaline conversion coating pretreatment process for aluminum alloys and was recently modified for use as an adhesion promoting layer on galvanized steel [42]. Hydrotalcite conversion coatings on aluminum alloys can be formed by simple immersion in alkaline lithium salt solution baths. The coating that forms is polycrystalline and deposits in thicknesses of several micrometers (Figure 6). Corrosion protection is highest on Cu-free alloys, though methods of increasing corrosion resistance on Cu-rich Al alloys have been reported. Depending on the alloy substrate and the details of the coating application, hydrotalcite conversion coatings will resist pitting in neutral salt spray exposure for 336 or more hours [43]. Post-coating treatments with cerium salt solutions have been developed to impart a form of self-healing to these coatings. [44]

The use of hydrotalcites to deliver corrosion inhibiting pigments in paints is a recent advance with at least two distinct approaches. Buchheit, et al. and

Williams, et al., have described the synthesis and application of hydrotalcites to deliver inorganic corrosion inhibitors [45, 46]. Sinko and Kendig recently filed a patent application on a process involving use of hydrotalcites to deliver organic inhibitor compounds [47].

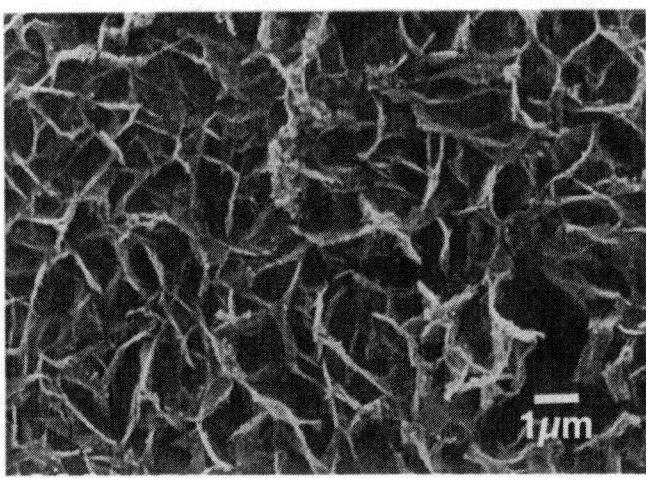

Figure 6. A hydrotalcite coating formed on an aluminum alloy substrate during immersion in a $LiNO_3$-based solution at pH 12 for 5 minutes.

The case of the aluminum-zinc hydroxide decavanadate hydrotalcite illustrates important aspects of corrosion inhibiting hydrotalcite pigments in coatings. This compound has been found to be an amphoteric exchanger that exchanges both decavanadate ($V_{10}O_{28}^{6-}$) and Zn^{2+}. Serial washing experiments intended to characterize the depth of the inhibitor reservoir show that inhibitor release is slow and exhaustion of the reservoir is difficult. In 100 hours of exposure to sodium chloride solutions (0.1 to 1.0 M) only 0.012 mmol/g vanadate and 0.059 mmol/g Zn^{2+} were released into solution. Indeed, exchange isotherms characterizing vanadate/chloride exchange and Zn^{2+}/Na^+ exchange showed an inhibitor affinity for the solid rather than the solution phase. The Gaines-Thompson equilibrium constants were determined to be 0.03 for vanadate/chloride exchange, and 1.10 for Na^+/Zn^{2+} exchange. These exchange characteristics are desirable for minimizing tendencies for osmotic blistering provided that sufficient inhibitor is released to achieve corrosion protection.

In potentiodynamic polarization of Al 2024-T3 aluminum in aerated sodium chloride solutions, vanadate is a good anodic inhibitor and a modest inhibitor of oxygen reduction. Figure 7 shows anodic polarization curves for Al 2024-T3 in a 0.124M NaCl solution to with additions of 0.1M sodium metavanadate or decavanadate corrosion inhibitors have been added. In anodic polarization, the pitting potential is increased by about 0.15V compared to the chloride-only case. Increases in the repassivation potential are not evident in these measurements.

Zinc is a good inhibitor of oxygen reduction on aluminum, but is not an inhibitor of anodic reactions. Figure 8 shows cathodic polarization curves for Al 2024-T3 in aerated 0.124 M NaCl solutions with additions of ZnCl. in amounts ranging from 0.05 to 0.25M. In the region where diffusion-limited oxygen reduction is observed in the polarization response, a decrease in the reaction rate of about an order of magnitude is observed upon addition of 0.25M $ZnCl_2$. The fact that the reaction remains mass transport limited suggests that Zn^{2+} is acting by blocking sites supporting cathodic reactions. These local sites are likely to be Cu-rich inclusions in the alloy.

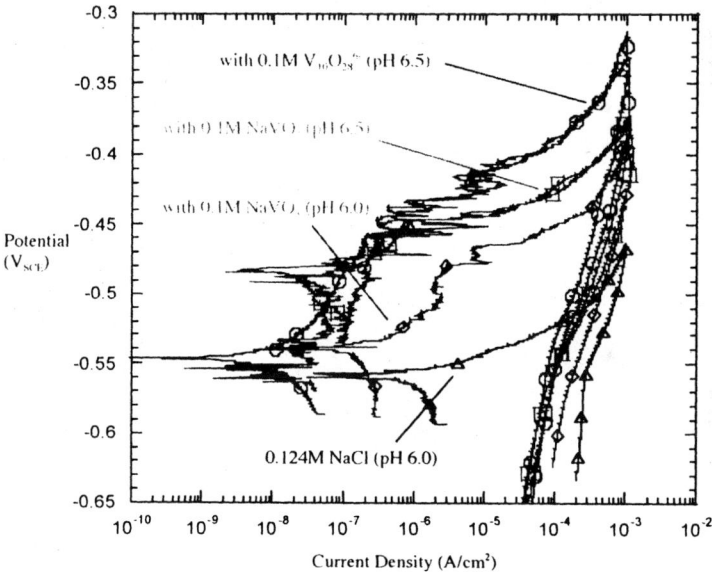

Figure 7. Anodic polarization curves of Al alloy 2024-T3 exposed to 0.124M NaCl at pH 6.0 to 6.5 with vanadate additions indicated in the figure.

These polarization results suggest that the mix of Zn^{2+} and vanadates is a potent inhibitor combination. However, as was made evident in the case of Ce inhibition, the extent to which polarization measurements accurately reflect inhibition of undercoating corrosion cells should always be considered carefully.

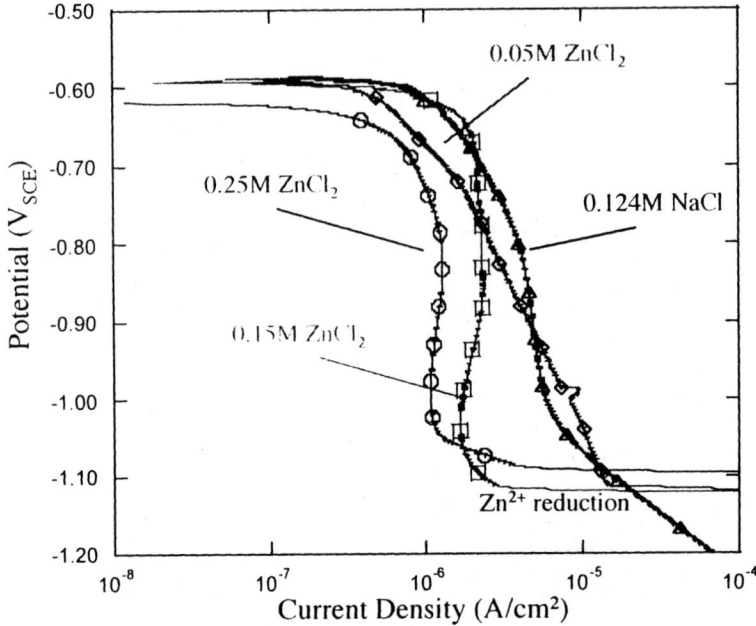

Figure 8. Cathodic polarization curves of Al alloy 2024-T3 exposed to $ZnCl_2$ and 0.124M NaCl (control).

The ability of hydrotalcite-pigmented coatings to release vanadate and Zn^{2+} inhibitors into solution and provide corrosion protection from afar has been demonstrated in simulated scratch cell experiments. A simulated scratch cell consists of two parallel aluminum coupons separated by about 5mm by an O-ring. This forms a small cell into which an aggressive electrolyte can be placed. One coupon is coated with a protective coating, while the other is left bare. A coating that releases inhibitor will do so into the cell electrolyte. If release is sufficient, inhibitors will be detected in solution and may interact with the bare surface to provide corrosion protection. In this arrangement the bare coupon serves as a simulated scratch that is to be protected by the inhibitor reservoir in the coating. This cell provides a means for testing key elements in the self-healing process: 1) release of inhibitor into solution, 2) interaction of the released inhibitor with the bare surface, and 3) increases in corrosion protection that can be detected electrochemically when the cell is outfitted with a counter and reference electrode.

Figure 9 shows optical photographs of simulated scratch cell coupons after exposure experiments. This sequence of photographs shows the coated and uncoated companion coupons from cells made with a 25 wt.% loading $SrCrO_4$

epoxy coating positive control, an experimental 25 wt.% hydrotalcite pigmented epoxy coating, and a negative control cell with two bare coupons. The electrolyte in these cells was 0.5M NaCl solution. The upper row of photographs shows the coated coupons and the lower row shows the uncoated "simulated scratch" coupons. Visual inspection shows staining and pitting on the negative control sample. The coupons from the SrCrO$_4$ and hydrotalcite coatings show no evidence of corrosion. Solution analysis showed a steady-state concentration of 0.2mM for CrO$_4^{2-}$ and 0.7mM for Sr^{2+}. In comparison, the concentration of vanadate decreased from 0.1 to 0.001 mM over 1000 hours of exposure while the Zn^{2+} concentration remained relatively steady at about 0.5mM. Figure 8 suggests that inhibitor concentrations at these levels are sufficient to preserve passivity for Al 2024-T3 in 0.5M NaCl solution. For the SrCrO$_4$ cell, in situ EIS measurements showed that higher total impedance values were sustained by the bare coupons in the positive control and experimental hydrotalcite coatings. Lower total impedance values that decreased with exposure time were observed for the negative control samples.

In neutral salt spray exposure, hydrotalcite pigmented epoxy coatings routinely protect against scribe corrosion for exposure durations in excess of

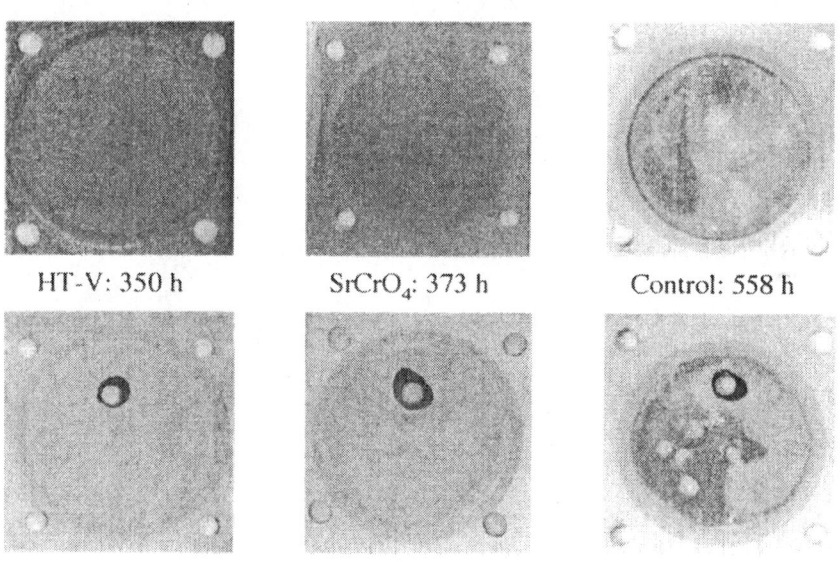

Figure 9. Optical photographs of simulated scratch cell faces after exposure to 0.5M NaCl solution for the times indicated in the figure. For the coated samples, the coated side of the cell is presented in the upper panel, the uncoated side is in the lower panel. (a) SrCrO$_4$-pigmented epoxy, a) Zn-Al-hydroxide decavanadate hydrotalcite pigment in epoxy, c) no coating (negative control).

1000 hours. Figure 10 shows an optical photograph of a scribed 25 wt.% loading hydrotalcite epoxy coatings applied to Al 2024-T3 after exposure to salt spray exposure for 1200 hours. The sample on the left was applied on top of a silane adhesion promoter, while the sample on the right was applied on a degreased and deoxidized surface where there was no attempt to promote adhesion. In both cases, the scribes remain shiny and unattacked at the conclusion of the exposure period. The sample applied on the degreased and deoxidized surface shows evidence of blistering that the silane pretreated sample does not. This suggests the need to promote adhesion to get the best performance from a hydrotalcite pigmented coating.

Sensing

The use of inorganic ion exchange compounds leads to the possibility of sensing water uptake and inhibitor exhaustion in organic coatings.

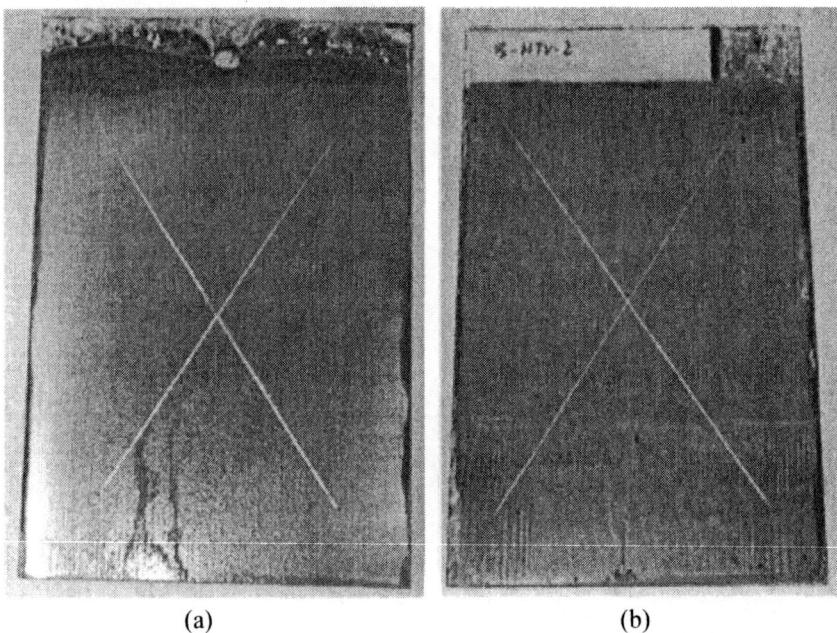

Figure 10. Optical photographs of scribed Zn-Al decavanadate hydrotalcite pigmented epoxy coatings (25 wt% loading) on 2024-T3 substrates after exposure to neutral salt spray exposure for 1008 hours. (a) surface pretreated with a silane adhesion promoter, (b) epoxy coating applied directly to a degreased and deoxidized surface.

Sensing water uptake in coatings

The hydrotalcite compound $Li_2[Al_2(OH)_6]_2 \cdot CO_3 \cdot nH_2O$ demonstrates a structural memory effect. When the compound is calcined at sufficiently high temperatures, it loses water and carbonate (as gaseous CO_2). At calcination temperatures at about 220°C, the compound loses crystallinity, which is readily detected by x-ray diffraction. When the poorly crystalline calcined product is exposed to an aqueous electrolyte, the compound rehydrates and takes up anions in a manner that restores its original structure and diffraction pattern with great fidelity. The diffraction patterns in Figure 11 illustrate the changes associated with calcination and rehydration. When calcined hydrotalcite powder is dispersed in an organic resin and applied to a metallic substrate as a coating, the rehydration transformation is retained and can be exploited for sensing water uptake in the coating during exposure to aqueous environments [48]. The rehydration transformation of calcined hydrotalcite can be tracked remotely by x-ray diffraction, and is indicated by an increase in the intensity of the diffraction pattern of the hydrated form of the hydrotalcite compound. The extent of the rehydration transformation can be tracked using the peak height ratio (PHR) of the (003) hydrotalcite diffraction peak, which is an indication of the basal plane spacing that is determined by the presence of exchangeable anions in the structure, to a diffraction peak from the substrate, which is used as an internal reference.

For the purposes of relating the PHR to coating water content, the volume fraction of water, X_v in the coating was estimated from coating capacitance data (from electrochemical impedance spectroscopy) using Brasher-Kingsbury equation. Both PHR (Figure 12) and X_v exhibit the three-stage "sigmoidal" curve shape, and regression analysis shows that X_v is linearly related to PHR (Figure 13) for an epoxy resin coating with about 10 wt.% calcined hydrotalcite pigment addition. Total coating thicknesses for coating used in this characterization were about 125 μm. With this model, XRD can be used to assess water uptake by the coating. Evidence suggests that substrate damage or coating/metal separation occurs at the end of the coating saturation period.

Sensing with Ce-bentonite Pigments

The Ce^{3+} cation is considerably larger than that of the sodium cation. Substitution of Ce^{3+} for Na^+ in the cation interlayer of bentonite produces a change in layer spacing along the c-axis of the crystal structure that is detectable by x-ray diffraction. The positions of the diffracted (100) peaks arising from the interlayer spacing determined by Na^+ ions and Ce^{3+} ions occurs between 4 and 8° 2-theta and the peaks themselves are separated by about 1.5 to 2.0°. Figure 14 shows x-ray diffraction patterns from epoxy coatings on Al 2024-T3 substrates containing 25 wt.% Ce-exchanged bentonite. These samples were exposed

126

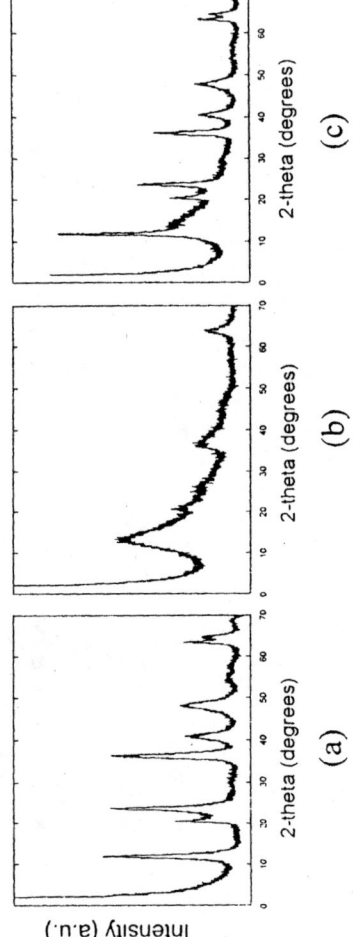

Figure 11. X-ray diffraction patterns of a Al-Li hydroxide carbonate hydrotalcite compound. (a) as synthesized, b) calcined in air at 220 °C for 2 hours, c) calcined and then immersed in 0.5M NaCl solution for 5 days. Note the coincidence in the peak positions for patterns a and c.

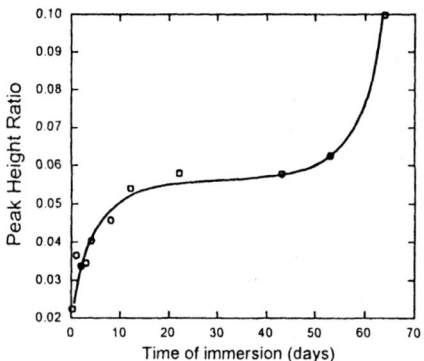

Figure 12. Variation in (003) hydrotalcite to (111) Al peak height ratio as a function of immersion time in 0.5M NaCl solution.

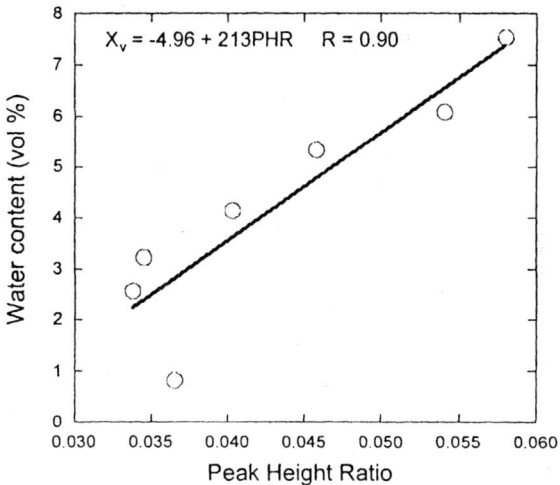

Figure 13. A plot of coating water content determined by electrochemical impedance spectroscopy and coating capacitance measurements and the 003 hydrotalcite to (111) Al peak height ratio. The line and its equation indicate a fit determined from a linear regression of the data set.

to neutral salt spray exposure and periodically removed (as indicated in Figure 14) for characterization by x-ray diffraction. The positions for the (100) basal plane peaks for Na and Ce-exchanged bentonite are indicated in the figure. During the course of exposure, the relative intensity of the peaks changes due to ion exchange. The patterns are virtually identical except for peaks associated with the interlayer cation. The extent of the exchange reaction can be characterized by measuring and comparing the intensities of the two (100) peaks. Figure 15 shows the variation in the ratio of the (100) peaks from Ce-bentonite to Nabentonite as a function of the square root of time for neutral salt spray exposure and exposure to an aqueous 0.5M NaCl solution during complete inundation. The variation in PHR suggests that the ion exchange process is subject to mass transport control independent of the exposure conditions examined.

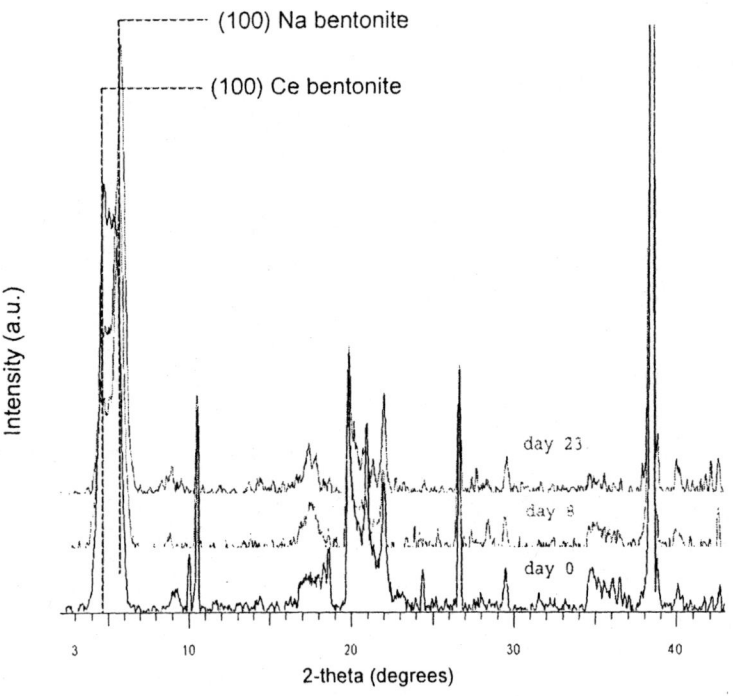

Figure 14. X-ray diffraction patterns from an epoxy coating on 2024-T3 Al loaded with 25% Ce-exchanged bentonite. Coated samples were exposed to neutral salt spray exposure for the times indicates on the plot. The positions of the peaks used to track Ce^{3+}-Na^+ ion exchange are indicated in the figure.

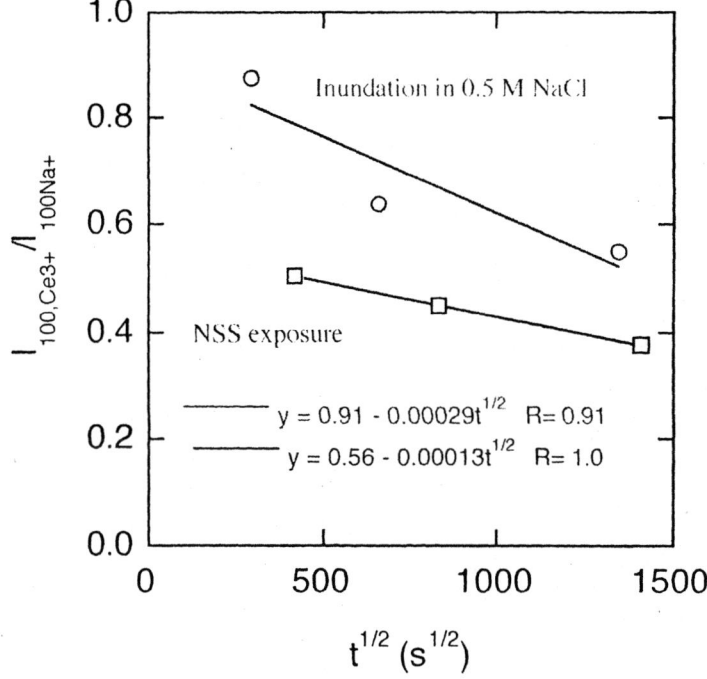

Figure 15. Variation in (100) Ce: (100) Na peak height ratio is consistent with diffusion-controlled Ce exchange.

Sensing with Hydrotalcites

An analogous form of inhibitor exhaustion sensing has been explored with an Al-Zn-decavanadate hydrotalcite-based pigment[45]. In this case, the objective was to sense exchange between chloride ion taken up from solution and decavanadate released into solution in hydrotalcite pigments added to an organic coating. Figure 16 shows diffraction patterns for an Al-Zn-chloride hydrotalcite and the Al-Zn-decavanadate hydrotalcite. These patterns are virtually identical to one another with the exception of the (003) basal plane peak in the chloride bearing compound and the (006) basal plane peak in the decavanadate compound. Sensing is then derived from relative intensity of these two peaks in the diffraction pattern of hydrotalcite pigment added to organic coatings. Figure 17 shows x-ray diffraction patterns for Al-Zn-decavanadate pigments exposed to a range of chloride bearing solutions for 22 days. Each pattern is from a sample that was exposed to solution, peaks indicating the presence of chloride and decavanadate in the structure are evident showing that the exchange process is occurring. The presence of chloride in the sample exposed to deionized (DI) water is due to chloride contamination released from the hydrotalcite pigment sample used in the experiment. The (003) chloride to (006) decavanadate peak height ratios in these samples are very small and present no systematic trend with solution chloride concentration. In experiments with Al-Zn-decavanadate pigments added to epoxy coatings and applied as coatings, diffraction patterns are obtainable, but changes in the peak height ratio are small and within the scatter of the measurements at least up to several hundred hours of salt spray or inundation exposure. The absence of a strong sensing response from this particular hydrotalcite compound is consistent with the limited extent of exchange discussed earlier. As a result, this compound is not expected to be a good short-term sensor. It may prove to be a better sensor of inhibitor exhaustion over long-term exposures, but experiments to support this have not yet been carried out.

Sensing Practicalities

Remote sensing using ion exchange pigments as presented here is dependent on the use x-ray diffraction as the method of interrogation. This interrogation method is subject to practical constraints if it is to be used in an

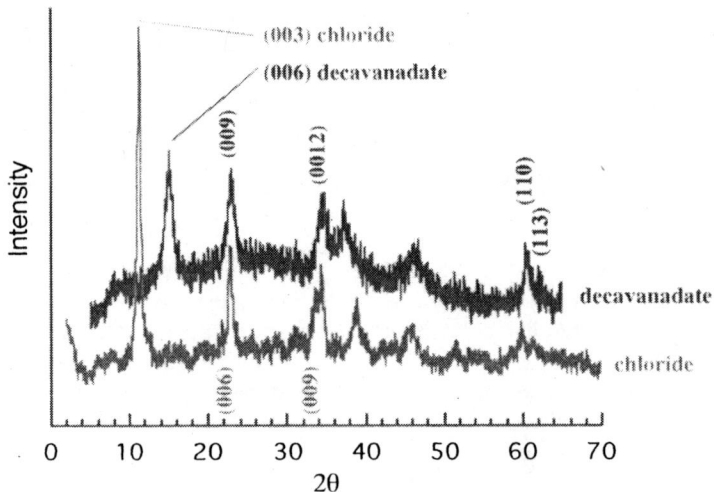

Figure 16. X-ray diffraction patterns for Al-Zn-chloride and Al-Zn-decavanadate hydrotalcite compounds showing the positions of the peaks associated with interlayer decavanadate and chloride.

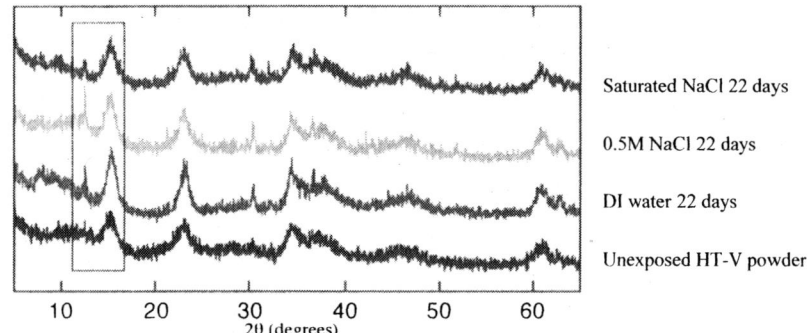

Figure 17. X-ray diffraction patterns for Al-Zn-decavanadate hydrotalcite compounds exposed to deionized water and chloride solutions. The peaks within the box are associated with interlayer chloride and decavanadate.

industrial setting. One issue is how to interrogate large surface areas rapidly and inexpensively. Laboratory-based powder diffraction-type measurements use monochromatic x-radiation in conjunction with a complex dynamic goniometer that enables a systematic collection of the diffraction pattern. How a laboratory x-ray diffraction measurement might be used to measure a large painted surface efficiently is difficult to imagine. However, it is possible to measure diffraction patterns rapidly using energy dispersive x-ray diffraction (EDXD), which utilizes a white radiation source [49, 50]. In this technique, the x-ray source-sample-detector geometry is fixed rather than adjustable as is the case in laboratory-based hardware. Large segments of the diffraction pattern can be obtained in a matter of seconds enabling the survey of large segments of surfaces area using a matrix of point-to-point measurements, each of which might be collected in a few seconds.

In order to obtain useful diffraction patterns it is necessary to get enough diffracted intensity from the pigment particles that are embedded within the coating system. These particles might be located at the base of the coating, or under layers of coating that are heavily pigmented with other x-ray-absorbing compounds. Simple mass absorption calculations can be used to determine the exiting intensity of x-rays based on knowledge of the effect mass absorption coefficients for coating components, the expected path length of x-rays through the coating system, and the energy of the x-ray source used. These calculations must be made on a case-by-case basis and are beyond the scope of the discussion here. However, the energy of Mo K_α radiation, a common industrial source of x-rays, is 17.5keV. This is sufficiently energetic to penetrate a nominal TiO_2-loaded coating to depth of several millimeters, which should be sufficient for many applications.

A final consideration for x-ray diffraction is the possibility of "pathological" overlap in the diffraction pattern of an ion exchange pigment and some other crystalline compound in the coating system. Overlaps of this type are those in which a diffraction peak of interest from an ion exchange compound overlaps with one from another component. This is also an issue that must be explored on a case-by-case basis.

Summary

The use of synthetic inorganic IECs is receiving new attention in the coating science and technology communities. Though efforts in this area are still limited, the potential may be significant. The use of these compounds enable coating designers the latitude to select from a broad range of inhibitors to meet a range of performance needs and manufacturing constraints.

Results reported in the literature so far show that useful levels of corrosion protection can be imparted by properly selected pigments. Mechanistic studies show that exptrapolation of electrochemical data on inhibition must be done with care–and especially with recognition of the local environments that develop in corrosion cells under organic coatings.

This new work shows that it is possible to impart new functionalities to coatings beyond corrosion protection. Examples presented include two types of sensing. The use of calcined hydrotalcite pigments leads to the ability to sense water uptake and x-ray diffraction methods can be used to track the progress of Ce^{3+}-Na^+ exchange in bentonite pigments to an extent that might enable a user to anticipate inhibitor exhaustion.

Acknowledgements

The authors would like to acknowledge the support of the Air Force Office of Scientific Research for support of work on hydrotalcite pigments at The Ohio State University conducted under contract no. F49620-01-1-0352.

References

1. Sinko, J., Prog. Organic Coat. **2001**, *42*, 267.
2. Helfferich, F., Ion Exchange. McGraw-Hill: New York, 1962; p.p 5.
3. sissoko, I.; Iyagba, E. T.; Sahai, R.; Biloen, P., J. Sol. St. Chem. **1985**, *60*, 283-288.
4. Grim, R. E., Clay Mineralogy. 2nd ed.; McGraw-Hill: New York, 1968; p p. 79.

5. Panda, K.; Das, J., J. Molecular Gatal. **2000**, *151*, 185-192.
6. Rives, V.; Uiibarni, M. A., Goord. Ghem. Rev. **1999**, *181*, 61-120.
7. Uiibarni, M. A.; Pavlovic, I.; Barriga, C.; Hermosin, M. C.; Gornejo, J., Applied Clay Sci. 2001, 18, 17-27.
8. Hussein, Z.; Zainal, Z.; Yaziz, 1.; Beng, T. C., J. Environ. Sd. Health 2001, A36, *565-573*.
9. Huang, F-C.; Lee, J.-F.; Lee, C-K.; Tseng, W.-N.; Juang, L.-G., J. Coil. Interf. Sci. 2002, 256, 360-366.
10. Meier, L. P.; Nuesch, R., J. Coil. Intef. Sci. 1999, 217, 77-85.
11. Heifferich, F., Ion Exchange. McGraw-Hill: New York, 1962; p p. 151.
12. Hutson, N. D.; Hoekstra, M. J.; Yang, R. T., Micro. Meso. Mater. 1999, 28, (28), *447-459*.
13. Zin, I. M.; Pokhmurskii, V. I.; Scantiebury, J. D.; Lyon, S. B., J. Eiectrochem. Soc. 2001, 148, B293-B298.
14. Zin, I. M.; Pokhmurskii, V. I.; Scantlebury, J. D.; Lyon, S. B., J. Electrochem. Soc. **2001**, *148*, B293-B298.
15. Zin, I. M.; Lyon, S. B.; Pokhmurskii, V. I., Corrosion Sci. **2003**, *45*, 777-788.
16. Zin, I. M.; Lyon, S. B.; Pokhmurskii, V. I.; Simmonds, M. C., Corrosion Eng. Sci. Tech. **2004**, *39*, 167-173.
17. Zin, I. M.; Lyon, S. B.; Hussain, A., Prog. Org. Coat. **2005**, *52*, 126-135.
18. Hinton, B. R. W.; Arnott, D. R.; Ryan, N. E., Metals Forum **1984**, *7*, 211-217.
19. Hinton, B. R. W.; Arnott, D. R., Microstructural Sci. **1989**, *17*, 311-320.
20. Arnott, D. R.; Hinton, B. R. W.; Ryan, N. E., Corrosion **1989**, *45*, 12-18.
21. Davenport, A. J.; Isaacs, H. S.; Kendig, M. W., Corrosion Sci. **1991**, *32*, 653-663.
22. Hinton, B. R. W., J. Alloys Cpds. **1992**, *180*, 15-25.
23. Aldykewicz, A. J.; Isaacs, H. S.; Davenport, A. J., J. Electrochem. Soc. **1995**, *142*, 3342-3350.
24. Bethencourt, M.; Botana, F. J.; Calvino, J. J.; Marcos, M.; Rodriguez-Chacon, M. A., Corrosion Sci. **1998**, *40*, 1803-1819.
25. Williams, G.; McMurray, H. N.; Worsley, D. A., J. Electrochem. Soc. **2002**, *149, (4)*, B154-B162.
26. Gavarini, S.; Guittet, M. J.; Trocellier, P.; Gautier-Soyer, M.; Carrot, F.; Matzen, G., J. Nucl. Mater. **2003**, *322*, 111-118.
27. Arnott, D. R.; Hinton, B. R. W.; Ryan, N. E., Mater. Perf. 1987, August, 42-47.
28. Bohm, S.; Greef, R.; McMurray, H. N.; Worsley, D. A., J. Electrochem. Soc. **2000**, *147*, 3268-
29. Bohm, S.; McMurray, H. N.; Powell, S. M.; Worsley, D. A., Werkst. Korros. **2001**, *52*, 896-903.
30. Arnott, D. R.; Ryan, N. E.; Hinton, B. W. R.; Sexton, B. A.; Hughes, A. E., Appl. Surf. Sci. **1985**, *22-23*, 236-251.

31. Aldykiewicz, A. J.; Isaacs, H. S.; Davenport, A. J., J. Electrochem. Soc. **1996**, *143*, 147-154.
32. Baes, C. F.; Mesmer, R. E., The Hydrolysis of Cations. Robert E. Knieger Pub. Co.: Malabar, FL, 1986; p p. 138.
33. Hinton, B. R. W.; Wilson, L. Effect fo Cerous Chloride Concentration on Corrosion Rate. 1988.
34. Mansfeld, F.; Lin, S.; Kim, S.; Shih, H., Electrochim. Acta **1989**, *34*, 1123.
35. Chrisanti, S.; Buchheit, R. G., ACS Abstr. **2004**, 228, 240-POLY Part 2 U360-U360.
36. Gui, J.; Dveine, T. M., Scripta Met. **1987**, *21*, 853-857.
37. Craig, J. G.; Newman, R. C.; Jarrett, M. R.; Holroyd, N. J. H., J. de Physique **1987**, *48*, C3-825 - C-833.
38. Rangel, C. M.; Travassos, M. A., Corrosion Sci. **1992**, *33*, 327-343.
39. Buchheit, R. G.; Wall, F. D.; Moran, J. P.; Stoner, G. E., Role of Anodic Dissolution in the Stress Corrosion Cracking of Al-Li-Cu Alloy 2090. In Proceedings of the First International Symposium on Environmental Effects on Advanced Materials, R.D. Kane, E., Ed. NACE: Houston, TX, 1992; pp 8-1.
40. Tanaka, S.; Isobe, Y.; Hine, F., Corrosion Eng. **1990**, *39*, 479-488.
41. Buchheit, R. G.; Bode, M. D.; Stoner, G. E., Corrosion **1994**, 205.
42. Buchheit, R. G.; Guan, H., J. Coat. Tech. Res. **2004**, *1*, 277-290.
43. Zhang, W.; Buchheit, R. G., Corrosion **2002**, *58,* 591.
44. Buchheit, R. G.; Mamidipally, S. B.; Schmutz, P.; Guan, H., Corrosion **2002**, *58,* 3-14.
45. Buchheit, R. G.; Guan, H.; Mahajanam, S.; Wong, F., Prog. Org. Coat. **2003**, *47*, 174.
46. Williams, G.; McMurray, H. N., Electrochem. Sol. St. Lett. **2003**, *6*, B9-B11.
47. Sinko, J.; Kendig, M. W. Pigment grade corrosion inhibitor host-guest compositions and procedure. Feb. 3, 2005, 2005.
48. Wong, F.; Buchheit, R. G., Progress in Organic Coatings **2004**, *51*, 91-102.
49. Ruud, C. O., X-ray Analysis and Advances in Portable Field Instrumentation. J. of Metals 1979, June, p. 10.
50. Sarrazin, P., A Miniature XRD/XRF Instrument for in-situ Characterization of Martian Soils and Rocks. J. Phys. IV **1998**, *8,* Pr5-465.

Chapter 9

Polyurethane–Polysiloxane Ceramer Coatings

Hai Ni[1,4], William J. Simonsick[2], and Mark D. Soucek[3,*]

[1]Polymers and Coatings Department, North Dakota State University, Fargo, ND 58105
[2]DuPont Marshall Lab, 3401 Grays Ferry Avenue, Philadelphia, PA 19146
[3]Polymer Engineering Academic Center, 250 South Forge Street, Room 221, Akron, OH 44325–0301
[4]Current address: BEHR Corporation, 3400 Segerstrom Avenue, Santa Ana, CA 92704

New low VOC (volatile organic compounds) polyurethane/oligosiloxane coatings were formulated using, 1, 6-hexamethylene diisocyanate (HDI) isocyanurate, alkoxysilane-functionalized HDI isocyanurate, tetraethyl orthosilicate (TEOS) oligomers and three cycloaliphatic polyesters. One series of polyester was synthesized using 1,4-cyclohexanedimethanol (CHDM) with 1,4-cyclohexanedicarboxylic acid (1,4-CHDA) and 1,3-cyclohexanedicarboxylic acid (1,3-CHDA). A second series of polyesters was synthesized using 2-butyl-2-ethyl-1,3-propanediol (BEPD) with 1,4-CHDA or 1,4-CHDA and 1,3-CHDA, respectively. The polyurethane provided the required mechanical properties and polysiloxane functioned as an adhesion promoter and a corrosion inhibitor.

The continuous organic polyurethane phase was coupled to the inorganic polysiloxane phase via the alkoxysilane-functionalized HDI isocyanurate. The general coatings, tensile, viscoelastic properties, and corrosion resistance were evaluated for the ceramer coatings as functions of the polyester and concentration of TEOS oligomers. The coatings properties were dominated by the organic phase. The corrosion resistance was found to be a function of TEOS with a maximum of corrosion resistance and coating properties at 10-15 wt % TEOS. The film morphology was investigated using the combination of SEM (Scanning Electron Microscope) and X-ray analysis. Phase separation was observed for ceramer coatings resulting in SiO_2 particles. The size of the pre-ceramic silicon-oxo-particles was controlled by the Tg of the organic phase and the concentration of TEOS oligomers.

Introduction

The aluminum substrates for aircraft are coated with many steps. First, the aluminum is pretreated with a chromate solution as a conversion layer to prevent corrosion. Following that, a 2-K epoxy-polyamine system with the chromate pigment is applied as a primer, in which the chromate pigment functions as corrosion inhibitor. The topcoat is typically a two-component polyurethane coating consisting of an HDI oligomer and a polyester polyol. In this system, the presence of chromate is carcinogenic. Many approaches have been carried out in this area to limit the usage of chromate.[1,2]

Our approach is to develop a polyurethane/polysiloxane ceramer self-priming coating or "Unicoat" system, which was designed on the basis of the traditional aircraft topcoat. In this system, polyurethane provides the general mechanical properties and polysiloxane functions as an adhesion promoter and corrosion inhibitor. Our previous studies of polyurea/polysiloxane ceramer coatings showed that the alkoxysilanes have a dramatic effect on the adhesion enhancement.[3] The polysiloxane formed by the sol-gel precursor, Tetraethyl orthosilicate (TEOS) oligomers, has a strong effect on the corrosion inhibition.[4]

To improve the UV-resistance for the traditional aircraft topcoat, a series of cycloaliphatic polyesters were also investigated in our previous study.[5] The cycloaliphatic polyesters can provide better flexibility and yellowing resistance compared to the phenyl structure of aromatic diacids in the traditional formulation.

To further simulate the traditional aircraft topcoat, the cycloaliphatic polyesters were added into the polyurea/polysiloxane ceramer coating system to formulate a polyurethane/polysiloxane ceramer coating system in this study. Two typical high-solid polyesters were synthesized from the diols, 1,4-cyclohexanedimethanol (1,4-CHDM) and 2-butyl-2-ethyl-1,3-propanediol (BEPD), and the diacids, 1,4-cyclohexanedicarboxylic acid (1,4-CHDA) and 1,3-cyclohexanedicarboxylic acid (1,3-CHDA). The esterification reaction is shown in Equation 1. The chemical structure of BEPD is shown in Scheme 1.

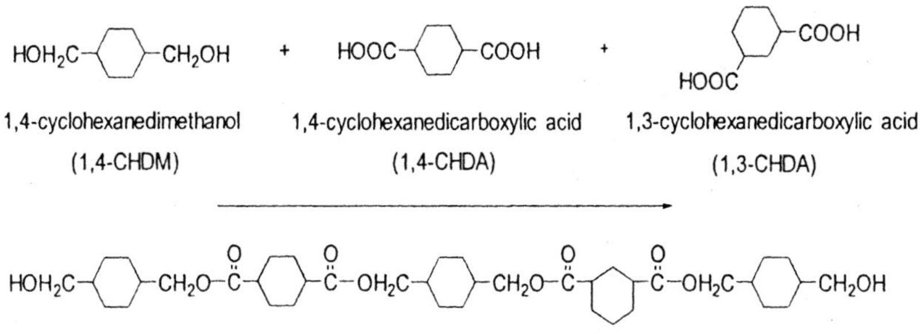

HOH₂C—⟨ ⟩—CH₂OH   +   HOOC—⟨ ⟩—COOH   +   HOOC—⟨ ⟩—COOH
                                                    |
                                                  HOOC

1,4-cyclohexanedimethanol   1,4-cyclohexanedicarboxylic acid   1,3-cyclohexanedicarboxylic acid
(1,4-CHDM)   (1,4-CHDA)   (1,3-CHDA)

$$HOH_2C-\bigcirc-CH_2O-\overset{O}{\overset{\|}{C}}-\bigcirc-\overset{O}{\overset{\|}{C}}-OH_2C-\bigcirc-CH_2O-\overset{O}{\overset{\|}{C}}-\bigcirc-\overset{O}{\overset{\|}{C}}-OH_2C-\bigcirc-CH_2OH \qquad (1)$$

$$\begin{array}{c} CH_2CH_3 \\ | \\ HO-H_2C-C-CH_2-OH \\ | \\ CH_2CH_2CH_2CH_3 \end{array}$$

2-butyl-2-ethyl-1,3-propanediol
(BEPD)

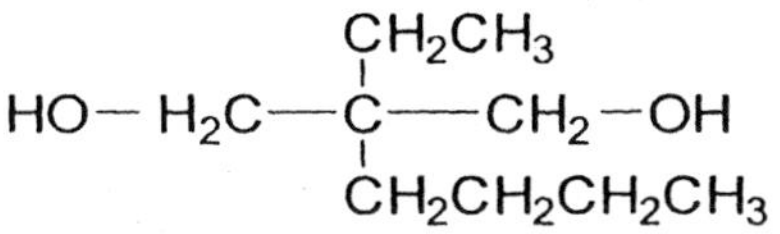

Scheme 1. Chemical structure of 2-butyl-2-ethyl-1,3-propanediol.

Experimental

The monofunctionalized HDI isocyanurate,[6] TEOS oligomers[4] and polyesters[7] were prepared as previously reported. The chemistry of the preparation of monofunctionalized isocyanurate is shown in Equation 2. The HDI isocyanurate (**1**) and monofunctionalized isocyanurate (**2**) were formulated with TEOS oligomers and polyesters as shown in Table 1.

The films were cast on aluminum panels (3003 H14, Q-Panel) with a wet thickness of 152.4 μm (6 mil) by a drawdown bar for the general mechanical measurement and with a wet thickness of 101.6 μm (4 mil) on the aluminum alloy 2024-T3 for the corrosion resistance test (salt spray). The films were set in a closed box with a stable humidity of 30% at 25°C for 96 hour until surface became tack-free, and then were put into a closed oven and further moisture cured at 88-90 °C under 100% humidity for 10 hours.

The Sample "Chromate Pretreatment" was pretreated with Alodine 1200 solution then coated with a 4 mil thickness of coating BEPD(4) in Table 1. The sample "Primer" was first coated with 2 mil thickness of two-component epoxy-

$$\underset{(1)}{\text{OCN-(H}_2\text{C})_6\text{-N-C-N-(CH}_2)_6\text{-NCO, isocyanurate with (CH}_2)_6\text{-NCO branch}} + \underset{(2)}{\text{NH}_2(\text{CH}_2)_3\text{-Si-(OC}_2\text{H}_5)_3}$$

$$\longrightarrow \text{OCN-(H}_2\text{C})_6\text{-[isocyanurate ring with (CH}_2)_6\text{-NCO]-(CH}_2)_6\text{-NHCONH(CH}_2)_3\text{-Si-(OC}_2\text{H}_5)_3$$

polyamide primer (Deft 02-Y-40) and then coated with 2 mil thickness of coating BEPD (4) after the primer was dried. The samples "Chromate Pretreatment" and "Primer" were cured according to the standard procedures for products 02-Y-40 (epoxy-polyamide primer) and 03-W-127A (polyurethane topcoat) from Deft Chemicals.

All mechanical tests were performed with standard procedures according to ASTM standards, including indentation hardness (Tukon hardness, ASTM D 1474-85), pencil hardness (ASTM D 3363-74), crosshatch adhesion (ASTM D 3359-87), pull-off adhesion (ASTM D 4541-85), and reverse impact resistance (ASTM D 2794-84). The prohesion was performed in a cyclic corrosion tester (Q-Panel) according to the ASTM standard G85-94. The test and evaluation were previously reported.[8] The surface morphology was investigated using Scanning Electron Microscopy (SEM) combined with Energy Dispersion X-ray Analysis (EDAX) with a JEOL JSM-6300 scanning electron microscope. The film was prepared on an epoxy plate and the sample sputter was coated with Au. Five random points were selected for the elemental analysis by X-ray.

Results and Discussion

A common ratio of NCO/OH, 1.1/1.0, was used for all formulations in Table 1. The general coating properties are shown in Table 2. The pencil hardness showed no significant difference between the formulations of either series. The difference can be observed comparing the formulations with different polyesters. Polyester with diol 1,4-CHDM possesses a rigid

Table 1. The Formulations of Coatings[a]

Coatings	1	2	Polyesters	TEOS Oligomers
CHDM(4/3)	36.7	0	CHDM(4/3): 63.3	0
CHDM(4/3)T0	19.9	25.0	CHDM(4/3): 55.1	0
CHDM(4/3)T2.5	19.4	24.3	CHDM(4/3): 53.8	2.5
CHDM(4/3)T10	17.9	22.4	CHDM(4/3): 49.7	10.0
BEPD(4)	36.7	0	BEPD(4): 63.3	0
BEPD(4)T0	19.9	25.0	BEPD(4): 55.1	0
BEPD(4)T2.5	19.4	24.3	BEPD(4): 53.8	2.5
BEPD(4)T10	17.9	22.4	BEPD(4): 49.7	10.0

[a] All the compositions are weight percentages and the reactants were diluted with 10 wt % acetone.

cyclohexane ring, providing films with harder properties compared to the branched diol, BEPD. The Tukon hardness shows that the hardness increased with the addition of 2 for both groups.

For the four formulations with CHDM(4/3), the Tukon hardness further increased with the initial addition of TEOS oligomers (2.5 wt %) and then decreased with increasing concentration of TEOS oligomers (10 wt %).

For the four formulations with BEPD, Tukon hardness decreased with the addition of TEOS oligomers. Similar to the polyurea, the effect of alkoxysilane on the adhesion improvement can also be observed for the polyurethane. The polyurethane CHDM(4/3) has a low pull-off adhesion, 143 lb_f/in^2, as well as a low crosshatch adhesion, 3B (The other seven formulations have the highest grade crosshatch adhesion, 5B). The adhesion was increased from 143 to 347 lb_f/in^2, a 143% increase with the addition of 2, and then increased to 467 lb_f/in^2, a 228% increase with addition of 10 wt % TEOS oligomers. The pull-off adhesion of four formulations with BEPD shows a similar variation tendency to the formulations with CHDM. Both the alkoxysilane group of 2 and TEOS oligomers has an effect on the adhesion improvement. The film CHDM(4/3) has lower impact resistance compared to other films, which may be caused by the poor adhesion. The reverse impact resistance of other formulations can not be distinguished due to the limitation of the scale.

Reverse impact resistance was greatly increased by the polyesters comparing the polyurethane/polysiloxane ceramers with the polyurea/

polysiloxane ceramers. The range of reverse impact resistance of polyurea/polysiloxane ceramers is from 10 to 35 lb·in.

Table 3 shows the tensile properties for the two series of films. The four films with polyester CHDM(4/3) have higher tensile strength and tensile modulus, and lower elongation-at-break than the four films with polyesters BEPD(4). The tensile properties are mainly controlled by the polyesters. In the four films with CHDM(4/3), the tensile strength increases from 39.3 to 54.2 MPa with the addition of alkoxysilane-functionalized isocyanurate (2) comparing the film CHDM(4/3)T0 with the film CHDM(4/3)T0. When TEOS oligomers were added into the formulation CHDM(4/3)T2.5, the tensile strength decreased from 54.2 to 45.8 MPa. This was attributed to the phase separation from the TEOS oligomers which produces defects in the film. When the TEOS oligomers concentration further increases to 10 wt %, the tensile strength decreases to 42.0 MPa. A similar trend as the CHDM polyurethane was observed for the formulations based on BEPD(4).

Table 2. General Mechanical Properties of the Films

Coatings	Pencil Hardness	Tukon Hardness (KHN)	Pull-off Adhesion (lb_f/in^2)	Reverse Impact (lb·in)
CHDM(4/3)	H/F	7.2±1.0	143±41	73±6
CHDM(4/3)T0	H/F	8.2±0.8	347±73	80+
CHDM(4/3)T2.5	H/F	9.5±0.5	346±69	80+
CHDM(4/3)T10	H/F	7.4±0.6	467±33	80+
BEPD(4)	HB/B	2.3±0.1	207±46	80+
BEPD(4)T0	HB/B	2.5±0.2	375±56	80+
BEPD(4)T2.5	HB/B	2.0±0.2	436±63	80+
BEPD(4)T10	HB/B	1.2±0.1	428±50	80+

The salt spray experiment was performed in a cyclic corrosion tester (Q-Panel) according to the ASTM standard G 85-94. The electrolyte called dilute Harrison solution is a solution of NaCl (0.05 wt %) and $(NH_4)_2SO_4$ (0.35 wt %) in water with a pH of 5.0-5.4. A cyclic fog/dry procedure was applied in the test, which consists of 1 hour fog at 25 °C followed by 1 hour dry-off at 35 °C. After 2400 hour exposure to salt spray, the bare aluminum panels (2024 T3) coated with BEPD (4)T2.5, BEPD(4)T10, chromate pretreatment and epoxy-polyamide primer was rated as 8, 8, 8, and 9 as shown in Figure 1 . Among the four samples, the epoxy-polyamide primer gave the highest value for corrosion

Table 3. Tensile Properties of the Films

Coatings	Tensile Strength (MPa)	Elongation-at-Break (%)	Tensile Modulus (MPa)
CHDM(4/3)	39.3±6.0	3.9±0.7	1003±54
CHDM(4/3)T0	54.2±4.7	5.1±0.5	1072±61
CHDM(4/3)T2.5	45.8±3.9	4.9±0.7	1018±64
CHDM(4/3)T10	42.0±5.3	4.2±0.7	1003±43
BEPD(4)	17.1±4.5	108.3±24.9	363±46
BEPD(4)T0	17.4±2.0	74.0±21.2	437±49
BEPD(4)T2.5	19.1±4.2	97.8±22.0	425±22
BEPD(4)T10	14.7±3.9	101.4±20.8	206±25

inhibition; however, none of the other systems utilized pigment. There are different inhibiting mechanisms for these samples. The mechanism of chromate pretreatment was proposed that the hexavalent chromium oxidized the aluminum forming a series of mixtures which passivate aluminum substrate. For the epoxy-polyamide primer, the pigment Strontium Chromate in the primer functions as corrosion inhibitor. It has proper leaching rate in the epoxy resin and migrates onto the exposed area of the aluminum substrate providing the corrosion protection.

The pigment can provide adequate inhibitor for a long period. As proposed in our previous study,[4] the TEOS ologomers may transport to the substrate and form a condensed barrier layer which functions as corrosion protection. The difference of corrosion inhibition between BEPD (4) T2.5 and BEPD (4) Tl0 was not distinguishable.

The organic/inorganic phases in the polyurethane/polysiloxane ceramer coating system were investigated using SEM. The phase separation was observed for both polyurethane coatings with polyester BEPD when the TEOS oligomers concentration reached 10 wt % as shown in Figure 2. A similar result of phase separation was also observed for the samples CHDM (4/3) T0, CHDM (4/3) T2.5, and CHDM (4/3) T10.

No phase separation was observed from the cross section of film BEPD (4) 10 and BEPD (4) T2.5. The HDI isocyanurate (**1**) and alkoxysilane-functionalized isocyanurate (**2**) have good compatibility and formed a homogeneous phase. Visually, the film of BEPD (4) T0 is transparent. No phase separation was observed within the SEM with the addition of 2.5 wt% TEOS oligomers into the system. When the TEOS oligomers increased to 10 wt% (BEPD (4) T10), spherical aggregates were observed in the cross section for the

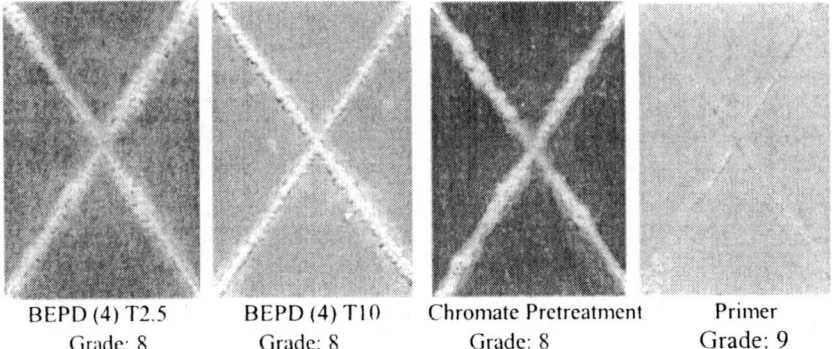

| BEPD (4) T2.5 | BEPD (4) T10 | Chromate Pretreatment | Primer |
| Grade: 8 | Grade: 8 | Grade: 8 | Grade: 9 |

Figure 1. The corrosion photographs of aluminum panels (2024 T3 bare) at the exposure time 2400 h.

ceramer coating as shown in Figure 2. An elemental analysis of silicon content was performed using EDAX for both the aggregate and continuous area. The silicon content by weight is 2.3-3.3 Si wt % for continuous phase. The silicon content for the aggregates was dependent on the selected position on the aggregate ranging from 6.0 to 37.2 Si wt%.

Conclusions

The polyurethane/polysiloxane ceramer coatings were prepared with the addition of polyesters into the polyurea/polysiloxane ceramer system. The impact resistance was increased by the polyesters. Similar to polyurea/polysiloxane ceramer coatings, the adhesion was also increased by the alkoxysilane group and TEOS oligomers. The hardness was determined by the polyesters. The films with BEPD are soft with a lower tensile strength, tensile modulus and lower hardness, and higher elongation-at-break as previously reported.[7]

The corrosion inhibition of polyurethane/polysiloxane ceramer coatings can compete with the chromate pretreatment. The polyurethane/polysiloxane ceramer coatings show the potential applications as a self-priming unicoat system for aircraft and also as high-performance topcoat, basecoat, or refinish for automotive.

(a): BEPD (4) T0

(b): BEPD (4) T2.5

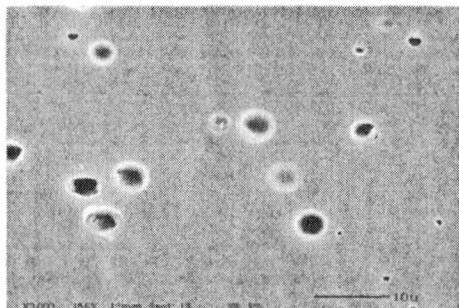
(c): BEPD (4) T10

Figure 2. SEM photograph of a cross section of ceramer coatings BEPD (4) T0, BEPD (4) T2.5 and BEPD (4) T10; magnification: 2000.

Acknowledgement

Financial support for this research from Air Force Office of Science Research grant (AFOSR/NL F49620-97-1-0218) is gratefully acknowledged. Dr. Bierwagen's group is acknowledged for the corrosion test and Dr. Glass' group for the viscosity measurement. Eastman Chemicals is acknowledged for the 1, 4-CHDA, 1, 3-CHDA, and CHDM, Bayer Corporation for HDI isocyanurate (Desmodur N-3300), and Deft Chemicals for the epoxy-polyamide primer (Deft 02-Y-40).

References

1. Holmes-Farley, S.D.; Yanyo, L. C. *J. Adhesion Sci. Technol.* **1991**, *5(2)*, 131.
2. Hegedus, C. R.; Pulley, D. F.; Eng, S. A. T.; Hirst, D. J. *J. Coat. Technol.* **1989**, *61(778)*, 31.
3. Ni, H.; Skaja, A.D.; Sailer, R.A.; Soucek, M.D. "Moisture-Curing Alkoxysilane-Functionalized Isocyanurate Coatings", *Macromol. Chem. Phys.* **2000**, *201*, 722.
4. Ni, H.; Simonsick, W.J.; Skaja, A.D.; Williams, J.P.; Soucek, M.D. *Prog. Organic Coat.* **2000**, *38(2)*, 97.
5. Ni, H.; Skaja, A.D.; Thiltgen, P.R; Soucek, M.D.; Simonsick, W.J.; Zhong, W. *Polym. Prepr.,* Am. Chem. Soc., Div. Polym. Chem. **1999**, *40(1).*
6. Ni, H.; Aaserud, D.J.; Simonsick, W.J. Jr.; Soucek, M.D. *Polymer,* **2000**, *41*, 57.
7. Ni, H.; Johnson, A.H.; Fries, M.L.; Skaja, A.D.; Worden, J.G.; Soucek, M.D. *Proceedings of the 78th Annual Meetings Technical Program of the FSCT, p.177,* Oct. 2000, Chicago, Illinois.
8. Ni, H.; Li, J.; He, L.; Skaja, A.D.; Fries, M.L.; Soucek, M.D. *Polym. Mater. Sci. Eng.,* Am. Chem. Soc., Div. Polym. Chem. **2000**, *83*, 303.

Specialty Coatings

Chapter 10

Effect of Selected Processing Parameters on the Key Properties of Solvent-Cast Polyimides for Optical Waveguides

Andrew J. Guenthner, K. R. Davis, L. Steinmetz, and J. M. Pentony

Code 4982, Naval Air Warfare Center, Department of the Navy,
China Lake, CA 93555-6106

The relationship between the key processing parameters and properties of interest for the fabrication of integrated optical devices are examined for four representative polyimide materials. Methods for estimating and controlling both the film thickness and refractive index profile of polyimide films (including experimental verification) that have proven successful in a variety of applications are presented and discussed.

Introduction

For five decades, polyimides have played an increasing role in high-performance polymer structures and coatings.[1] The widespread use of polyimides results not only from their excellent combination of thermal, electrical, and optical properties, but also from the high degree of flexibility afforded by the chemical synthesis procedures. Polyimide copolymers have been synthesized from a vast array of starting materials, including, for instance, crown ethers[2] and polyhedral oligomeric silsesquioxanes.[3] As a result, extensive tuning of the properties of interest and the incorporation of highly specialized chemical functionalities are possible in a single material that features excellent thermal stability.

Although polyimides have been used extensively as structural materials, a large number of applications for polyimide coatings also exist. In addition to the more traditional microelectronics applications, such as interlayer dielectrics, there has been substantial interest in the use of polyimide coatings for integrated optics applications.[4-10] In general, polymers for integrated optics must be highly stable, in order to survive processing while (in the case of active devices) providing protection for delicate nonlinear optical chromophores. At the same time, they must have electrical and optical properties that may be readily adjusted in order to fine-tune device performance without the need to develop a new material "from scratch" with each new device. Thus, in many ways polyimides are ideally suited for use in integrated optics.

The fabrication of an optical waveguide device typically involves the use of solution-cast polyimide coatings a few microns in thickness, with substrates that include metal-patterned semiconductors and patterned polymers, including previously deposited layers of the same or other polyimides. For instance, an optical waveguiding structure may be formed by deposition of a (relatively) low refractive index polyimide on a patterned semiconductor, followed by deposition of a higher index polyimide. After patterning of the high-index polyimide, another layer of the low-index polyimide might then be deposited on the patterned high-index layer. A typical deposition sequence for creating an optical waveguide from a set of polyimides is depicted in Figure 1.

The performance of integrated optical devices is normally predicted and controlled by defining each material as a uniform geometric region of a given complex refractive index. Because the devices typically require extremely low levels of scattering and absorption in order to perform adequately, the assumption of material uniformity on the scale of interest is usually justified, provided that the devices being modeled are of practical significance. In such

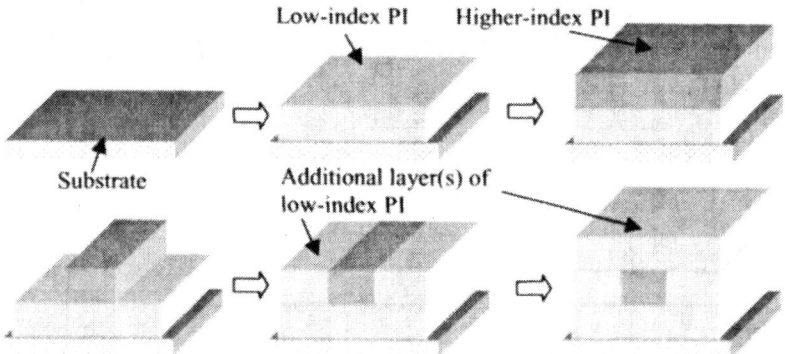

Figure 1. Schematic of waveguide formation using low and high refractive index polyimides (PI). After deposition and patterning, light in the high-index region (core) is confined by surrounding regions of lower index (cladding).

cases the imaginary part of the refractive index is also typically quite small. Thus, in order to design integrated optical devices with good performance, the key properties are the refractive index (both in-plane (TE) and out-of-plane (TM) at a given wavelength) along with the geometrical dimensions of the particular set of coatings that will be the end result of the fabrication process.

The chapter that follows presents models (including experimental verification) for determining these key properties, with a discussion of the underlying structure-processing-property relationships. The examination of the models is limited to processes in which the polyimide coating is formed as a film via spin casting from solution with a subsequent thermal treatment. Although other processes (such as subsequent etching) may also impact the film properties, the high glass transition temperature (>200°C) of most polyimides ensures that, provided a sufficient degree of imidization has taken place, these effects are secondary, with the index and geometry basically "frozen in" once the high temperature cycle is complete. Moreover, the effects of processes such as wet etching and development on polyimides and/or their precursors have been explored in detail elsewhere.[1] Although the data presented here is limited to polyimides, the general principles apply equally well to other rigid polymers with a high glass transition temperature that are formed via spin casting.

Materials and Methods

The polyimides used in this study include Pyralin 5878G, a commercially available poly(amic acid) obtained as a solution from HD Microsystems, and three polyimides synthesized at the Naval Air Warfare Center in China Lake, California. The first of these, designated Polyimide A, is a copolymer made from approximately 35 mol% s-biphenyl dianhydride (s-BPDA), 15 mol% hexafluoroisopropidyl dianhydride (6FDA), 37 mol% methylene bis(2-methyl-4-aminocyclohexane) (the commercial curative RF24), and 13 mol% 4,4'-oxydianiline (ODA). The combination of fluorinated and aliphatic groups makes this polyimide exceptionally soluble and results in low orientation. Polyimide B is a copolymer made from roughly 25 mol% BPDA, 25 mol% 6FDA, 25 mol% 2,2'-bis(trifluoromethyl)-4,4'diaminobiphenyl (TFDB), and 25 mol% ODA. It provides moderate orientation with good solubility. Polyimide C is a copolymer made from 50 mol% TMEG (the diester condensate of ethylene glycol and 2 mol trimellitic dianhydride) 45 mol% 2,5-bis(4-aminophenyl)-1,3,5-oxadiazole (BOAZ), and 5 mol% 1,4-diamino-2-phenol, and provides high orientation and good adhesion. Polyimides A and B were used in the pre-imidized form, while Polyimide C was used in the amic acid form. Details of the synthesis of these polyimides, with additional property data, have been reported elsewhere.[11-13]

Film thickness values were measured using a Dektak III stylus profilometer. FTIR spectra were recorded on a Nicolet 710 spectrometer using the average of

32 scans, with samples cast onto salt plates. Film weights were recorded using a Mettler AE series analytical balance, with a resolution of 0.01 mg, using the average of three separate measurements for each sample. The refractive index profile of the samples was determined by a prism coupling technique using 1310 nm radiation, with a verified accuracy of 0.001, as described elsewhere.[11]

Film Geometry

The key geometrical parameter for polymer films formed by spin casting is the field thickness of the film, which refers to the average thickness of the film formed over a smooth region of the substrate far away from any substrate topographical features or boundaries. Although somewhat arbitrary, a good working definition of "smooth" and "far away" would involve a height difference of no more than 0.01 times the intended film thickness extending over an area spanning 100 times the intended film thickness in any direction. An exception is the region of non-uniform thickness near the outermost limits of wetting during spin casting, or other places where large quantities of casting solution may accumulate. Depending on solution viscosity and surface tension, this region typically extends between 1 and 5 mm from the boundary.

The field thickness of the film depends on a number of key material and processing variables. These include the solution shear viscosity, solids concentration, and to a minor extent, surface tension. The key parameters of the spinning process that affect the field thickness are spin speed, spin time, and the rate of solvent evaporation. These parameters have been used in a number of previously developed models of the spin casting process in order to aid in the prediction of the field thickness.[14-16]

During the spin casting process, centrifugal forces push the liquid solution out to the edges of the substrate, where (except in the case of low boiling solvents), it forms into drops and is slung off. The rate of outward transport is strongly dependent on the thickness of the solution layer, causing the layer to thin rapidly at first, followed by a period in which further thinning is accomplished mainly through solvent evaporation. These observations agree with the data shown in Figure 2, in which the thickness of a dried film of Pyralin 5878G is plotted as a function of spin time for a spin speed of 2000 rpm, and with a baking protocol of 70°C for 20 minutes, 150°C for 70 min, 180°C for 60 min, and 240°C for 60 min. Although the thickness of the wet film on the spinner may still vary with time after 60 seconds, the fact that the dried films are of nearly identical thickness shows that any thinning is due to evaporation.

For a given polyimide solution, the sensitivity of field thickness to spin time may be greatly reduced by choosing a spin time that is sufficiently long. As a result, for a given solution and casting apparatus, the only variable that remains to be controlled is the spin speed. (Evaporation may also be controlled, but often is not reproducible from one spin coater to the next.) Thus, the field thickness of polyimide films is typically controlled during film formation by adjusting the spin speed on the basis of an empirically determined spin speed to thickness relationship.

Since the empirical relationship may be different for each different solution used, determining these relationships can become a time-consuming process. However, as can be seen from Figure 3, the relationship obeys a power law for both Polyimide A and Pyralin 5878G (spin time 30 seconds). Moreover, for both the 15 wt% and 24 wt% solutions of Polyimide A, the power law exponent is the same. For most polyimides, a power law exponent around –0.8 is typical, with a range from –0.5 to –1.0 observed. These films were spun for 30 seconds and baked according to the schedule previously given for Pyralin 5878G.

With the expectation that a power law relationship will be obeyed, it is possible to construct (and validate) a useful thickness versus spin speed curve with data from just three or four test samples collected over a wide range of spin speeds using logarithmic intervals. Generally speaking, we have been able to successfully predict field thickness values to about 0.05 micron using this procedure.

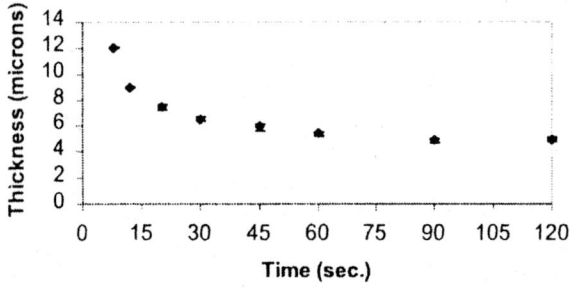

Figure 2. Field thickness as a function of spin time for Pyralin 5878G solution

Refractive Index

For most polymer films the refractive index, n, is not isotropic but depends on direction. For films formed via spin casting, the refractive index tensor typically has two principle values, denoted n_{TE} (in-plane) and n_{TM} (out-of-plane). These two quantities may be expressed in terms of two more fundamental quantities, the average refractive index, n_0, and the birefringence Δn, via the equations $n_{TE} = n_0 + 1/3\ \Delta n$ and $n_{TM} = n_0 - 2/3\ \Delta n$. The quantities n_0 and Δn are fundamental in the sense that n_0 should depend on chemical structure but not orientation, whereas Δn should be the product of the intrinsic birefringence of the polymer and its orientational order parameter. As a result, for chemically stable polymers undergoing no changes in phase, only Δn should be affected by processing. Because of the difference in the factors controlling n_0 and Δn, it is

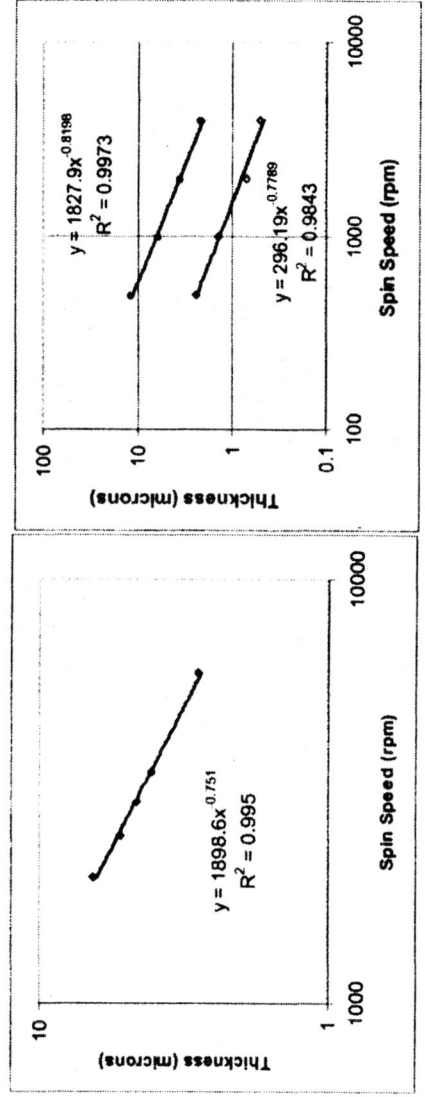

Figure 3. Field thickness versus spin speed for (a) Pyralin 5878G, and (b) Polyimide A at 15 wt% (lower) and 24 wt% (upper) in 1,1,2,2-tetrachloroethane

important to have available, 1) a means of predicting n_0 as a function of chemical composition, and 2) procedural means for fine-tuning the index profile via adjustment of Δn. With these in hand, it is possible to work around the difficult problem of *a priori* prediction of Δn, especially when Δn is very small.

For polyimides, correlations such as the ones developed by van Krevelen[17] and Bicerano[18] for the refractive index at visible wavelengths provide at least qualitatively good predictions. With minor modifications, the correlation of Bicerano has been used to predict the average refractive index of amorphous polyimides at near infrared wavelengths quantitatively.[11] Table 1 lists the predicted and observed index of Polyimides A-C and Pyralin 5878G. Note that the predicted average index of Polyimide C is somewhat lower than observed, perhaps due to enhanced charge transport afforded by the oxadiazole group.[13]

For solution-cast polyimide films, there are several generally recognized means of adjusting orientation, and therefore birefringence, via processing. These include alteration of the polymer-solvent interaction parameter,[19] adjustment of low-temperature baking conditions,[20] changes in imidization temperature, use of pre- versus post-imidized spin casting solutions,[22] thermal treatment of free standing or adhered films, mechanical stretching, tailoring of chemical composition and molecular weight, adjustment of heating rates during drying, and, for films that crystallize, controlling the extent of crystallization.[1]

For the amorphous polyimides used in integrated optics, however, one particularly useful means of controlling birefringence in solution-cast polymer

Table I. Refractive Index Profiles for Polyimides A-C and Pyralin 5878G

Sample	n_{TE}	n_{TM}	n_0	no (Pred)	Δn
Pralin5878G	1.673	1.615	1.654	1.655	0.058
PolimideA	1.589	1.565	1.581	1.578	0.024
PolimideB	1.613	1.574	1.600	1.599	0.039
Polyimide C	1.664	1.616	1.648	1.634	0.048

films is by altering the process of gel collapse. As mentioned earlier, spin casting and subsequent baking remove enough solvent to solidify the film into a gel, in which the polymer molecules may move locally, rotate and become oriented, but exhibit limited long-range translational mobility, as in any rubbery polymer solid. As evaporation proceeds, the polymer chains can therefore move locally to fill the voids left by the solvent, resulting in a net motion of the polymer perpendicular to the plane of the film. However, without the ability to move relative to each other, the only means of achieving the desired net motion is for the chains to assume a net planar orientation.

Although sophisticated analytical models of orientation development exist,[23,24] a simple qualitative explanation of gel collapse is sufficient to capture the general features of the process. During gel collapse, the planar orientation of the polymer chains distorts the molecules from their equilibrium configuration (random coils for amorphous materials). The result is an entropic restoring force that attempts to eliminate the planar orientation. Thus, geometric and mass transport factors (via the need to maintain a constant density) determine the rate of increase in orientation, while the rotational mobility of the chains (aided greatly by the presence of solvent) determines the rate at which orientation may decrease. In addition to solvent content, the rotational mobility of the polymer chains is determined to a great extent by the temperature of the system relative to the glass transition temperature (Tg), with the mobility above Tg being orders of magnitude higher than the mobility below Tg.

For fully dried rigid polymer, however, the rotational mobility of the chains above Tg is still quite low. Thus, after evaporation is mostly completed, the rotational mobility will be low no matter what the temperature. Moreover, the rate of shrinkage will also be low, since little solvent remains to generate shrinkage. Consequently, gel collapse can alter the orientation of the film mainly in the early stages of baking, which are usually done at the lowest temperatures. Therefore, if increased orientation is desired, the temperature of the initial bake can be lowered relative to the Tg of the wet film. Note that the Tg itself is highly concentration dependent, and although means of estimating it for a wet polymer film are very approximate,[18] it can fall below ambient temperatures for wet films containing a low concentration of polymer.

As a result, even for rigid polymers with high (dry) Tg values, initial baking temperatures near to room temperature may be required to stay well below the Tg of the wet film if a high level of orientation is desired. At these low temperatures, the evaporation of solvent can be quite slow, necessitating long drying times. As the film continues to dry, the Tg will increase, allowing the baking temperature to be increased while maintaining low mobility. The process can be continued until the film is fully dried. On the other hand, if a reduction in birefringence is desired, then the best approach would be to elevate the baking temperature to above the Tg of the wet film as soon as possible in the drying process, and to delay moving below the Tg as long as possible.

For polyimides cast in the poly(amic acid) form, imidization during baking at high temperature induces further shrinkage of the film due to the release of volatiles (usually water) that are a by-product of the reaction. This shrinkage enhances gel collapse, further orienting the film. In fact, releasing larger molecules during imidization (thus increasing the degree of shrinkage) results in even higher orientation.[25] At very high temperatures, changes in packing or "cure" reactions further increase the density of polyimide films.[26] As mentioned earlier, if solvent is mostly absent during these events, then the rotational mobility is likely to remain too low to counteract the increases in orientation.

Investigations of the Effect of Film Processing on Refractive Index Profile

Polyimide B was dissolved at 12 wt% in veratrole (b.p. 206°C), filtered to 0.2 microns, and spun at 1000 rpm at spin times ranging from 10 to 60 s. The films were subsequently baked at 70°C for 20 minutes, 150°C for 70 min, 180°C for 60 min, and 240°C for 60 min. At various points during baking, the films were weighed. An additional film was spun on a salt plate at 1000 rpm for 30 seconds and subjected to the same baking protocol, with FTIR spectra collected at the same point during the bake that the other films were weighed. The refractive index profile was then measured after baking was complete.

As seen from Table 2, the refractive index profile of the polymer changes only modestly with increased spin time. According to the data in Table 3, no significant amount of solvent is lost during spin casting of the polyimide solution. As indicated by Table 3 and Figure 4 (the FTIR spectra), however, the vast majority of solvent is lost in 20 minutes of baking at 70°C. For the longer spin times, the loss occurs mainly during the early stages of baking, whereas for the shorter spin times, the loss is more gradual. This difference is readily explained by noting the difference in thickness that accompanies the different spin times (see Table 2). For thinner films, the same flux of solvent molecules will result in more of an increase in concentration over time. It is worth noting, however, that according to Table 3, the total flux of solvent decreases with

Table 2. Refractive Index Profile of Polyimide B Versus Spin Time

Film ID	Time (1000 rpm)	Thickness	n_{TE}	n_{TM}	n_0	Δn
B-1	10 sec	5.5	1.613	1.573	1.600	0.040
B-2	15 sec	5.2	1.613	1.574	1.600	0.039
B-3[a]	30 sec	2.4	1.615	1.571	1.600	0.044
B-4[a]	45 sec	2.8	1.615	1.570	1.600	0.045
B-5[a]	60 sec	2.3	1.615	1.569	1.600	0.046

[a] Index measured on a thicker part of the film. The thickness at the center of the film was too thin for index measurement.

increased spin time, since at high polymer concentration, the drying process will become diffusion limited, slowing evaporation substantially. This phenomenon may explain why no sample is able to reach more than 80-90% polymer at 70°C.

In order to understand the process in terms of gel collapse theory, it is important to have at least an approximate idea of how the glass transition temperature of the film varies with solvent content. Although only very rough general models exist, the predicted Tg versus volume fraction of solvent is shown in Figure 5, based on the Tg of the dry polymer obtained via DSC, the supplier's listed melting point of the solvent, and theoretical equations given by

Table 3. Weight Fraction of Polyimide B in Films at Various Points During the Baking Cycle as a Function of Spin Time

Point in Drying Cycle	Original Spin Time sec				
	10	15	30	45	60
wet	11%	11%	11%	11%	11%
after 5 min. at 70°C	34%	28%	57%	n/a	89%
after 10 min. at 70°C	77%	77%	83%	81%	90%
after 20 min. at 70°C	84%	81%	83%	85%	91%
after 70 min. at 150°C[a]	94%	94%	96%	99%	96%
after 60 min. at 180°C[a]	98%	97%	98%	100%	99%
after 60 min. at 240°C[a]	99%	100%	98%	101%	100%

[a] also after completing all previous steps at lower temperatures.

Bicerano.[18] The model predicts that during drying at 70°C, the film will fall below its glass transition when reaching about 70% solids by volume (about 75% by weight). Thus, the gel will most likely form at 70°C, and since most solvent loss takes place at 70°C, the orientation of the film should develop at 70°C as well.

The best opportunity for loss in orientation due to relaxation would then be the point at which the temperature is raised to 150°C. With shorter spin times (and thus thicker films), more solvent remains in the film at this point, as indicated in Table 3, thus films spun under this condition will come closer to (or perhaps surpass) the Tg than those spun at longer times. After the 150°C step,

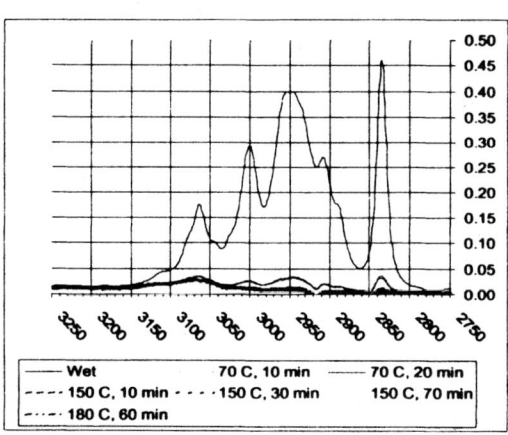

Figure 4. C-H stretch region of FTIR spectra of Polyimide B at various points in the drying process. The horizontal axis is the wavenumber in cm^{-1}.

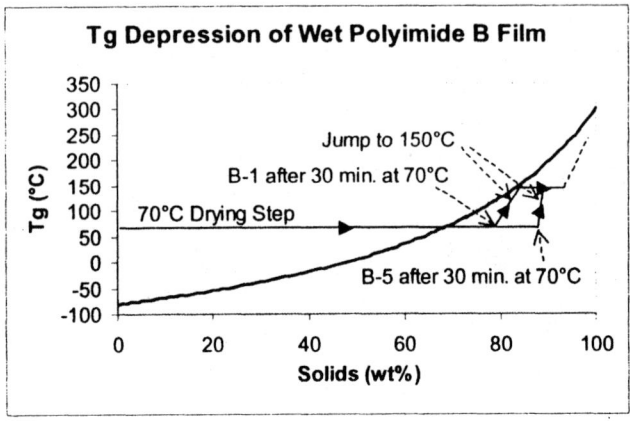

Figure 5. Estimated Tg versus Solids Content (thick line) for Polyimide B / veratrole wet films. The thin lines with solid arrowheads show the trajectory of the baking protocol for samples B-1 and B-5.

the films are dry enough that any subsequent gel collapse will be of limited extent, and should be about the same for all the spinning conditions studied. These concepts are illustrated schematically in Figure 5, where the temperature of the system as a function of the amount of solvent remaining is shown as a multi-step trajectory for the shortest and longest spinning times. The trajectories show clearly that the shorter spin time will pass closest to (or go farthest past) the Tg. Note that this is so no matter what the actual Tg value is, provided that the actual dependence on solvent volume fraction has the same general monotonic shape shown in the figure, a very reasonable assumption.

Thus it appears that the effect of spin time on birefringence is an indirect consequence of the difference in the amount of solvent lost at 70°C (the more solvent lost, the higher the birefringence). In order to test this hypothesis more directly, another set of similar data was collected for Polyimide C. Polyimide C was cast in the poly(amic acid) form by preparing an 18 wt% solution in NMP, followed by the same filtration, substrate cleaning, and spin casting procedure outlined previously. In this case, the spinning conditions were the same for all samples: 600 rpm for 30 seconds, resulting in films that were all 2.5 +/- 0.1 microns thick. A four-stage drying cycle was used, the first stage involved a variable time (5-120 minutes) at 70°C, the second and third stages 60 minutes at 180°C and 240°C, respectively, for all samples, and the fourth stage a variable time (up to 120 minutes) at 280°C. Thus the process variables for comparison purposes are the baking time at 70°C and the baking time at 280°C.

Unlike Polyimide B, the film not only undergoes solvent loss, but also imidization during baking. According to FTIR scans of the film casting solution spun on a salt plate and dried for 120 minutes at 70°C and 120 minutes at 280°C, the imidization (indicated by the 1378 cm^{-1} band shown in Figure 6) is very low after the 70°C step, but is mostly complete after the 180°C step. However, some imidization continues at 240°C, and at 280°C there are subtle shifts in the spectrum. These changes cannot be due to solvent loss, or even to substantial imidization, as the weight loss table (Table 4) shows no change in the weight once the 240°C step is complete.

Table 4 does indicate that, as with Polyimide B, longer baking times at 70°C result in more solvent loss. Although the exact Tg of the poly(amic acid) is difficult to predict, especially since NMP and poly(amic acids) are known to form strongly interacting complexes below 150°C, it will nevertheless be true that, as with Polyimide B, the heating immediately after the 70°C step will bring the sample with the most remaining solvent closest to (or farthest above) the Tg. Given how much of the solvent evaporates in the first 5 minutes at 70°C, it is likely that the formation of a stable acid! NMP complex is what limits the extent of evaporation. Further support for this explanation comes from the fact that a 1:1 molar ratio of NMP to acid groups in the amic acid form of Polyimide C would represent about a 76 wt% solids mixture.

In Table 5, the refractive index profile of Polyimide C is displayed as a function of the two baking time variables (time at 70°C and at 280°C). As expected from the preceding discussion, the birefringence increases significantly as the baking time (and extent of solvent evaporation) at 70°C increases. In addition, for both the shortest and longest baking times at 70°C, there is a further increase in birefringence with increased baking time at 280°C. As mentioned earlier, no significant solvent loss or imidization takes place that could explain further increases in anisotropy. It is worth noting, however, that the effects of baking time at 70°C and 280°C have markedly different effects on the relative values of the n_{TE} and n_{TM} components of refractive index. Increased baking time at 70°C for Polyimide C (like the increased spin time for Polyimide B) results in a relatively large decrease in n_{TM} with only a slight increase in n^{TE}, while increased baking time at 280°C results in a relatively large increase in n_{TE} with a much smaller increase in n_{TM}.

As previously mentioned, changes in orientation should not affect the average refractive index. For both the effect of spin time on Polyimide B and the effect of baking at 70°C on Polyimide C, this condition comes close to being met, although especially in the latter case the value of n_{TE} is too low. It may be that, as the baking time at 70°C is increased and the polymer is hindered from relaxing, further loss of solvent results in excess free volume or porosity. Since samples of both polyimides remained completely transparent, even when viewed under an optical microscope, any porosity must be confined to a length scale considerably less than 1 micron.

Conversely, the changes in Polyimide C with increased baking time at 280°C result in an increasing average refractive index over time. Both the corresponding FTIR and weight loss data, however, indicate that significant chemical changes are not taking place. Thus, the most likely explanation is an

Table 4. Weight Fraction of Polyimide C in Films at Various Points During the Baking Cycle as a Function of Baking Time at 70°C

Point in Drying Cycle	Total Baking Time at 70 °C (minutes)			
	5	30	60	120
wet	18%	17%	17%	18%
after completing 70°C stage	65%	68%	69%	73%
after 60 min. at 180°C[a]	93%	93%	94%	91%
after 60 min. at 240°C[a]	98%	99%	101%	100%
after 30 min at 280°C[a]	98%	99%	100%	99%

[a] also after completing all previous steps at lower temperatures.

Table 5. Refractive Index Profile of Polyimide C Versus Spin Time

Film ID	Bake Time at 70 °C	Bake Time at 280 °C	n_{TE}	n_{TM}	n_0	Δn
C-1	5 min.	0 min.	1.664	1.621	1.650	0.043
C-1	5 min.	30 min.	1.669	1.618	1.652	0.051
C-1	5 min.	120 min.	1.673	1.612	1.652	0.061
C-2	30 min.	0 min.	1.664	1.616	1.648	0.049
C-3	60 min.	0 min.	1.666	1.613	1.648	0.053
C-4	120 min	0 min.	1.665	1.612	1.647	0.053
C-4	120 min	30 min.	1.676	1.606	1.653	0.070
C-4	120 min.	120 min.	1.675	1.607	1.652	0.068

[a] index measured on a thicker part of the film. The thickness at the center of the film was too thin for index measurement.

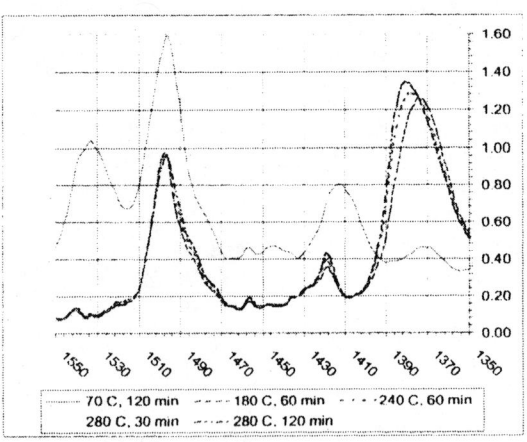

Figure 6. FTIR spectra of Polyimide C films after completion of various points in the drying process. The horizontal axis is the wavenumber in cm^{-1}.

increase in the density of the polymer. Two potential mechanisms that would contribute to an increase in density are physical aging and ordering. The films of Polyimide C used to obtain the data in Table 5 undergo at least part of the solidification process below Tg, and a reduced density is evident when longer baking times at 70°C are used, as described earlier. Thus, during the 280°C baking step, when the Tg is exceeded, some physical aging effects (i.e. density equilibration via removal of excess free volume) would be expected. Moreover, the average index of films C-1 and C-4 appears to approach the same final value, as would be expected if the samples were approaching an equilibrium density. Interestingly, however, physical aging would not normally be expected to lead to an increase in birefringence. However, it may be that the removal of void-like regions allows for increased planar alignment in the films.

An alternative explanation for the increase in birefringence would be the formation of a crystalline (or perhaps liquid crystalline) phase. However, in that case, one would expect the changes in average index and birefringence to be correlated, whereas for sample C-1 the change in birefringence takes place without a corresponding increase in the average index. In addition, the films remained completely transparent during the baking process at 280°C.

The behavior of Polyimide C during baking has important practical consequences. As Table 5 clearly shows, increases in the baking time at 70°C result in a decrease in n_{TM} with little change in n_{TE}, due to the combination of increased orientation with presumably decreased density. On the other hand, increases in the baking time at 280°C result in relatively large changes in n_{TE} with smaller changes in n_{TM}, due to presumably increased density and increased orientation. Thus, it is possible, by judicious choice of baking protocol, to control the n_{TE} and n_{TM} values of a polymer such as Polyimide C independently. Moreover, it is possible, by making appropriate adjustments to the baking times at low and high temperature, to adjust the values of n_{TE} and n_{TM} for the purpose of maintaining or fine-tuning waveguide properties. These adjustments have the advantage of not altering other key properties such as the film geometry (which would be the case if a variable such as spin time were used).

Although data has been presented for only two particular polyimides, we have observed similar trends in more than half a dozen different polyimides made in our laboratory, and have been able make adjustments to the TE and TM index values of polyimides rapidly and reliably whenever needed using these principles. For instance, it is possible to adjust the n_{TM} value of Pyralin 5878G between about 1.605 and 1.62 depending on the baking protocol used, while maintaining n_{TE} at 1.675 +/- 0.02. As a result, using a combination of comonomer selection (for obtaining an approximate index profile), baking protocol adjustment (for fine-tuning the index profile), and test sample production (to establish geometrical film quality parameters), the development of a polyimide that conforms to pre-determined specifications for optical waveguide production is now a routine matter that may be accomplished in a few days under normal circumstances. Consequently, the full potential of polyimides as materials that may be tailored efficiently and affordably for use in a wide variety of optical waveguide devices is being realized.

Conclusions

The development of polymer-based optical waveguides depends critically on the availability of highly stable materials that may be "tuned" readily to exhibit a wide range of properties. We have examined polyimides as a versatile class of materials that is ideally suited to fulfilling these requirements. For the design of waveguides, the key properties of interest include the field thickness and the refractive index profile of the film layers that constitute the device. We have shown that production of three to four test films made by variation of spin speeds is the most reliable means of estimating and controlling the field thickness via the establishment of an empirical power law relationship. We have also demonstrated the ability to predict and control the index profile of the films (at near infrared wavelengths) using a combination of topological correlations for the average index and variation in comonomer ratios for coarse control over birefringence. The ability to make more precise adjustments rests on an understanding of the processes of gel collapse and the dynamics of orientational order during the baking process. Using these concepts, we have developed a method for separately controlling the TE and TM index of polyimide films via adjustment of the baking protocol that has proven successful for a wide range of materials.

Acknowledgement

The authors wish to acknowledge the work of Drs. Peter Zarras, Stephen Fallis, Matthew Davis, and Andrew Chafin of the Naval Air Warfare Center at China Lake, CA for synthesizing the polyimides, and the efforts of Eric Webster and Rebecca Trimm at Redstone Arsenal in Huntsville, AL for providing additional film processing data.

References

1. Polyimides: Fundamentals and Applications; Ghosh, M. K.; Mittal, K. L., Eds.; Marcel Dekker: New York, 1996.
2. Zarras, P. et al. *Polymer News* **2001**, *26*, 335.
3. Phillips, S. H.; Haddad, T. S.; Tomczak, S. J. *Curr. Opinion Solid State Mater. Sci.* **2004**, *8*, 21.
4. Reuter, R.; Franke, H.; Feger, C. *Appl. Optics* **1988**, *27*, 4565.

5. Matsuura, T. et al. *Macromolecules* **1992**, *25*, 3540.
6. Ma, H.; Jen, A. K.-Y.; Dalton, L. R. *Adv. Mater.* **2002**, *14*, 1339.
7. Singer, K. D. et al. *Optoelectronic integrated circuits, Proceedings of the Conference,* San Jose, CA; 12-14 Feb. 199, pp. 326.
8. Hornak, L. A. et al. *Miniaturized systems with micro-optics and micromechanics II; Proceedings of the Meeting,* San Jose, CA; 10-12 Feb. 1997, pp. 124-135.
9. Robitaille, L.; Callender, C. L.; Noad, J. P. *Electrical, optical, and magnetic properties of organic solid state materials III, Proceedings of the MRS Symposium,* Boston, MA; 27 Nov.- 1 Dec. 1995, pp. 305-310.
10. Selvaraj, R.; Lin, H. T.; Mcdonald, J. F. *J. Lightwave Tech.* **1988**, *6*, 1034.
11. Guenthner, A. J., et al. in *Linear and Nonlinear Optics of Organic Materials II* Eich, M.; Kuzyk, M. G.; Eds.; Proceedings of SPIE Vol. 4798, SPIE: Bellingham, WA, 2002, pp. 184-194.
12. Lindsay, G. A., et al. *Polym. Mater. Sci. Eng.* **2003**, *88*, 170.
13. Lindsay et al., Polyimides Having an Unusually Low and Tunable Electrical Resistivity Useful for Electrical and Optical Applications. U.S. Patent 6,861,497, 2005.
14. Flack, W. W. et al., *J. Appl. Phys.* **1984**, *58*, 1199.
15. Meyerhofer, D. *J. Appl. Phys.* **1978**, *49*, 3393.
16. Strawhecker, K. E. et al. *Macromolecules* **2001**, *34*, 4669.
17. van Krevelen, D. W. *Computational Modeling of Polymers* Bicerano, J., Ed.; Marcel Dekker: New York, 1992.
18. Bicerano, J. *Prediction of Polymer Properties, 3rd ed.;* Marcel Dekker: New York, 2002.
19. Li, B.; He, T.; Ding, M. *Polymer* **1997**, *36*, 6413.
20. Ree, M. et al. *J. Polym. Sci. Part B Polym. Phys.* **1998**, *36*, 1261.
21. Lin, L.; Bidstrup, S. A. *J. Appl. Polym. Sci.* **1993**, *49*, 1277.
22. Russell, T. P; Swalen, J. D. *J. Polym. Sci. Polym. Phys. Ed.* **1983**, *21*, 1745.
23. Bornside, D. E.; Macosko, C. W.; Scriven, L. E. *J. Imaging Tech.* **1987**, *13*, 122.
24. Dabral, M. et al. *J. Polym. Sci. Part B Polym. Phys.* **2001**, *39*, 1824.
25. Takeichi, T.; Zuo, M.; Hasegawa, M., *J. Polym. Sci. Part B Polym. Phys.* **2001**, *39*, 3011.
26. Morino, S.; Hone, K. *Polymer Journal* **1999**, *31*, 707.

Chapter 11

Sensor Coatings: Responsive Coatings for the Detection, Identification, and Removal of Surfaceborne Plutonium and Uranium

H. Neil Gray[1] and Betty Jorgensen[2]

[1]Department of Chemistry, The University of Texas at Tyler, 3900 University Boulevard, Tyler, TX 75799
[2]Materials Science and Technology Division, Mail Stop E–549, Los Alamos National Laboratory, Los Alamos, NM 87545

> Strippable coatings that are capable of both detecting and removing surface-borne nuclear and heavy metal contaminants have been developed. When applied to a contaminated surface, the coatings display responsive behavior; areas of contamination are indicated by a color change. As the coatings dry, the contaminants are drawn into and fixed within the polymer matrix. Subsequent removal of the coating with entrapped contaminants results in surface decontamination. Here we report the development and investigation of a sensor coating for uranium and plutonium.

The development of materials for the detection, identification, and sequestration of hazardous chemical and biological contaminants in the environment is a significant element of many applications in environmental remediation and national security. The presence of such contaminants on structural and other fabricated surfaces is a significant concern as they can be transferred to workers by casual contact and, if disturbed, easily made airborne. Of particular interest to nuclear facilities, and some national laboratories, are the gloveboxes, walls, and tools contaminated by radioactive isotopes. Additionally, attacks against the United States by terrorists using unconventional weapons such as radiological, chemical, or biological dispersion devices can lead to surfaces contaminated with a variety of hazardous species. The rapid and sensitive detection, speciation, and removal of chemical and biological weapon agents (CBWs) dispersed by such devices is critical. Any effective decontamination scheme must have, as an integral part, a plan for surface decontamination. Our work has focused toward the realization of such procedures via the development of responsive, strippable, decontaminating coatings.

Strippable coatings are usually polymeric solutions or dispersions that can be applied to a surface by brushing or spraying. Upon curing or drying these coatings form elastic films that can be easily peeled from the surface in large sheets. During the drying process, the contaminants are drawn into and mechanically trapped within the coating matrix. Subsequent removal of the coating with entrapped contaminants results in some degree of surface decontamination. Although conventional strippable coatings have demonstrated some success at several nuclear facilities and accident sites (*1-8*), they still have many associated problems and limitations.

One problem is that conventional coatings are often toxic themselves, containing carcinogenic solvents and chelators, and other materials such as ammonia that make application dangerous or unpleasant. Additionally, conventional coatings offer no indication of the presence or identity of contaminants, cannot be reused or recycled, and typically do not decontaminate well with a single application (*8*).

Many methods have been reported for the detection and identification of biological (*9-15*), chemical (*16-37*), and radiological (*38-46*) contaminants and warfare agents. Most methods require onsite sample collection and preparation, and subsequent extensive offsite analyses using expensive and sophisticated instrumentation. In many cases the collection and transport of such hazardous materials can be dangerous and time consuming, often leading to delays in obtaining results. The development of sensors and correspondingly simple procedures for the rapid onsite analysis of CBWs without the need to collect, prepare, or transport material is of great importance. Specifically, sensors for the real-time detection and identification of CBWs on surfaces such as walls, floors, windows, pavement, furnishings, aircraft, vehicles, and military equipment would make an immediate and significant contribution to national security.

Although the heart of this work was the development of sensor coatings with focused remediation utility, we believe it has established an emerging and potentially transformative research area in which rapid and innovative advances in sensor materials are likely to occur. The impact of expanding this work could lead to novel sensor technologies based on sensor coatings for biological, chemical, and radiological analytes. Such sensor coatings would not only have the potential to detect and identify specific warfare agents, but to sequester them as well. Moreover, the coatings are water soluble, so entrapped contaminants could be transported within the coating to a laboratory facility for subsequent recovery and analysis.

Here we present the results observed in our laboratory from the development and study of water-based, contaminant-sensing, decontaminating coatings for the detection and removal of plutonium and uranium from a variety of fabricated surfaces.

Results and Discussion

Typical coating compositions consisted of aqueous solutions of film forming polymers, chelators, indicators, and plasticizers. Some compositions contained additional additives, such as defoaming agents and preservatives. When applied to a contaminated surface, contaminants were indicated by a color change. Our best coating for uranium and plutonium not only detected these species but could distinguish between them by turning purple with uranium and red with plutonium. The coatings have an extended drying period (24 h), allowing the contaminants ample time to migrate into and become fixed within the polymer matrix. Stripping the coating with entrapped contaminants results in some degree of surface decontamination (Figure 1). To mask interferences of the colorimetric response, a chelating masking agent was added to the coating composition that formed stable complexes with many elements but less stable complexes with the contaminants to be detected (Figure 2).

The polymer compositions studied for uranium/plutonium sensing coatings included aqueous solutions of poly(vinyl alcohol) (PVA), poly(vinyl pyrrolidone) (PVP), poly(ethylene oxide) (PEO), poly(vinyl amine) (PVAm), and poly(ethenyl formamide) (PEF). The specific coating components and concentration ranges studied are shown in Table I. Coating compositions containing PEO formed flexible, strong films but did not detect the contaminants well. While an observable color change occurred soon after treating a uranium or plutonium contaminated surface with a PEO based coating, it quickly faded. Coating compositions containing PVAm were unable to detect the presence of contaminant via a color change. This is likely due to competition between the indicator and PVAm for complexation with the metal contaminant. Compositions containing PEF tended to have a low wet tack, which resulted in coatings that ran when painted on vertical surfaces. In addition, coatings

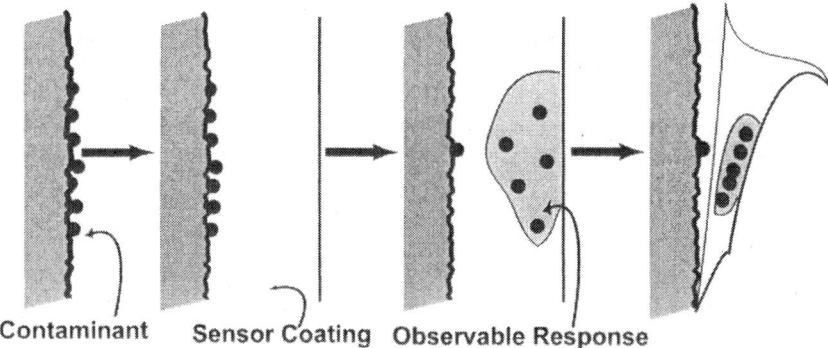

Figure 1. *General process for the detection and removal of contaminants from surfaces using sensor coatings.*

containing PEF tended to foam severely when applied using a power sprayer. These studies indicated that, of the systems investigated, PVA, or a blend of PVA and PVP, provided the best polymer foundation for the coatings used in the decontamination of uranium and plutonium contaminated surfaces.

Poly(vinyl alcohol) is a non-toxic, water-soluble material that is a good film former. Our PVA films and coatings did not require a curing cycle as film

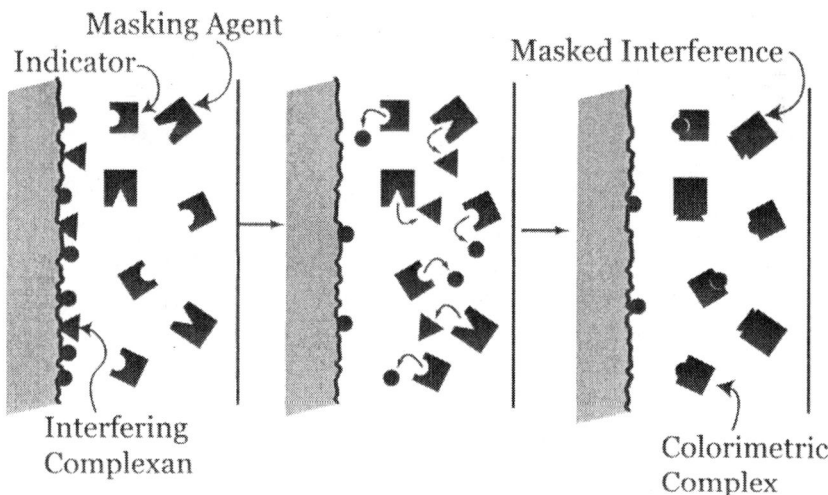

Figure 2. *Masking complexans which interfere with the colorimetric response by the addition of masking agents.*

Table I. Coating Compositions Investigated for their Potential at Sensing and Removing Uranium and Plutonium Contaminants from Surfaces.[a]

Polymeric Composition	Indicator Composition	Masking Chelator	Plasticizer
PVA(5-25)	Br-PADAP[b] (3x10^{-3})	EDTA[d] (0-1)	Glycerin (0-10)
		DTPA[e] (0-1)	Glycerin (0-10)
	CP-III[c] (3x10^{-3})	EDTA (0-1)	Glycerin (0-5)
		DTPA (0-1)	Glycerin (0-5)
PEO (5.5)	Br-PADAP (3x10^{-3})	EDTA (0-1)	N/A
	CP-III (3x10^{-3})	EDTA (0-1)	N/A
PVP (10-32)	Br-PADAP (3x10^{-3})	EDTA (0-1)	Glycerin (0-4)
		DTPA (0-1)	Glycerin (0-4)
	CP-III (3x10^{-3})	EDTA (0-1)	Glycerin (0-4)
		DTPA (0-1)	Glycerin (0-4)
PVA (10-18) and PVP (2-15)	Br-PADAP (3x10^{-3})	EDTA (0-1)	Glycerin (0-4)
		DTPA (0-1)	Glycerin (0-4)
		EGTA[f] (0-1)	Glycerin (0-4)
		TTHA[g] (0-1)	Glycerin (0-4)
PVA (10-18) and PVP (2-15)	CP-III (3x10^{-3})	EDTA (0-1)	Glycerin (0-4)
		DTPA (0-1)	Glycerin (0-4)
		EGTA (0-1)	Glycerin (0-4)
		TTHA (0-1)	Glycerin (0-4)
PVA (10-18) and PEO (5-8)	Br-PADAP (3x10^{-3})	EDTA (0-1)	Glycerin (4)
		DTPA (0-1)	Glycerin (4)
PVA (10-18) and PVAm (5-8)	Br-PADAP (3x10^{-3})	EDTA (0-1)	Glycerin (4)
		DTPA (0-1)	Glycerin (4)
PVA (10-15) and PEF (4-8)	Br-PADAP (3x10^{-3})	EDTA (0.5)	Glycerin (4)

[a] Units are weight percent in aqueous solution. [b] 2-(5-bromo-2-pyridylazo)-5-diethylaminophenol. [c] Chlorophosphonazo-III. [d] Ethylenediaminetetraacetic acid disodium salt. [e] Diethylenetriaminepenta-acetic acid. [f] Ethyleneglycol-2-(aminoethyl)tetra-acetic acid. [g] Triethylenetetraminehexa-acetic acid.

SOURCE: Adapted from Reference 8. Copyright 2001 American Chemical Society.

formation occurred via the simple evaporation of water. The PVA grades used were obtained commercially as DuPont™ Elvanol®, which is made by the alcoholysis of polyvinyl acetate (PVAc). The grade of each PVA is determined by the percent hydrolysis and corresponding residual acetate groups. Both cold- and hot-water soluble films were prepared from Elvanol by varying the grade, imparting greater versatility into its use in our strippable coatings. A wide range of films with varying physical characteristics such as tensile strength, elongation, wet tack, and dry adhesion, were produced via the addition of plasticizers, and other polymers. PVA was also shown to work very well with a variety of indicators in the detection of both uranium and plutonium.

Poly(vinyl pyrrolidone) is a nontoxic, water-soluble material that was added to enhance the properties of PVA. The addition of PVP imparted high initial-tack, strength, and hardness to PVA. It also reduced the tendency of the PVA films to curl. Furthermore, PVP increased the adhesion of PVA on the surfaces studied. It was also found to increase the grease resistance of our coating.

The preferred coating composition contained 12 wt% PVA (partially hydrolyzed) and 7 wt% PVP in water as the coating foundation, 4 wt% glycerin as the plasticizer, 0.5 wt% EDTA as the masking agent, and 3×10^{-3} wt% Br-PADAP as the colorimetric indicator. This coating composition exhibited different color changes for each contaminant (orange to purple for uranium, and orange to red for plutonium) and was extremely effective at removing varying levels of both contaminants from different types of surfaces. This composition was used to obtain all the data reported here.

Decontamination Studies

To measure the effectiveness of our sensor coatings, a variety of uranium and plutonium contaminated planchets were prepared and decontaminated. The planchets were weighed before and after contamination. Each contaminated planchet was analyzed via α-scintillation counting, and then treated with our sensing strippable coating. All coating applications were allowed to dry at least 24 h before removing. The coatings were stripped and the planchets were again analyzed via α-scintillation counting. Decontamination factors (DF) were calculated using DF = α_1/α_2, where α_1 = [α-count before decontamination] and α_2 = [α-count after decontamination]. The ability of the sensor coating to decontaminate several different types of surfaces contaminated with varying amounts of contaminant was studied; the results are displayed in Table II.

The decontamination factors obtained for uranium on all the surfaces studied were very high, and were observed to decrease as the amount of contaminant on the surface increased. This observation is likely due to the rate of contaminant diffusion into the coating lagging behind the drying time; in other words, it is likely that the coating dried before the larger amount of uranium was drawn into it.

Table II. Decontamination Factors for the Decontamination of a Variety of Uranium Contaminated Planchets Using Sensor Coating.

Surface	Coating	U (mg)^a	Pu (mg)^a	DF^b
Glass	Sensor Coating	6.7	0	1529
Glass	Sensor Coating	32.4	0	1395
stainless steel	Sensor Coating	6.4	0	1451
stainless steel	Sensor Coating	31.3	0	1220
stainless steel	Sensor Coating	48	0	813
stainless steel	ALARA^c	28.7	0	276
painted Al	Sensor Coating	17.1	0	Complete^d
painted Al	Sensor Coating	41.3	0	959
painted cement	Sensor Coating	15.3	0	524
painted cement	Sensor Coating	38.3	0	418
Al	Sensor Coating	17.8	0	646
Ni	Sensor Coating	9.3	0	487
stainless steel	Sensor Coating	0	50.1	147
stainless steel	ALARA	0	50.0	2.16
stainless steel	Stripcoat TLC^e	0	49.8	12.6

^{a} Each area of contamination was approximately 0.78 cm^2.
^{b} DF = α_1/α_2, α_1 = α-count before decontamination, α_2 = α-count after decontamination.
^{c} Alara is a latex-based commercially available strippable coating, and is shown here for comparison.
^{d} No detectable contaminant remained after decontamination.
^{e} Stripcoat TLC is a commercially available strippable coating.
SOURCE: Adapted from Reference 8. Copyright 2001 American Chemical Society.

For comparison, a commercially available strippable coating, ALARA DECON 114B, was carried through the same decontamination procedure described above (Table II). The analysis indicated that the commercial coating was only 25% as effective in uranium decontamination as the sensor coating we developed. Moreover, unlike the sensor coating, the commercial coating did not have the ability to detect the presence of the contaminant.

Do to the danger of working with plutonium, the majority of our work focused on uranium; nonetheless, a small study involving plutonium was completed. Our sensor coating was not as effective in plutonium removal as it was for uranium (Table II), but it was still much more efficient than commercially available strippable coatings. Two commercially available strippable coatings, ALARA DECON 114B, and StripCoat TLC, were tested for their ability to decontaminate plutonium contaminated stainless steel planchets. The results for these coatings are shown as the last two entries in Table II. As can be seen from the DF values, the sensor coating is far superior to the commercial coatings studied for the decontamination of plutonium from stainless steel.

Selectivity and Sensitivity of the Colorimetric Response

To mask interferences of the colorimetric response, a chelating masking agent was added to the coating composition which forms stable complexes with many elements but less so with the contaminants to be detected (Figure 2). The abilities of several chelating agents to mask the response of interfering metals were investigated. To determine the effectiveness of a specific masking agent, a stainless steel planchet was contaminated with one of the metals shown in Table III. For simplicity, only metals shown to interfere in the absence of a masking agent are presented. The contaminated surface was treated with sensor coating containing a potential masking chelator (0.5% by mass) and observed for several hours. If there was no colorimetric response, the metal was assumed to have been masked by the chelating agent. Each observation was verified by 7 or more replicate experiments. Subsequent studies were completed to determine if each masking agent itself had any negative effects on the detection of uranium, and to insure that uranium could still be detected in the presence of the masked interferences. These results are also summarized in Table III. After a tremendous amount of work, including solution studies, it was determined that EDTA was the most effective masking agent for this particular coating. This decision was based on both masking effectiveness and cost.

The visible detection limit for the coatings, i.e. the smallest amount of contaminant that will cause a *visible* color change, is very important. This limit determines how small of an amount of contaminant will actually be detected by a worker on-site. The determination of this limit for our uranium/plutonium coating in the detection of the uranyl cation was quite simple. Surfaces were contaminated with specific amounts of uranium contaminant, as uranyl nitrate, and then treated with the smart coating. If a visible color change was observed it was noted, then a smaller amount of contaminant was tried. This process continued until the amount of uranium used resulted in no color change. Using this procedure the detection limit of uranium on glass was determined to be approximately $0.55 \mu g/cm^2$, indicating that the coating is very sensitive to small amounts of uranium.

Table III. A Study of Potential Masking Agents for Our Uranium/Plutonium Decontaminating Coatings.

Potential Interference	Amount* (mg)	Complex With Indicator	Masked by EDTA	Masked by DTPA	Masked by EGTA	Masked by TTHA
Zr	5	YES	YES	YES	NO	YES
Zr	10	YES	YES	NO	NO	YES
Hf	5	YES	YES	YES	NO	YES
Th	5	YES	YES	NO	NO	NO
Th	10	YES	YES	NO	NO	NO
Y	10	YES	YES	NO	NO	NO
La	10	YES	YES	NO	NO	NO
Gd	5	YES	NO	NO	NO	NO
Gd	10	YES	NO	NO	NO	NO
Er	10	YES	NO	NO	NO	NO
Pb(II)	15	Minor	YES	YES	NO	YES
Cu(II)	15	Minor	YES	YES	NO	YES
Ti(IV)	15	Minor	NO	YES	NO	YES
W(IV)	15	Minor	YES	YES	NO	YES
Mo(VI)	15	Minor	YES	YES	NO	YES
Bi	15	Minor	YES	YES	YES	YES
Fe(III)	15	Minor	YES	YES	NO	YES
Sn(IV)	15	Minor	YES	YES	NO	YES

* Each area of contamination was approximately 0.78 cm^2.
SOURCE: Adapted from Reference 8. Copyright 2001 American Chemical Society.

Field-testing of the Sensor Coating

The effectiveness of our sensor coating was field-tested at the Waste Characterization, Reduction and Repackaging Facility (WCRRF) at Los Alamos National Laboratory in Los Alamos, NM. This facility repackages and stores nuclear contaminated gloveboxes. Part of this process involves the destruction (cutting up), compacting, and repackaging of contaminated gloveboxes in a special 50' x 30' x 15' compartmental stainless steel structure. The interior of this structure is highly contaminated with both uranium and plutonium. Workers seldom enter, and when they do it is for very short periods of time (<5 minutes). Moreover, when entering they must wear layers of protective clothing and breathe through a self-contained breathing apparatus.

Several 2' x 2' plutonium/uranium contaminated areas inside the WCRRF were treated with the sensor coating. As the coated areas began to dry, color changes were observed where uranium and plutonium contamination was present. The coatings were peeled from the surfaces the following day. Swipes of the surfaces were taken prior to and after treatment, and were used to determine DF values for the three treated areas. The average DF for the three decontaminated areas was 179. Considering the extreme levels of surface contamination present within the WCRRF, this value is very respectable.

Experimental

Materials

All water was doubly distilled using an automated distillation apparatus. Medium viscosity, 88% hydrolyzed Polyvinyl alcohol (PVA) was obtained as Elvanol 52-22 from DuPont. Poly(vinyl pyrrolidone) (PVP), $Mw \approx 360,000$, was obtained from Scientific Polymer Products, Incorporated. Glycerin, ethylenediaminetetraacetic acid disodium salt (EDTA), ethanol (absolute), and 2-(5-bromo-2-pyridylazo)-5-diethylaminophenol (Br-PADAP) were obtained from Aldrich. Both uranium and plutonium were used as their oxides in 0.5 M nitric acid at the Isotope and Nuclear Chemistry Facility at Los Alamos National Laboratory.

Safety Considerations

Both uranium and plutonium are significant health hazards and should only be used after proper training. These materials should only be handled and stored at appropriately designed facilities.

Uranium and plutonium poisoning may damage eyes, lungs, kidneys, skin, liver, lymphatics, blood and bone marrow. Both may cause significant radiation damage. Long term exposure to soluble uranium and plutonium salts has been reported to cause an increase in cancer of the lymphatic and blood-forming tissue of man.

General Coating Preparation.

In a 2-L beaker, 765 g of distilled water were stirred mechanically using a Teflon-bladed mechanical stirrer. To the water were added 120 g of PVA in 10-g aliquots with good stirring. After all the PVA had dissolved, 70 g of PVP were added with good stirring. The polymer blend was stirred for 20 min. to ensure proper mixing. To the polymer solution were added in order, 40 g of glycerin, 5 g of EDTA, and 0.03 g of Br-PADAP (in 1 mL absolute ethanol), followed by 30 min of additional stirring. The orange, viscous coating mixture was defoamed by allowing it to sit in the beaker for 2 h without stirring, then was transferred to a clean polyethylene bottle for storage.

Surface Decontamination in the Laboratory.

In a typical decontamination procedure, uranium or plutonium contaminated surfaces were prepared by evaporating a known amount of a 0.1 M UO_2^{2+} in 0.1 M HNO_3 solution onto a variety of planchets. The planchets were weighed before and after contamination. Each contaminated planchet was analyzed via alpha-scintillation counting, and then treated with the sensing strippable coating. All coatings were allowed to dry at least 24 hours before removing. Typical drying conditions consisted of 35% relative humidity at 20°C. The coatings were stripped, and the planchets were again analyzed via alpha-scintillation counting. Using the count-rates before and after decontamination, decontamination factors (DF) were calculated as described above.

Conclusion

The sensor coatings we have developed actually address several environmental concerns. Unlike other strippable coatings they are completely safe water-based materials. They are capable of decontaminating contaminated surfaces better than conventional coatings, and they demonstrate some level of responsive behavior by indicating the areas of contamination. This is important, because contaminated portions of the coating can be separated from uncontaminated areas and disposed of or treated accordingly. Furthermore, the design of the coatings is such that they have the potential of being redissolved, purified, and reused.

We believe the sensor coatings we have developed establish an emerging and potentially transformative research area in which rapid and innovative advances in sensor materials are likely to occur. The impact of expanding this work could lead to novel sensor technologies based on sensor coatings for biological, chemical, and radiological analytes. Such sensor coatings would not only have the potential to detect and identify specific warfare agents, but to sequester them as well. The coatings are water soluble, so entrapped contaminants could be transported in the coating to a laboratory facility, where they could be recovered and further analyzed. From a broader point of view, sensor-coating technologies will likely make significant contributions to other areas, including: environmental remediation, worker safety, optical sensor design and development, and responsive polymeric materials.

Acknowledgements

The authors gratefully acknowledge The Welch Foundation Departmental Research Grant Program (Grant No. BP-0037), and The United States Department of Energy for their financial support of this research.

References

1. Bargues, S.; Favier, F.; Pascal, J-L.; Lecourt, J-P.; Damerval, F. *PCT Int. Appl.* **1999**, EP 928489, 51 pp.
2. Arnaud, A.; Coen, S.; Jouve, A.; Lelaidier, M.; Perichaud, A. *J. Radioanal. Nucl. Chem.* **1995**, *201(3)*, 213-223.
3. Gopinathan, C.; Balan, T. P. Decontamination of Surfaces by the use of Peelable Gel Membranes. RadTech Asia '91, Conf. Proc. (1991), 71-76. Publisher: RadTech Int. North Am., Northbrook, Ill.
4. Pandur, J. *Muanyag Gumi* **1990**, *27(6)*, 172-174.
5. Kondo, S.; Moriya, T.; Kariya, Y.; Karya, Y. *Jpn. Kokai Tokkyo Koho* 1986, JP 61269095, 8 pp.
6. Bernaola, O. A.; Filevich, A. *Health Phys.* **1970**, *19(5)*, 685-687.
7. Sugalski, A.; Jagielski, E. J. Experiences with Latex-Type Materials as Decontaminating Agents at the Knolls Atomic Power Laboratory. Proc. Conf. Remote Syst. Technol. **1967**, *15*, 294-295.
8. Gray, H. N.; Jorgensen, B.; McClaugherty, D.L.; Kippenberger, A. *Ind. Eng. Chem. Res.* **2001**, *40*, 3540-3546.

9. Ji, J.; Schanzle, A.; Tabacco, B. *Anal. Chem.* **2004**, *76*, 1411-1418.
10. Botzung-Appert, E.; Monnier,V.; Ha Duong, T.; Pansu, R.; Ibanez, A. *Chem. Mater.* **2004**, *16*, 1609-1611.
11. Kievit, O.; Busker, R. W.; Kientz, Ch. E.; Marijnissen, J. C. M. *VDI-Berichte* **2003**, *1772*, 111-116.
12. Dickert, F. L.; Hayden, O. *Anal. Chem.* **2002**, *74(6)*, 1302-1306.
13. Hayden, O.; Dickert, F. L. *Adv. Mater.* **2001**, *13*, 1480-1483.
14. Chuang, H.; Macuch, P.; Tabacco, M. B. *Anal. Chem.* **2001**, *73*, 462-466.
15. Chang, A. C.; Gillespie, J. B.; Tabacco, M. B. *Anal. Chem.* **2001**, *73*, 467-470.
16. Sadik, O.; Land, Jr., W.; Wanekaya, A.; Uematsu, M; Embrechts, M.; Wong, L.; Leibensperger, D.; Volykin, A. *J. Chem. Inf Comput. Sci.* **2004**, *44*, 499-507.
17. Muir, Bob; Slater, Ben J.; Cooper, David B.; Timperley, Christopher M. *J. Chrom., A* **2004**, *1028(2)*, 313-320.
18. Yang, Y.; Ji, H.; Thundat, T. *J. Am. Chem. Soc.* **2003**, *125*, 1124-1125.
19. Ciucu A. A.; Negulescu, C.; Baldwin, R. P. *Biosens. Bioelectron.* **2003**, *18*, 303.
20. D'Agostino, Paul A.; Hancock, James R.; Chenier, Claude L. *European Journal of Mass Spectrometry* **2003**, *9(6)*, 609-618.
21. Wang, J; Pumera, M.; Chatrathi, M. P.; Escarpa, A.; Musameh, M. *Anal. Chem.* **2002**, *74*, 1187-1191.
22. Karousos, N. G.; Aouabdi, S.; Way, A. S.; Reddy, S. M. *Anal. Chim. Acta* **2002**, *469*, 189.
23. Wang, J.; Krause, R.; Block, K.; Musameh, M.; Mulchandani, A.; Mulchandani, P.; Chen, W.; Schoning, M. *J. Anal. Chim. Acta* **2002**, *469*, 197-203.
24. Thurow, K.; Koch, A.; Stoll, N.; Haney, C. A. NATO Science Series, 1: Disarmament Technologies (2002), 37(Environmental Aspects of Converting CW Facilities to Peaceful Purposes), 123-138.
25. Sadik O. A.; Land W.; Wang J. T. *Electroanalysis* **2003**, *15*, 1149.
26. Wang, J.; Chatrathi, M. P.; Mulchandani, A.; Chen, W. *Anal. Chem.* **2001**, *73*, 1804.
27. Mulchandani, P.; Chen, W.; Mulchandani, A. *EnViron. Sci. Technol.* **2001**, *35*, 2562.
28. Mulchandani, P.; Chen, W.; Mulchandani, A.; Wang J.; Chen, L. *Biosens. Bioelectron.* **2001**, *16*, 433.
29. Sadik, O. A.; Ngundi, M. M.; Yan, F. *Biotechnology and Bioprocess Engineering* **2000**, *5*, 407.
30. Mulchandani, A.; Mulchandani, A.; Chen, W. *Anal. Chem.* **1999**, *71*, 2246.

31. Mulchandani, A.; Kaneva, I.; Chen, W. *Anal. Chem.* **1998**, *70*, 5042.
32. Everett, W. R.; Rechnitz, G. A. *Anal. Chem.* **1998**, *70*, 807.
33. Abad, J. M.; Pariente, F.; Hernandez, Abruna, H. D.; Lorenzo, E. *Anal. Chem.* **1998**, *70*, 2848.
34. Hadd, A. G.; Raymond, D. E.; Halliwell, J. W.; Jacobson, S. C.; Ramsey, M. *Anal. Chem.* **1997**, *69*, 3407.
35. Sadik, O. A.; Van Emon J. M. *Biosens. Bioelectron.* **1996**, *11*, 9.
36. Sherma, J. *Anal. Chem.* **1995**, *67*, R1-R20.
37. Sherma, J. *Anal. Chem.* **1993**, *65*, R40-R54.
38. Wilson, Richard E.; Hu, Yung-Jin; Nitsche, Heino *AIP Conference Proceedings* (2003), 673(Plutonium Futures--The Science), 111-112.
39. Roane, J. E.; DeVol, T. A *Anal. Chem.* **2002**, *74*, 5629-5634.
40. Greenwell, R. A.; Addleman, R. S.; Crawford, B. A.; Mech, S. J.; Troyer, G. L. *Fiber and Integrated Optics* **1992**, *11(2)*, 141-150.
41. Boxall, C.; Port, S.; Shackleford, S. *PCT Int. Appl.* **2002**, Patent No. WO 2002001214, 44 pp.
42. Hauser, W.; Goetz, R.; Klenze, R.; Kim, J. I. Proceedings of SPIE-The International Society for Optical Engineering (1996), 2778(Pt. 2, 17th Congress of the International Commission for Optics, 1996, Pt. 2), 528-599.
43. Reboul, Scoff H.; Borai, Emad H.; Fjeld, Robert A. *Analytical and Bioanalytical Chemistry* **2002**, *374(6)*, 1096-1100.
44. Kamenev, A. G.; Efimov, I. A.; Chubinskii-Nadezhdin, I. V. *Atomnaya Energiya* **1996**, *80(1)*, 43-47.
45. Crain, Jeffrey S.; Mikesell, Barbara L. *Applied Spectroscopy* **1992**, *46(10)*, 1498-1502.
46. Lu, Qin; Callahan, John H.; Collins, Greg E. *Chem. Comm. (Cambridge)* **2000**, *19*, 1913-1914.

Chapter 12

High-Performance UV-Cured Acrylic Coatings

Christian Decker

Département de Photochimie Générale (UMR 7525 - CNRS), Ecole Nationale Supérieure de Chimie de Mulhouse, 3 rue Werner 68200 Mulhouse, France

Highly crosslinked polymers have been rapidly produced at ambient temperature by light-induced polymerization of multifunctionalized monomers and oligomers. The reaction was monitored in real time by infrared spectroscopy and shown to proceed extensively within a fraction of a second for acrylate-based resins containing highly efficient photo-initiators (diphenoxybenzophenone and α-aminophenyl-ketone) and a very reactive monoacrylate carbamate monomer. Dual-cure systems that combine photopolymerization and thermal curing have been developed to achieve an effective hardening of resins in areas receiving no UV light (shadows areas, thick pigmented coatings). The physico-chemical properties of the photocured polymer were modulated in a large range by a proper selection of the chemical structure of the oligomer and of the monomer functionality. High-modulus crosslinked polymers have been produced for coatings applications to improve the surface properties of various substrates, in particular their resistance to scratching, moisture and weathering. The UV-curing technology proved to be an environment-friendly process which transforms rapidly a solvent-free resin into a highly resistant coating with minimum energy consumption and no VOC emission.

Light-induced polymerization of multifunctional monomers, also called UV-radiation curing, is commonly recognized as the most effective process to transform quasi-instantly a solvent-free liquid resin into a highly resistant polymeric material. The subject has been extensively investigated in the past few years with respect to both the polymerization kinetics and mechanism, and the properties of different types of UV-cured polymers (acrylates, epoxides, vinyl ethers, thiol/polyene,...)[1-9]. Because of its distinct advantages regarding processing and product performance, this environment-friendly technology has found a wide range of applications, in particular in the coating industry for the surface protection of a variety of materials (plastics, wood, paper, metals). Remarkable progress has been made during the past decade regarding the processing stage, with the design of highly reactive UV-curable formulations and of well-suited light sources, as well as the service properties of the final product, which can be adjusted to fit the requirements imposed for the considered application (coatings, adhesives, composites, etc...). Such UV-curable resins usually consist of acrylate monomers and oligomers, associated to radical-type photoinitiators. The viscoelastic properties of the cured coatings can be finely controlled through the chemical structure of the oligomer and the monomer functionality, depending on the type of substrate used (flexible or rigid). Most of the research efforts in UV-curing chemistry have been devoted to the development of new photoinitiators, monomers and functionalized oligomers, specially designed for improving the performance of both the processing stage (speed and cure extent) and the coating properties (abrasion and scratch resistance). In this contribution, we report some of the progress recently made in the development of UV-curable acrylic resins with respect to both the resin formulation and the characteristics of the final product. Special attention will be given to the kinetic aspect because the properties of UV-cured coatings are largely depending on a good understanding and control of the manifold reactions occurring during such ultrafast molecule to material transformation.

Experimental

The UV-curable resins contained three basic components: a photoinitiator which generates free radicals upon UV-exposure, an acrylate end-capped oligomer having different chemical structures (polyurethane, polyether, polyester,...) which will determine the properties of the UV-cured polymer, and an acrylate monomer used as reactive diluent to adjust the resin viscosity. The following compounds were used for the formulation of UV-curable resins: Irgacure 2959, Irgacure 651, Irgacure 819, Irgacure 369 (Ciba SC), benzophenone (BP), and 4,4'diphenoxybenzophenone (DPBP) as photoinitiators, Ebecryl 600 (UCB), Ebecryl 284 (UCB), Laromer 8987 (BASF) and Laromer 8949 (BASF) as telechelic acrylate oligomers, hexanedioldiacrylate (HDDA from UCB) and a carbamate (Acticryl CL-960 from SNPE) as reactive diluents. The chemical formulas of the compounds used are given in Chart 1.

The formulation was coated on a BaF_2 crystal or on a glass plate at a typical thickness of 25 μm or 50 μm, respectively, by means of a calibrated wire-

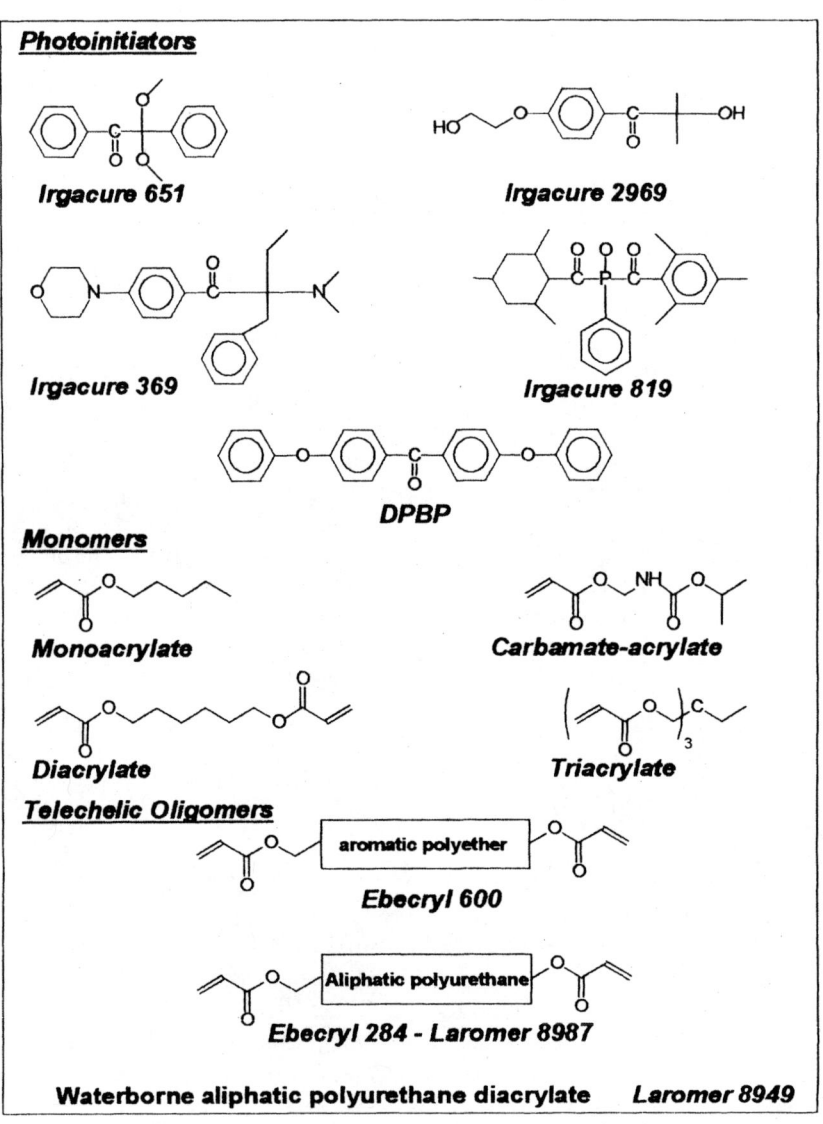

Chart 1. Chemical formulas of the compounds used

wound bar. The sample was exposed for a short time to the UV-radiation of a medium pressure mercury lamp, at a light intensity ranging from 15 to 600 mW cm^{-2}. The polymerization was followed in real time by infrared spectroscopy through the decrease of the IR bands characteristic of the acrylate double bond (810 cm^{-1} and 1410 cm^{-1}). Monomer conversion was determined from the relative decrease of the IR band after a given exposure. The Persoz hardness of the UV-cured coating was evaluated by monitoring the damping of the oscillations of a pendulum placed onto a coated glass plate. Persoz hardness values are ranging from 50 s for soft elastomers up to 400 s for very hard and glassy materials. The scratch resistance, expressed in grams, was evaluated according to the Taber procedure. The weathering resistance of some UV-cured samples coated onto a BaF$_2$ crystal was tested in an accelerated QUV-A weatherometer operated under wet cycle conditions (8 h UV exposure at 70°C, 4 h in the dark at 50°C with water condensation).

Results and Discussion

Photocuring of acrylate-based resins

Acrylate-based resins are widely used today for producing UV-cured coatings because of the great reactivity of the acrylate double bond and the large choice of monomers and oligomers available. The latter differ by their chemical structures, which can be polyesters, polyurethanes, polyethers or polysiloxanes, and by their molecular weight, which ranges typically between 500 and 2000 g. During the light-induced liquid to solid phase change, the viscosity increases rapidly and will cause the polymerization to slow down progressively, up to a complete stop when vitrification occurs. Replacing the di- or triacrylate reactive diluent by a monoacrylate will increase the molecular mobility and lead to a more complete but slower polymerization, as shown by the conversion versus time profiles recorded by real-time infrared spectroscopy for a polyurethane-acrylate (PUA) coating (Figure 1). Fast and complete curing was achieved by using as reactive diluent a monoacrylate carbamate (Acticryl CL-960). The reason of this much-enhanced reactivity is still unknown, but it seems to be related to a lower termination rate constant, as well as to the presence of labile hydrogens which favors chain transfer reactions. This would account for the fact that this monomer bearing a single acrylate function becomes completely insoluble upon UV exposure. Similar results were obtained by introducing a cyclic carbonate in the structure of a monoacrylate monomer, with a 5 fold decrease of the ratio of the termination and propagation rate constant (k_t/k_p). Another distinct advantage of this monomer used as reactive diluent is that it imparts hardness and flexibility to the UV-cured polymers, thus making them more resistant to abrasion, scratching and shocks.

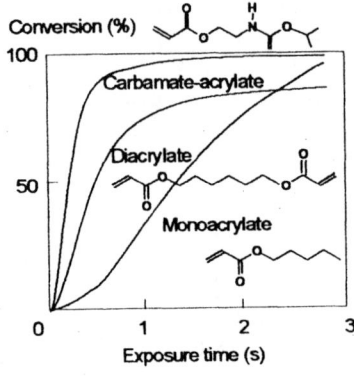
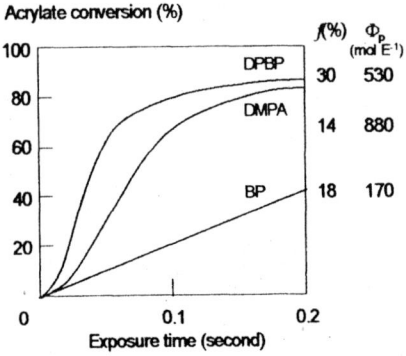

Figure 1. Influence of the reactive acrylic diluent on the photopolymerization of a polyurethane-acrylate coating. [Irgacure 651] = 3 wt%; $I = 40$ mW cm^{-2}; $l = 24$ μm

Figure 2. Influence of the photoinitiator (2 wt%) on the UV-curing of a polyurethane-acrylate in air. $I = 500$ mW cm^{-2}; $l = 25$ μm. [MDEA] = 5wt%; f: fraction of UV-light absorbed; Φ_p: quantum yield of polymerization

The constant search for evermore photosensitive resins has led to the development not only of highly reactive monomers but also of very efficient photoinitiators. In this respect, it is worth mentioning the outstanding performance of a novel photoinitiator, diphenoxybenzophenone (DPBP), which was found to initiate very effectively the photopolymerization of acrylate resins, when associated to a hydrogen donor molecule like methyldiethanolamine (MDEA). Figure 2 compares the polymerization profiles, obtained by infrared spectroscopy, of a polyurethane acrylate (PUA) resin containing 30 wt% HDDA upon exposure to intense UV-radiation (500 mW cm^{-2}) in the presence of various photoinitiators (2 wt%) benzophenone (BP), dimethoxyacetophenone (DMPA) and DPBP. The greater efficiency of DPBP. compared to the unsubstituted benzophenone, was attributed to both a stronger absorption in the UV range (by a factor of 2) and a larger production of initiating radicals (by a factor of 3, based on quantum yield evaluation).

Another advantage of DPBP is to achieve both a surface cure (like BP) and a deep-through cure (like DMPA), thus leading to a marked decrease of the residual unsaturation content of the tack-free film (Figure 3). The more complete curing of the deep-lying layers produces not only a harder polymer (Figure 3), but it also improves the adhesion of the UV-cured coating onto the support. Very hard and glassy polymers were obtained by replacing the flexible polyurethane chain by a stiff polyphenoxy chain, the Persoz hardness of the

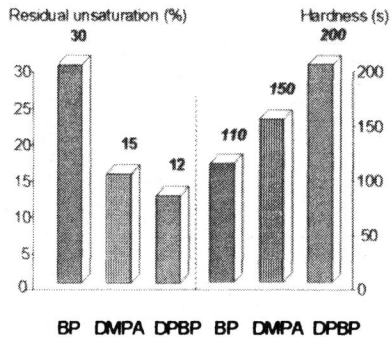
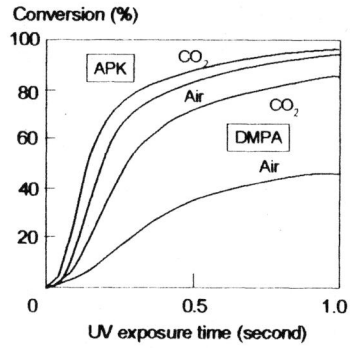

Figure 3. Performance analysis of radical-type photoinitiators in UV-curing of a PUA coating [HDDA] = 30 wt%. UV dose: 0.5 J cm^{-2}

Figure 4. Influence of oxygen on the photopolymerization of a polyurethane-acrylate containing 3 wt% photoinitiator. $I = 30$ mW cm^{-2} $l = 7$ μm.

coating UV-cured with DPBP + MDEA reaching then values superior to 320 s. It should be mentioned that the addition of a co-initiator (MDEA) is no more necessary when the acrylate resin contains labile hydrogen atoms, thus avoiding the plasticizing effect of this compound. As a non-toxic and low cost compound (DPBP is an intermediate product in the manufacturing of polyetherketones), this novel photoinitiator should prove most valuable in UV-curing applications where high-speed and extensive polymerization are of prime importance.

The inhibitory effect of atmospheric oxygen on the radical-induced polymerization of acrylic resins remains one of the critical issues, specially in the coating industry. Different approaches have been used to overcome this unwanted reaction. We have recently found that O_2 inhibition can be effectively suppressed by a proper choice of photoinitiator [10] The best performance was obtained with acylphosphine oxides (Irgacure 819) and amino-phenylketones APK (Irgacure 369), both from Ciba SC. It can be seen in Figure 4 that the photopolymerization of a 6 μm thick PUA coating (neat Laromer 8987 from BASF), exposed to low intensity UV-radiation (30 mW cm^{-2}), proceeds nearly as fast in air as in a CO_2 atmosphere by using APK as photoimtiator, but 3 times as slow with DMPA. In the latter case, the gap between the two curves was greatly reduced by the addition of MDEA which consumes the dissolved oxygen by undergoing a chain peroxidation [10]. Tack-free coatings were thus obtained, like with the DPBP photoinitiator, even by performing the curing under low light intensity in the presence of air.

Waterbased systems are increasingly used in the coating industry, because of ever more severe regulations on VOC emission. They can be advantageously combined with the UV-curing technology to speed up the hardening of the dried coating [11]. As polymerization occurs slowly in a solid material of low molecular mobility, it is recommended to perform the UV-irradiation on the hot sample emerging from the drying oven, so as to achieve a more complete cure. Figure 5 shows the decay curves of the acrylate double bond for a waterbome polyurethane-acrylate coating (Laromer 8949) exposed to UV-light at 25°C and 80°C. As expected, the light-induced crosslinking polymerization is making the coating harder and very resistant to chemicals and organic solvents, specially when the UV-curing is performed at 80°C, due to a more complete polymerization (Figure 5). Waterbased UV-curable resins have the distinct advantage of being much less sensitive to oxygen inhibition than liquid acrylic resins, because of the slower diffusion of atmospheric oxygen into a solid film.

When UV-cured coatings are used to protect three-dimensional objects having complex shapes, e.g., a car body, some remote areas may not be exposed to UV-radiation and will remain uncured. The same problem arises with thick pigmented coatings which cannot be deep-through cured because of the limited penetration of the incident light in opaque materials. To address this issue, dual-cure systems have been developed that combine photopolymerization of acrylates and thermal curing of isocyanates with alcohols [12,13].

Infrared spectroscopy proved to be well suited to study quantitatively the two curing reactions, as it allows one to monitor accurately the chemical modifications induced in thin films by heat or light, namely the disappearance of

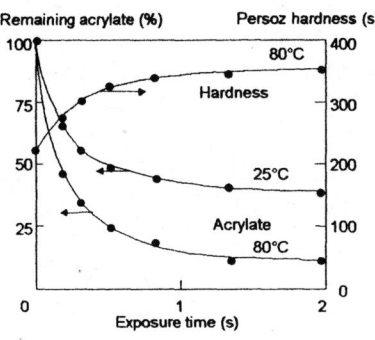
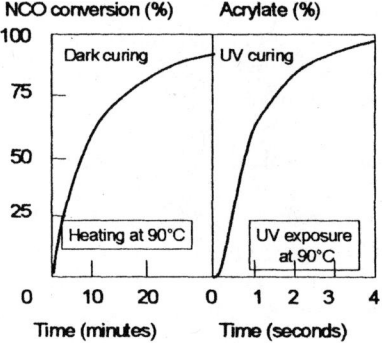

Figure 5. Polymerization and hardening of a waterborne PUA coating upon UV exposure. [Irgacure 2959] = 3 wt%; l : 25 µm; I = 500 mW cm^{-2}

Figure 6. Thermal and photochemical curing at 90°C of a mixture of NCO and OH functionalized diacrylate oligomers. [Lucirin TPO] = 1wt%

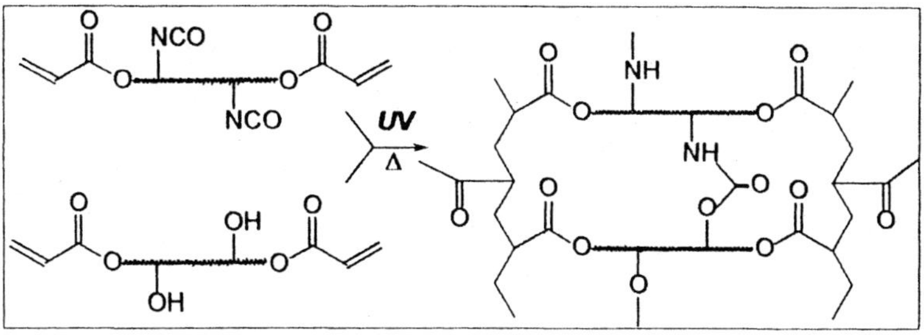

Chart 2. Dual curing of isocyanate and hydroxyl functionalized diacrylate oligomers upon heating and UV-exposure

the characteristic IR band of the isocyanate group at 2271 cm^{-1} and of the acrylate double bond at 1410 cm^{-1}, respectively. The best performance was achieved by combining an isocyanate acrylate oligomer (Laromer LR9000 from BASF) and a hydroxyl functionalized diacrylate oligomer. The dual-curing process is represented schematically in Chart 2. With this formulation, thermal curing occurs already at moderate temperature (90°C), with formation of a tack-free coating in the non-illuminated areas within 20 minutes. Figure 6 shows the polymerization profiles obtained upon heating at 90°C, followed by a short UV exposure of the hot sample, Lucirin TPO (1 wt%) being used as photoinitiator. To speed up the thermal curing, which is the rate limiting step, amines were used as hydrogen donor to form urea linkages by reaction with the isocyanate groups. With methyldiethanol-amine (17 wt%), the polyaddition reaction proceeds already at ambient temperature, with a 50% loss of NCO groups after 20 minutes and total consumption after 2 hours. But the potlife of this two-pack formulations is then quite short, thus requiring special processing conditions. The concentration of the amine has to be finely adjusted to get the best compromise between reactivity and potlife time.

The hardening of an acrylic resin in the non-illuminated areas can also be achieved by heating in the presence of a peroxide, the free radicals thus generated initiating the polymerization of the acrylate double bonds. With 1 wt% benzoyl peroxide, a UV-curable polyurethane-acrylate resin turned into a hard solid after a 20 min heating at 140°C, with a Persoz hardness value of 270 s. The great advantage of this purely acrylate resin is that it does not contain isocyanate groups and is a one- pack formulation with a long potlife. But to proceed rapidly and extensively, it required heating temperatures of at least 120°C that are ill-suited for heat-sensitive substrates. A 2 cm thick PUA sample containing 5 wt% carbon black has been fully cured by applying this procedure (95% conversion, Persoz hardness of 263 s). The addition to this acrylic resin of a tertiary amine (1 wt%), like dimethyl-p-toluidine or N-phenyldiethanolamine, leads to a rapid gelation, even at ambient temperature.

This two-component system has thus been successfully used as a glue to bond together within minutes different types of materials (glass, plastics, metals, wood). The processing is particularly easy to perform, one part of the assembly being coated with the resin containing benzoyl peroxide and the other part with the resin containing the amine, before pressing the two elements together during one minute. If necessary, the setting time can be shortened by a moderate heating with a hair drier.

Properties of UV-cured coatings

The physico-chemical and viscoelastic properties of ISV-cured coatings have to meet precise specifications, depending on the considered end-uses. This is usually achieved by a proper design of both the resin formulation and the processing conditions. The performance of UV-cured coatings, with respect to their mechanical behavior and their resistance to chemicals, moisture and weathering, will actually define the particular industrial applications where they are the best suited and have a chance to successfully compete with conventional solvent-based coatings.

Aliphatic polyurethane-acrylates (PUA) combined with monoacrylates generate upon UV-curing soft elastomeric coatings well suited for the surface protection of flexible supports (paper, plastic and metal foils, fabrics,...). By contrast, polyester-acrylate, aromatic polyether-acrylate and PUA combined with di- or triacrylates yield hard and tough coatings that are more appropriate for the protection of rigid substrates requiring a good resistance to scratching. Figure 7 shows some typical hardening profiles obtained by following the Persoz hardness of PUA coatings exposed to UV radiation (see monomer structure in Chart 1). The great influence of the functionality of the monomer used as reactive diluent is clearly apparent. A judicious combination of these monomers allows one to obtain a coating showing the precise hardness required for the considered applications. One of the distinct feature of waterbased UV-cured polyurethane-acrylate coatings [14], as well as those containing the monoacrylate carbamate [15], is that they combine hardness and flexibility, as shown in Figure 8, thus making such materials quite resistant to abrasion, scratching and shocks.

One of the key properties of organic coatings used to protect different types of materials (metals, plastics, wood, paper) is their resistance to scratching. In this respect, ISV-cured acrylic coatings were found to outperform most of the thermoset acrylate coatings, like those used as automotive finishes [16,17]. The

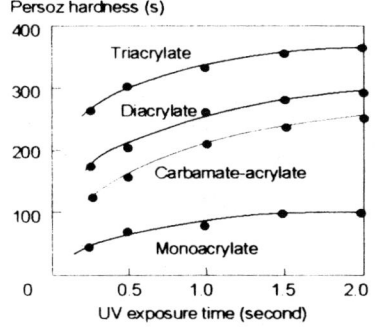
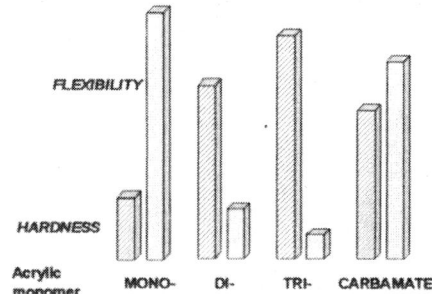

Figure 7. Influence of the reactive diluent (50 wt%) on the hardening of a UV-curable polyurethane-acrylate coating I = 500 mW cm^{-2}

Figure 8. Influence of the acrylic monomer used as reactive diluent (40 wt%) on the mechanical properties of a UV-cured polyurethane-acrylate coating (pendulum hardness and mandrel flexibility).

addition of silica nanoparticles causes a significant improvement of the scratch resistance measured by either the Taber test (load applied on a diamond tip) or the Scotch-brite test (loss of transparency after 200 double rubs), as shown in Figure 9, in full agreement with previous observations[18].

Because of their high crosslink density (up to 4 mol kg^{-1}), UV-cured acrylic polymers are insoluble in organic solvents and quite resistant to most chemicals and pollutants. A PVC plate coated on both faces with a UV-cured polyesteracrylate was thus rendered perfectly resistant to tetrahydrofuran. By using triacrylate reactive diluents, the stain resistance of UV-cured acrylic coatings was substantially enhanced [19]. One of the known limitations of UV-cured acrylic polymers in their moisture sensitivity, which leads to a softening of coatings placed in a humid atmosphere. The water uptake, followed quantitatively by IR spectroscopy, is yet fully reversible in dry conditions and does not affect its durability, by contrast to the melamine-acrylate thermosets used as automotive finishes [20], as shown in Figure 10. After a 2 month exposure at 70°C in a 100% humid atmosphere, the thermosetting acrylate coating was found to have lost 60% of its ether crosslinks, compared to a loss of only 8% of the urethane groups for the photoset PUA coating. The water permeability was significantly reduced for UV-cured clay/acrylic nanocomposites [21,22] as well as by addition of a fluorinated acrylate monomer (1 wt%) which migrates toward the surface and imparts to the UV-cured

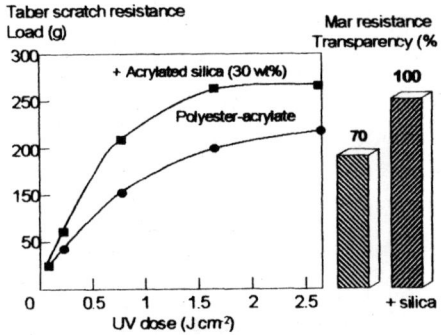 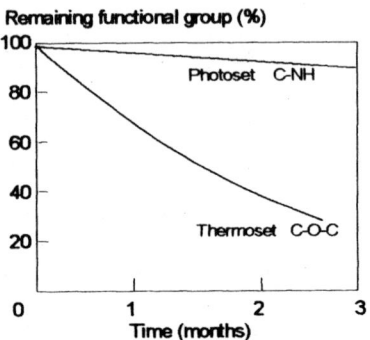

Figure 9. Influence of silica nanoparticles on the scratch resistance and mar resistance of a UV-cured polyester-acrylate coating

Figure 10. Moisture resistance of stabilized melamine-acrylate thermoset and polyurethane-acrylate photoset coatings placed in a 100% humid atmosphere at 70°C.

coating a strong hydrophobic character (contact angle of 1200 for a water droplet). The resulting decrease in die surface energy (30 mJ cm^{-2}) makes such coatings particularly well suited for anti-fouling applications.[2]

Specially designed aliphatic polyurethane-acrylate (PUA) resins, once cured by UV-irradiation, have been found to be very resistant to photodegradation when properly stabilized[23,24] In the presence of a UV absorber (2 wt% Tinuvin 400) and a HALS radical scavenger (1 wt% Tinuvin 292), such PUA coatings remained clear and glossy after 5000 h of wet cycle QUV-A weathering, and after as much as 8 years of outdoor exposure in Florida, without delamination from their metallic support, thus ensuring an effective and durable protection. Even waterbased UV-cured PUA coatings were found to exhibit an outstanding weathering resistance [25,26], as illustrated by the minor chemical modifications observed in a 40 μm thick coating after 10.000 h QUV-A ageing (Figure 11). A quite remarkable feature is that more than half of the UV absorber is still present after such a long exposure, thus ensuring a longlasting screen effect against sunlight. Well stabilized UV-cured PUA coatings have been successfully used to improve the exterior durability of various organic materials, like for instance to reduce the discoloration of clear PVC [27] as well as the loss of gloss and photobleaching of painted metals. They were found to outperform the thermosetting melamine-acrylate clearcoats currently used as automotive finishes, with respect to their resistance to scratching, moisture and weathering [28].

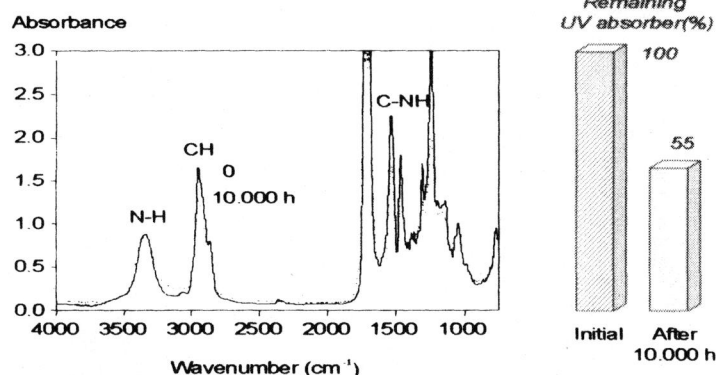

Figure 11. Infrared spectra of a stabilized UV-cured waterborne PUA coating before and after 10,000 h wet cycle QUV-A ageing Laromer ®LR8949+2% Irgacure 2959 (2% Tinuvin 400 + 1% Tinuvin 292, before (——) and after (----) 10,000 h QUV-A weathering

Conclusion

Photoinitiated polymerization of multifunctional monomers and oligomers remains one of the most effective methods to generate quasi-instantly highly crosslinked polymer materials. A considerable amount of research and development work has been devoted to this "green" technology which has found its major openings in the coating industry and in the graphic arts. We have reported here some of the progress recently made in several areas to further improve the performance of UV-cured acrylic polymers and widen their range of applications. Special attention has been given to the design and development of:

- highly efficient photoinitiators allowing to overcome the inhibitory effect of atmospheric oxygen on radical-type polymerization:
- very reactive acrylate monomers to be used as diluents to achieve a fast and complete curing;
- dual-curable resins combining thermal and photochemical polymerization to realize an effective hardening in the non-illuminated areas, in particular for thick pigmented samples;
- UV-cured coatings showing superior performance with respect to their surface properties and their weathering resistance, thus allowing to improve the exterior durability of various materials.

The production of protective coatings by UV-radiation curing is a great step toward the assigned objective of meeting the growing constraints of the international regulations to preserve the environment. hi this respect, this technology happens to meet the first of the grand challenges for chemists and chemical engineers issued by the National Academies in their recent Report entitled "Beyond the Molecular Frontier" [29], namely "Learn how to synthetize and manufacture any new substance that can have scientific or practical interest, using compact synthetic schemes and processes with high selectively for the desired product, and with low energy consumption and benign environmental effects in the process". Further research on photocrosslinking polymerization is still needed to improve both the resin performance and the final properties of such UV-cured polymers, thus broadening the field of applications of this environment-friendly technology. It should also be focused on the mechanistic aspects to gain a better understanding and control of the manifold and complex processes involved in ultrafast photocuring reactions.

Acknowledgements

The author wishes to thank his coworkers: Laurent Keller, Ivana Lorinczova, Frédéric Masson, Katia Studer, Estelle Weber-Kochl and Khalid Zaliouily.

References

1. Decker, C.; *Progr. Polym. Sci.*, **1996**, *21*, 593.
2. Oldring, K.T. (ed), *"Chemistry and Technology of UV and EB Formulation for Coatings, Inks and Paints"*, Vol I to VII, Wiley/SITA London 1997-1998.
3. Roffey, C.; *"Photogeneration of Reactive Species for UV-curing"*, Wiley, New-York. 1997.
4. Scranton, A.B.; Bowman, C.N.; Peiffer, R.W.; (eds), *"Photopolymerization Fundamentals and Applications"*, ACS Symp. Ser. 673, Amer. Chem. Soc., Washington 2003.
5. Davidson, S.; Exploring the Science, Technology and Applications of UV and EB Curing, SITA Techno., London, 1999.
6. Decker, C.; *Macromol. Rapid. Com.* **2002**, *23*, 1067.
7. Belfield, K.D.; Crivello, J.V.; (eds), *"Photoinitiated Polymerization"*, ACS Symp. Ser. 847, Amer. Chem. Soc. Washington 2003.
8. Decker, C.; *Surf Coat. Intern.* **2005**, *88B1*, 9.
9. Decker, C.; *Polymer News,* **2005**, *30*, 34.
10. Studer, K.; Decker, C.; Beck, F.; Schwalm R., *Prog. Org. Coat.* **2003**, *48*, 92 and 101

11. Masson, F.; Decker, C.; Jaworek, T.; Schwalm R., *Progr. Org. Coat.* **2000**, *39*, 115.
12. Decker, C.; Masson, F.; Schwalm, R.; *Macromol. Mater. Eng.* **2003**, *288*, 17.
13. Studer, K.; Decker, C.; Beck, E.; Schwalm, R.; *Europ. Polym. J.* **2005**, *41*, 157
14. Schwalm, R.; Haussling, L.; Reich W.; Beck, E.; Enenkel, P.; Menzel, K.; *Progr. Org. Coat.* **1997**, *32*, 191.
15. Moussa, K.; Decker, C.; *J. Polym. Sci. Polym. Chem. Ed.* **1993**, *31*, 2203.
16. Schwalm, R.; *"Farbe und Lack"*, **2000**, *106(4)*, 58
17. Schwalm, R.; Beck, E.; Pfare, A.; *Europ. Coat. J.* **2003**, *2*, 39.
18. Can, V.; La Ferté, O.; Eranian, A.; *Proc. Rad. Tech Europe,* 2003, p.1119
19. Zahouily, K.; Blachere, G.; Decker, C.; *Proc. Rad. Tech Europe (Lyon,),* 1997, p.266
20. Nguyen, T.; Martin, J.; Byrd, E.; Embree, N.; *Polym. Prep.* **2001**, *42(1)*, 420.
21. Decker, C.; Keller, L.; Zahouily, K.; Benfarhi S., *Proc. Rad. Tech Conf, Yokohama,* **2003**, p.516.
22. Keller, L.; Decker, C.; Zahouily, K.; Benfarhi, S.; Le-Meins, J.M.; Miehe-Brendle, J.; *Polymer,* **2004**, *45*, 7437.
23. Decker, C.; Moussa, K.; Bendaikha, T.; *J .Polym. Sci. Polym. Chem.* **1991**, *29*, 739.
24. Decker, C.; Zahouily, K.; Polym. Degr. Stab. **1999**, *64*, 293.
25. Decker, C.; Masson, F.; Schwalm, R.; *Polym.Degr.Stab.* **2004**, *83*, 309.
26. Decker, C.; Lorinzcova, I.; *J. Coat. Technol. Research,* **2004**, *1(4)*, 247.
27. Decker, C.; in *"Chemical Reactions on Polymers",* Benham, J.; Kinstle, J.; (eds), ACS Symp.Ser. 364, *Amer. Chem. Soc.,* Washington **1988**, p.201.
28. Decker, C.; Zahouily, K.; Valet, A.; *J. Coat. Technol.,* **2002**, *74*, 87.
24. National Academies Report. *Chemical and Engineering News,* 3 March 2003, p.39.

Chapter 13

Polyamide-11 Powder Coatings: Exceptional Resistance to Cavitation Erosion

T. Page McAndrew[1], Marc Audenaert[2], Jerry Petersheim[3], Dana Garcia[4], and Thomas Richards[4]

[1]Technical Polymers R&D, Arkema Inc., Research Center,
900 First Avenue, King of Prussia, PA 19406
[2]Technical Polymers R&D, Arkema S.A., Centre d'Etude de Recherche &
Développment, Route du Rilsan, Serquigny, 27470, France
[3]Technical Polymers Business Development International,
2000 Market Street, Philadelphia, PA 19103
[4]Analytical and Systems Research, Arkema Inc., Research Center,
900 First Avenue, King of Prussia, PA 19406

A first application of polyamides in the 1930's was replacements for metals. Seventy years later, a similar application is discovered for Polyamide-11 – cavitation resistant coatings that enable replacement of high-cost metals in demanding water applications. Polyamide-11 powder coatings are shown to have exceptional resistance to cavitation erosion – substantially better than other polymer coatings such as epoxies, and better even than stainless steels. Gravimetric, photographic, and spectrophotometric data are presented.

Introduction

Polyamide-11

Of the hundreds of polyamides developed since the 1930's, only one excels as a coating – Polyamide-11. See Figure 1.

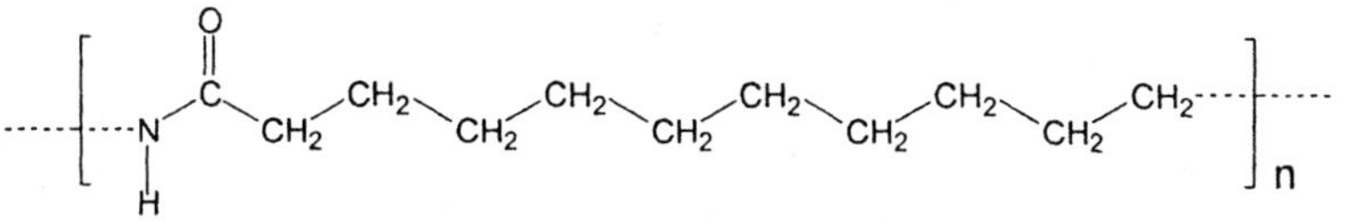

Figure 1. Polyamide-11 [poly(11-aminoundecanoic acid)]. Polyamide-11 is also called nylon-11.

To understand this, see Figure 2, which compares Polyamide-11 to Polyamide-6 and Polyamide-6/6 – commercially the two most important polyamides. (1) Polyamide-11 has features that make it very well-suited to coatings:

- melting point is low (185°C) – enabling convenient processing
- water absorption is low (approximately 2% by weight) – enabling good corrosion protection (absence of water inhibits the corrosion reaction) – see Equation 1
- impact resistance is high (approximately 50 kJ/m^2)

$$2Fe + O_2 + 2 H_2O \Leftrightarrow 2 Fe^{2+} + 4 (OH)^- \quad E° = +1.26 \text{ volts} \tag{1}$$

Polyamide-11 is unique in another aspect – it has a natural source. The monomer, 11-aminoundecanoic acid, derives from castor oil. (2)

Rilsan® Fine Powders

Arkema Inc. makes Rilsan® Fine Powders – thermoplastic powder coating products comprising Polyamide-11. Powders are applied by ordinary powder coating methods – electrostatic spray, fluidized bed dipping, and thermal spray. Worldwide, they have been used for over 40 years in water industry

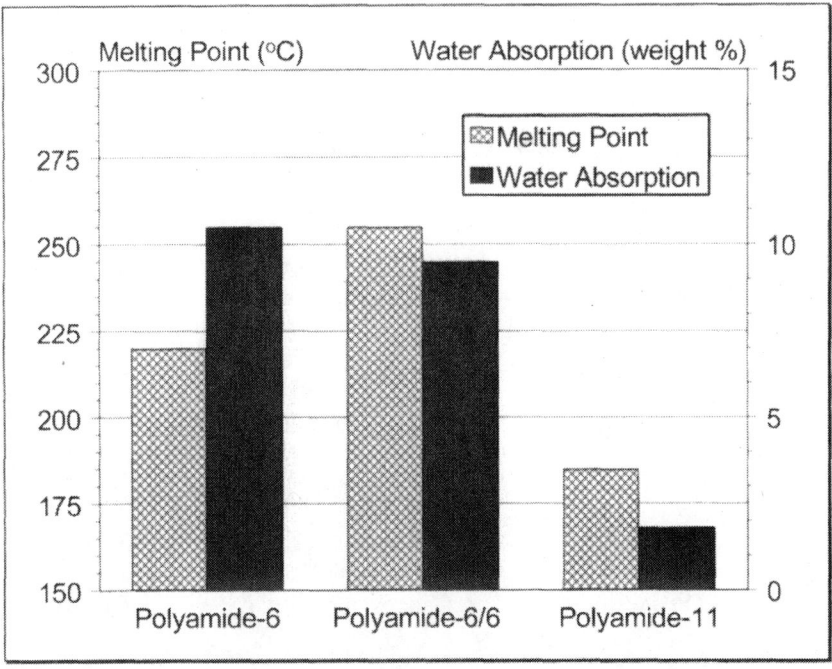

Figure 2. Comparison of polyamides.

applications, such as coatings for pumps, valves, couplings, and pipes. This utility is based upon excellent resistance to corrosion, abrasion, impact, and acidic/basic water-treatment chemicals. In the United States, Rilsan® Fine Powders are approved by several regulating agencies:

- American Water Works Association – ANSI/AWWA C-224 – *Nylon-11-Based Polyamide Coating System for the Interior and Exterior of Steel Water Pipe, Connections, Fittings and Special Sections*
- Underwriters Laboratories – Standard 1091 – *Butterfly Valves for Fire Protection Service*
- NSF (National Sanitation Foundation) International – NSF/ANSI Standard 61 – *Drinking Water System Components – Health Effects*

Recently a discovery was made that enhanced the utility of Rilsan® Fine Powder coatings in the water industry – exceptional resistance to cavitation erosion.

Cavitation Erosion

In high-velocity water, low-pressure regions form, where water vaporizes to form bubbles. When these bubbles move to high-pressure regions, they collapse. This is called *cavitation*. This collapse generates shock waves that erode surfaces nearby. This is called *cavitation erosion*. Cavitation erosion commonly causes premature failure of hydraulic equipment such as pumps, valves, pipes, turbines, and propellers. (3, 4)

Recent work examined the cavitation erosion of metals employed in municipal water systems – stainless steel, cast iron, bronze, brass, and copper – and concluded that stainless steel was the most resistant, by far. (5) Because of cost, use of high-performance materials like stainless steel usually is not practical. A protective coating that resists cavitation erosion would be of enormous value – improving equipment lifetimes and enabling use of lower-cost metals.

As described presently, the cavitation erosion resistance of Rilsan® Fine Powder coatings far exceeds that of other polymer coatings, and exceeds even that of stainless steels. This enables coated carbon steel (much lower cost) to be used instead of stainless steel where cavitation erosion is an issue.

Experimental

Sample Preparation

Coatings were formed on carbon steel panels (approximately 3 mm thick) at Arkema Inc. facilities in King of Prussia (PA) and Serquigny (France). Panels were prepared by abrasive blast cleaning, followed by washing with either fresh liquid toluene or trichloroethylene vapor. Powder coating materials were used as received, and applied according to manufacturer's specifications.

Regarding comparative metals: carbon steel panels (Type S-36) were from Q-Panel Lab Products (Cleveland, OH); steel alloy (CA-6-NM), stainless steel (316-L), ductile cast iron (65-45-12), brass (85-5-5-5, CDA 836) and phosphorus bronze (CDA 510) were from Metal Samples Company (Munford, AL).

Samples were 0.5 inch squares — cut from interior areas of panels to avoid any edge effect. Most samples were mounted as is. However, because of thinness, samples of carbon steel and bronze were mounted on a support piece of steel with epoxy.

Measurement of Cavitation Erosion

Cavitation erosion was measured at KTA Tator, Inc. (Pittsburgh, PA), using a method based on ASTM G-32-98 *(Standard Test Method for Cavitation Erosion Using Vibratory Apparatus)*. The ultrasonic probe was a Sonics and Materials, Inc. (Newton, CT) Model VC-501 – which produced 500 watts oscillating at 20 kHz. The diameter of the horn tip was 0.5 inch and the amplitude of the oscillation was 62 microns (peak-to-peak displacement). The distance between the probe tip and the sample surface was approximately 1 mm. The volume of immersion water (distilled/deionized) was approximately 1 liter. Samples were placed approximately 1.2 cm below the water surface. Water temperature was maintained between 22°C and 27°C. Water contained a proprietary (to KTA Tator, Inc.) corrosion inhibitor to prevent weight changes due to corrosion of exposed metal surfaces.

Exposure to cavitation was performed continuously for a fixed time. Then the sample was removed, dried in a dessicator and weighed. After this, exposure continued. Loss (microns) was calculated according Equation 2:

$$Loss = \frac{10^4 \times (weight\ loss\ in\ grams)}{(sample\ area\ in\ cm^2) \times (specific\ gravity\ in\ grams/cm^3)} \quad (2)$$

Sample area was 1.6 cm² (i.e., 0.5 inch²). Values of loss versus time are shown in Figures 3 and 4. Each value is an average of three (from three different panels). Optical imaging was performed with a Nikon® SMZ800 stereo microscope (magnification approximately 10X) using reflected light and an Optronics® digital camera. ATR infrared spectroscopy was performed with a Thermo Electron Corporation Nicolet® Nexus® FT-IR. Resolution was 4 cm⁻¹.

Results and Discussion

Values of loss versus time are shown in Figures 3 and 4. Photographs before and after testing are shown in Figures 5, 6 and 7. From Figure 3 it is seen that Rilsan® Fine Powder coatings show far better performance than the thermoset epoxy products, Scotchkote® 206N (3M) and Resicoat® R5 (Akzo-Nobel, Inc.). Photographic data support this. From Figure 5 observe that Rilsan® Fine Powder products undergo only a small loss of gloss, whereas both thermoset epoxy products undergo substantial pitting – to the point of exposing metal. These results are not surprising. Crosslinked coatings, such as epoxies,

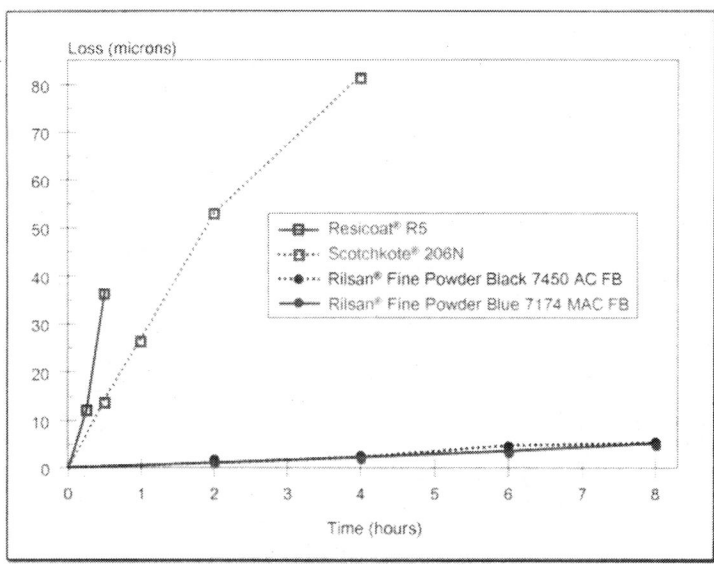

Figure 3. Loss data for coatings.

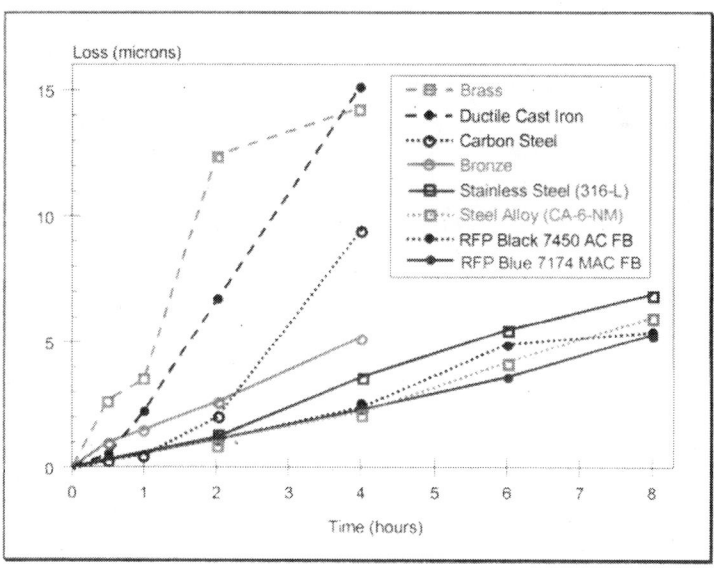

Figure 4. Loss data for coatings and metals.

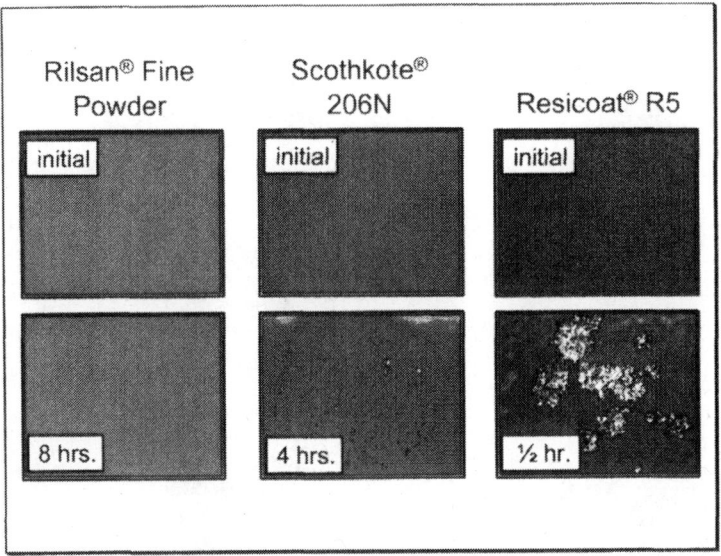

Figure 5. Photographs of coatings.

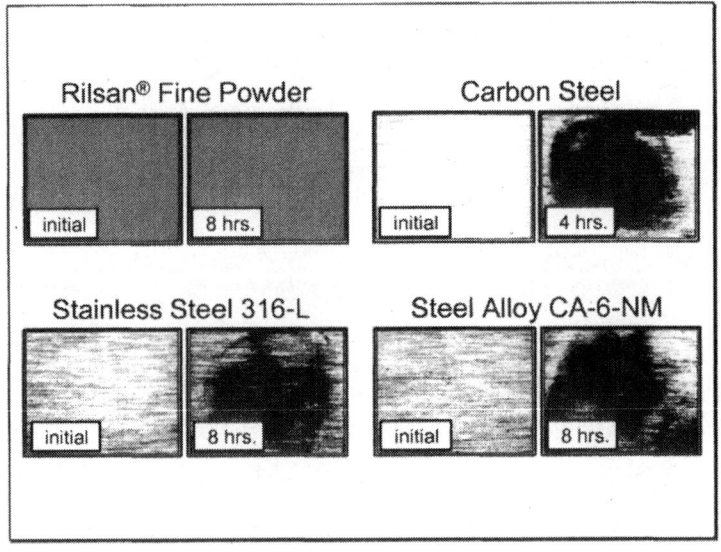

Figure 6. Photographs of coatings and metals.

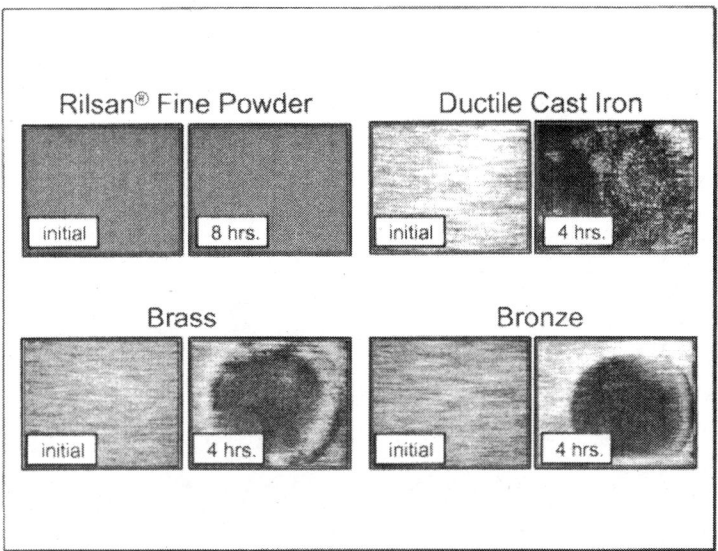

Figure 7. Photographs of coatings and metals.

would not be expected to withstand very well the impact of cavitation-generated shock waves. The data shown in Figure 4 are quite interesting. Rilsan® Fine Powder coatings show substantially better performance than brass, bronze, ductile cast iron, carbon steel, and stainless steel 316-L. These data are supported by photographic data shown in Figures 6 and 7. Concerning weight loss data, Rilsan® Fine Powder coatings perform approximately the same as Steel Alloy (CA-6-NM). However, photographic data show clearly that Steel Alloy (CA-6-NM) undergoes a visual change, whereas Rilsan® Fine Powder coatings undergo almost none. Thus the cavitation erosion resistance of Rilsan® Fine Powder coatings would be regarded as better than that of Steel Alloy (CA-6-NM).

Some work was done to understand the process of material loss that occurs during cavitation erosion. Infrared spectra are shown in Figures 8, 9, and 10, for the coatings, Rilsan® Fine Powder Black 7450 AC FB, Scotchkote® 206N, and Resicoat® R5, respectively. Rilsan® Fine Powder Black 7450 AC FB exhibits virtually no change after exposure to cavitation. The very small loss shown in Figure 4 appears to result only from a mechanical process. By contrast for Scotchkote® 206N and Resicoat® R5, there are substantial changes in the infrared spectra. There are notable increases in the areas of 3,400 cm^{-1} and 2,900 cm^{-1}, corresponding to –OH and sp^3 C-H, respectively. Exposure to cavitation

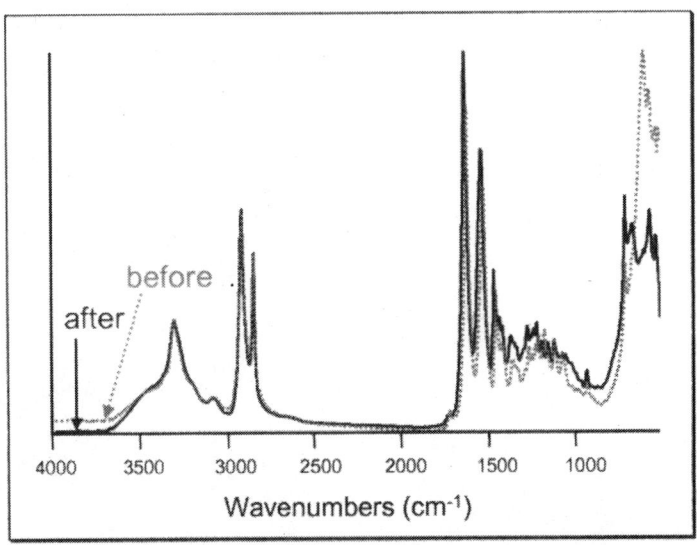

Figure 8. ATR infrared spectra of Rislan® Fine Powder Black 7450 AC FB coating. Shown are before and after 8 hours exposure to cavitation.

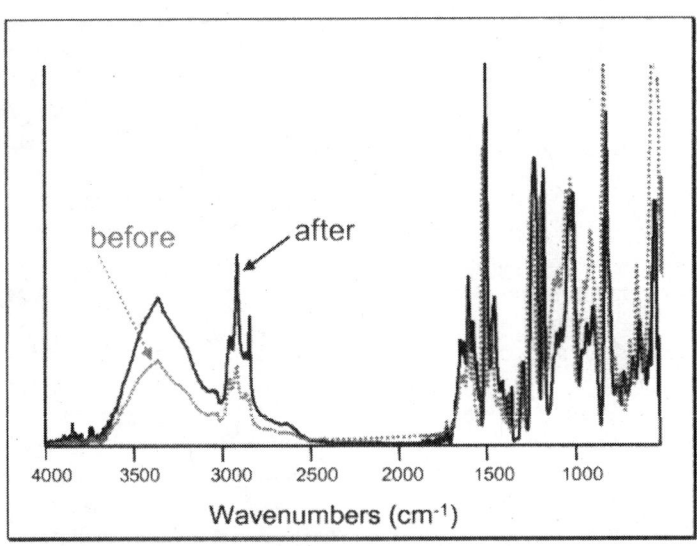

Figure 9. ATR infrared spectra of Scotchkote® 206N coating. Shown are before and after 4 hours exposure to cavitation.

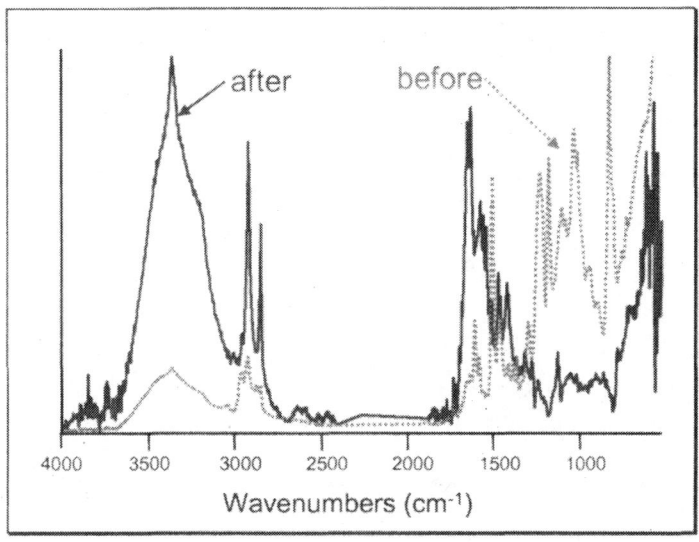

Figure 10. ATR infrared spectra of Resicoat® R5 coating. Shown are before and after 0.5 hour exposure to cavitation.

clearly caused a chemical change, possibly oxidation and chain scission, contributing to the very rapid loss of coating observed.

Conclusions

Rilsan® Fine Powder coatings have exceptional resistance to cavitation erosion – substantially better than other coatings and metals used in water industry applications – most notably stainless steels. For applications where cavitation erosion is a concern, Rilsan® Fine Powders enable replacement of high-cost metals like stainless steels with lower-cost coated metals, like coated carbon steel. Seventy years after the discovery of polyamides, another *metal-replacement* application has been identified.

Acknowledgements

Measurement of cavitation erosion at KTA Tator, Inc. was performed under the direction of William D. Corbett, Technical Services Administrator. Thanks are extended to Christophe Valentin (Powders R&D Department, Arkema S.A.,

Serquigny, FR) for preparation of some of the coatings. Philip L. Drooks of Metropolitan Water District of Southern California (La Verne, CA) is acknowledged for highlighting the criticality of the cavitation erosion phenomenon, and advice on materials important to the water industry.

References

1. *Nylon Plastics Handbook;* Kohan, M., Ed.; Hanser/Gardner Publications, Inc.: Cincinnati, OH, 1995: pp 576-582.
2. Apgar, G.B. and Koskoski, M.J. in *High Performance Polymers. Their Origin and Development;* R.B. Seymour and G.S. Kirshenbaum, Eds.; Elsevier Science Publishing Co., St. Louis, MO, 1986, pp 55-65.
3. Lecoffre, Y.; *Cavitation;* A.A. Balkema Publishers, Brookfield, VT, 1999, pp 244-291.
4. Rahmeyer, W.J. Journal AWWA, May **1981**, 270-274.
5. Chan, W.M., Cheng, F.T., and Chow, W.K. Journal AWWA, **2002**, *94 (8)*, 76-84.

Chapter 14

Using VOC-Exempt Solvents in Coatings: Performance, Productivity, and Lower Environmental Impact

Daniel B. Pourreau

Lyondell Chemical Company, 3801 Westchester Pike, Newtown Square, PA 19093

Introduction

For decades, regulatory agencies in most industrialized nations have enacted more and more stringent volatile organic compounds (VOC) regulations to reduce the solvent content in coatings. This effort is motivated, in part, by the desire to reduce emissions of solvents that are precursors to tropospheric ozone or smog. These VOC regulations have also spurred the growth of low-VOC technologies, such as waterborne, powder, high solids, and energy-curable coatings. While these technology advances are welcome and are excellent choices for some coating applications, they too have their limitations and environmental impacts.

Despite these environmental regulations and new coating technologies, solvent-borne coatings remain the second largest coating category, after waterborne coatings, for several reasons. First, industrial solvents are inexpensive and are excellent viscosity reducers for coating resins and additives. They are used at every stage of the coating process, from resin production to gun cleanup:

- Resin synthesis
- Coating formulation
- Surface cleaning and degreasing

- Thinning
- Application equipment cleanup

In coating formulations, they perform several important functions, including:

- Dispersing pigments
- Compatibilizing resins, additives and pigments
- Reducing viscosity for application
- Leveling the coating on the surface
- Evaporating quickly and consistently, regardless of environmental conditions

Solvents, therefore, play an important role in the formulation, storage, application, drying, appearance, performance and durability of coatings and inks.

Despite their shortcomings, solvent-borne coatings are still unmatched in their economy, ease of use, versatility, and performance. Therefore, solvent producers have been developing new solvents with lower health and environmental impacts to replace the ones still in use today.

Regulators are also beginning to see the benefit of encouraging this type of substitution. For example, the California Air Resources Board (CARB) has enacted a reactivity-based policy for aerosol coatings[i]. This approach is also being considered by the U.S. Environmental Protection Agency (EPA) as a possible future VOC compliance option instead of mass-based VOC content limits[ii].

The Science Behind VOC Regulations

Because most solvents evaporate and react in polluted urban atmospheres to produce tropospheric ozone, they are regulated as VOCs[iii]. Ozone is a respiratory irritant and is regulated as a criteria pollutant in the United States and Europe. Canada has also enacted strict VOC regulations patterned after U.S. regulations[iv]. Europe also regulates VOCs, but their regulations are based on volatility instead of photochemical reactivity. This approach is less likely to produce results because it ignores the vast difference in ozone-forming potentials between solvents and does not encourage substitution of less reactive solvents for more reactive ones.

In the United States and Canada, solvents that are shown to produce less ozone than ethane are considered negligibly photochemically reactive and are exempt from VOC regulations. Not all solvents do have the same impact on ozone (see Figure 1). Some, like acetone and tertiary butyl acetate (TBAC), produce very little ozone, whereas others, like toluene and xylene, produce many times their own weight in ozone when emitted. Many halogenated solvents have very low photochemical reactivities and are VOC-exempt. However, they are also relatively persistent in the environment. Chlorofluorocarbons (CFCs) and hydrochlorofluorocarbons (HCFCs), for example, have very long atmospheric lifetimes and have been shown to deplete stratospheric ozone[v]. Some CFCs and their substitutes are also believed to contribute to global warming[vi].

When comparing photochemical reactivities to that of ethane, the U.S. EPA has historically used two benchmarks. For VOCs whose reactivity is well below ethane, it is sufficient to consider their kinetic reactivities. The rate constant for hydrogen abstraction from the VOC by atmospheric OH radicals, or k_{OH}, expressed in cm^3/molecule•second is compared to that of ethane:

$$R\text{-}H + {}\cdot OH \xrightarrow{k_{OH}} R\cdot + H_2O \tag{1}$$

This reaction is the main initiation step for ozone formation for most VOCs. Another initiation pathway is direct photolysis to form a free radical. After the VOC radical is formed, it undergoes a series of decomposition steps, producing by-products that can also interact with ambient NO to form NO_2. Photolytic decomposition of NO_2 produces singlet oxygen which then reacts with oxygen to form ozone. The VOC oxidation by-product, $RO\cdot$, can then undergo further reactions to produce more ozone or terminate.

$$R\cdot + O_2 \rightarrow RO_2\cdot \tag{2}$$

$$RO_2\cdot + NO \rightarrow RO\cdot + NO_2 \tag{3}$$

$$NO_2 + O_2 \xrightarrow{h\upsilon} NO + O_3 \tag{4}$$

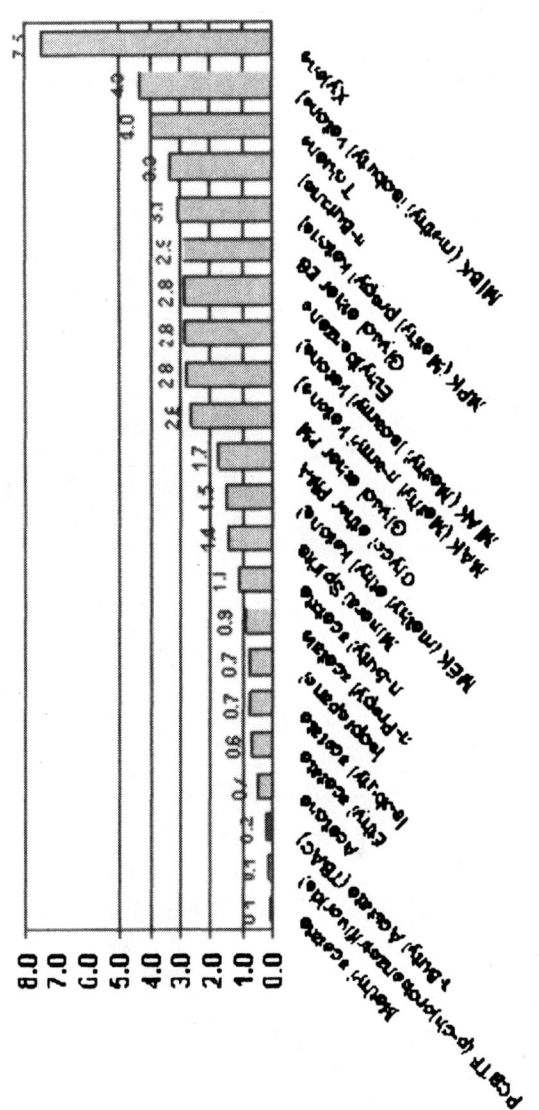

Figure 1. Ozone Forming Potential of Common Coating Solvents

Hence, depending on the VOC decomposition mechanism and environmental conditions, each mole of VOC might form less than one mole of ozone, exactly one mole, or several moles of ozone.

To account for this vast difference in ozone forming potential, VOCs whose kinetic reactivities are close to that of ethane are compared based on their "mechanistic" reactivities. Mechanistic reactivities are most often expressed as grams ozone/grams VOC to reflect the nonstoichiometric nature of ozone formation. Several state-of-the-art models are available for estimating mechanistic reactivity, but the most commonly used by the EPA is the SAPRC 99 model, developed by Dr. William Carter[vii]. The mechanistic reactivity conditions historically used for regulatory purposes is the one measured by adding a small amount of a VOC to a standard VOC mixture under optimal NO_x and photolytic conditions to form the maximum amount of ozone, which is called the maximum incremental reactivity (MIR). The MIRs of common coating solvents are compared in Figure 1.

Comparing VOC-Exempt Solvents

Acetone was the first compound exempted from VOC regulations based on MIR data. It is an excellent viscosity reducer for a variety of coating resins, but has several properties that make it less than ideal for formulating coatings. First, its low flash point makes it a severe fire hazard. Second, it has a high evaporation rate and is hygroscopic, which can result in the appearance of defects known as "blushing" and "solvent popping." Blushing is the appearance of a whitish haze in a clearcoat when moisture condenses in the coating due to evaporative cooling. Solvent popping appears as visible pinholes in the coating caused by the solvent escaping after the viscosity has increased to a point where the coating no longer flows (see Figure 2). Finally, acetone is a very strong solvent that can redissolve undercoats, causing the color to "bleed" into clear topcoats or even lift from the substrate.

Since acetone was exempted, solvent producers have petitioned the EPA to exempt several other solvents based on MJR. Of those, only volatile methyl siloxanes (VMS), parachlorobenzotrifluoride (PCBTF), methyl acetate, and TBAC have been exempted. VMS and methyl acetate have found limited use in coatings. VMS is relatively expensive and is not a very effective viscosity

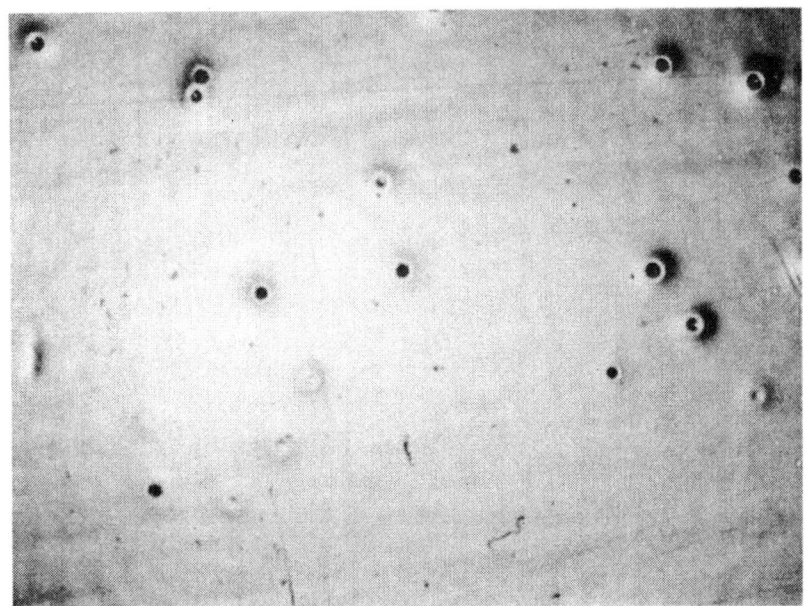

Figure 2. Solvent Popping in a 2K Urethane Clearcoat Using Surface Magnification (15x) and Coaxial Lighting

reducer for common coating resins. Methyl acetate has properties similar to acetone, but is more expensive. Despite its added cost, acetone's higher electrical resistivity and flash point make it a safer solvent more suitable for electrostatic spray applications.

PCBTF is currently used in a variety of compliant coatings, but is likely to be replaced by newly exempted TBAC in applications for which flammability is not an issue. The chief complaints from coating formulators regarding PCBTF are its high cost, unpleasant odor, and high density (11.2 lb/gallon). The high density of PCBTF compounds its high cost because coating solvents are usually purchased by the pound but coatings are sold by the gallon. The properties of these exempt solvents are listed in Table 1.

All VOC-exempt solvents have low ozone forming potentials (MIRs). Methyl acetate is the least reactive of the four, and acetone is the most reactive. PCBTF is about half as reactive as TBAC on a weight basis but almost the same on a volume basis. Therefore, substituting PCBTF for TBAC is not expected to

Table I. Physical Properties of Exempt Solvents for Coatings

VOC Exempt Solvent	Acetone	Methyl Acetate	TBAC	PCBTF
CAS number	67-64-1	79-20-9	540-88-5	98-56-6
Evaporation rate, n-BuAc = 1	6.3	6.2	2.8	0.9
Boiling point, °C	56	56	98	139
Flash point, °C	-4	9	42	109
lb/gal@25 °C	6.55	7.78	7.24	11.2
Surface tension, Dynes/cm^2	23.3	25.8	22.4	25
Electrical resistivity, Mohms	<0.4	0.4	23.8	
Solubility in water, %	100%	23%	3%	0%
MIR, g ozone/g	0.43	0.07	0.20	0.11
MIR, Kg ozone/gal	1.28	0.25	0.66	0.57
Hansen H	3.4	3.7	2.9	1.9
Hansen D	7.6	7.6	7.0	8.8
Hansen P	5.1	3.5	1.7	2.9
Hansen rotal	9.8	9.2	7.7	8.6

greatly increase ozone formation, and replacing acetone should decrease it by almost 50%.

Because of its intermediate evaporation rate, broad solvency for coating resins, and relatively low cost, TBAC is expected to find widespread use as coating solvent, thinner, and cleanup solvent. The use of acetone, methyl acetate, and PCBTF in coatings has been previously described. This chapter will focus on the use of TBAC as a VOC compliance tool in high-performance solvent-borne coatings.

Selecting a Solvent Blend

Solvent evaporation rate is a critical property that can affect the drying time of the coating or ink, as well as the film appearance and durability. Formulations typically contain a blend of fast, medium, and slow solvents to optimize film drying time, properties, and appearance. A typical air-dry formulation contains the following combination of solvents:

Solvent	Evaporation Rate vs n-BuAc	Wt %
Fast	> 3.0	10%
Medium	0.8 to 3.0	80%
Slow	< 0.8	10%

Because TBAC is a medium-fast solvent (2.8 vs. n-butyl acetate (nBuAc)), a fast solvent is not required, and more slow solvent might be needed especially if the coating is baked after application. For alkyd baking enamels, for example, a 70/30 blend of TBAC and a slow solvent, such as PM propionate[viii], is a good starting point to minimize solvent popping and other appearance defects.

Another important consideration in selecting a solvent blend is the activity of the solvents. Solvents can also be classified as active, latent, and diluent based on their ability to reduce the resin's viscosity. The activity of a solvent depends on the resin type, but in general, ketones, esters, and other oxygenated solvents are good viscosity reducers and are active solvents for most coating resins. Alcohols and aromatic hydrocarbons are typically latent solvents with intermediate solvency for resins, whereas aliphatic hydrocarbons are usually low-cost diluents. The exception is long oil alkyds for which aliphatic hydrocarbons are effective viscosity reducers and are, therefore, active solvents.

A conventional solvent blend contains approximately 40% active or latent solvents and approximately 60% diluent. However, VOC content limits are changing this conventional wisdom. Indeed, as the VOC content limits are reduced, less diluent and latent solvents can be used in the formulation because each gram of solvent must be used to reduce the coating viscosity. Also, the slow solvent must be an active solvent for the resin in question, or the film appearance could be compromised as the solvent blend evaporates and leaves behind the resin and a diluent for which it has little affinity.

For example, consider the sample reformulation in Table 2. The "old" formulation is a conventional solvent blend for a two-component acrylic urethane coating. It contains 60% inexpensive toluene, 10% fast methyl ethyl ketone (MEK) solvent, 20% n-BuAc, a medium-speed active solvent, and 10% methyl n-amyl ketone (MAK), a slow, active solvent.

Table 2. Solvent Blend Reformulation for HAP and VOC Compliance

Solvent	Old	New	Speed
MEK	10%		Fast
TBAC		71%	Medium-fast
Toluene	60%		Medium
n-BuAc	20%	9%	Medium-slow
MAK	10%	20%	Slow
Total	100%	100%	
ASTM evaporation rate, seconds	340	340	
VOC content, grams/liter	855	247	
HAP content, grams/liter	599	0	
MIR Kg ozone/liter	3.48	0.95	

This solvent blend has a VOC content of 855 grams per liter and a hazardous air pollutant (HAP) content of almost 600 grams per liter. To reduce the VOC content of this blend to less than 250 grams per liter, 71% of exempt TBAC must be used and the formulation must be rebalanced to match the evaporation rate of the old blend. Fortunately, software is now available to help make this rebalancing fairly easy.

Figure 3 shows the evaporation rate profile for these two solvent blends. Because n-BuAc and MAK are the last solvents to leave the film in both cases, the coating properties and appearance are likely to be nearly identical.

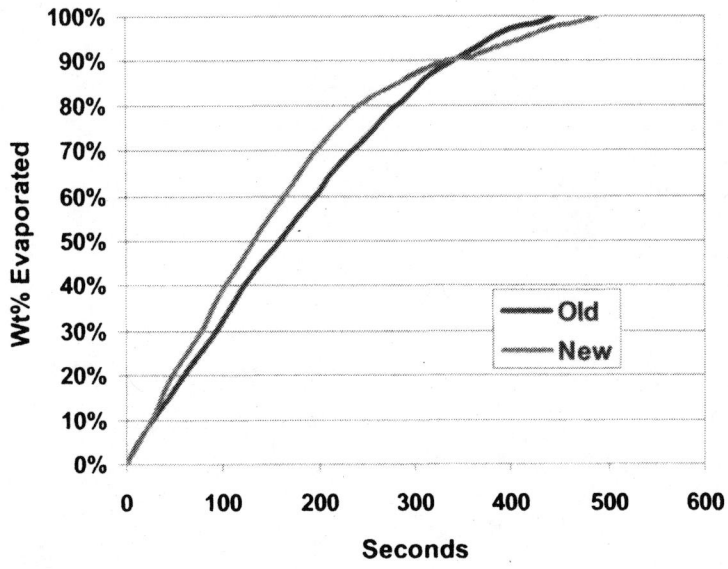

Figure 3. Calculated Evaporation Profiles for Two Solvent Blends

Replacing toluene and MEK in this formulation has the added benefit of reducing the HAP content to zero. The reduction in ozone forming potential (MIR) is 73%, which exceeds the VOC reduction benefit of 71%. In other words, the environmental benefit of this reformulation is slightly higher than its regulatory benefit.

Solvency Properties

Solvency power is another critical consideration when selecting a solvent. This property reflects a solvent's ability to compatibilize the formulation components and reduce the coating or ink viscosity so it can be applied to the substrate. Solvency properties are difficult to predict or model, but some general principles have predictive value, such as "like dissolves like." For example, hydrocarbon resins should have good solubility in hydrocarbon solvents.

This basic concept was further expanded and quantified by Charles Hansenlx. According to Hansen, solvency is a function of three basic parameters:

- Dispersion (Hd): The ability to disperse, a non-polar parameter related to molecular size. The smaller the size, the higher the dispersion power.

- Polar (Hp): Related to the dipole moment or the polarity of the solvent.

- Hydrogen bonding (Hh): The ability to accept or donate hydrogen bonds.

Although this concept can be used to predict the ability of a solvent to dissolve resins, it is most useful for comparing solvents because the Hansen parameters of small molecules are easy to calculate. It is particularly useful to plot the polar and H-bonding parameters and use the dispersion parameter as the diameter of the spot. This process helps visualize which solvents are most similar and helps identify solvent combinations that could replace other solvents. For example, Figure 4 graphically illustrates that the solubility properties of toluene are closer to those of TBAC and PCBTF than acetone or methyl acetate.

However, the best way to determine solvency is to measure the viscosity of the resin in several solvents at a few solids levels. Many techniques are available for measuring viscosity, from simple cup and bubble methods to the more sophisticated Brookfield viscometry. It is essential to compare the solution viscosities at the same solids level because solution viscosity increases exponentially with solids content.

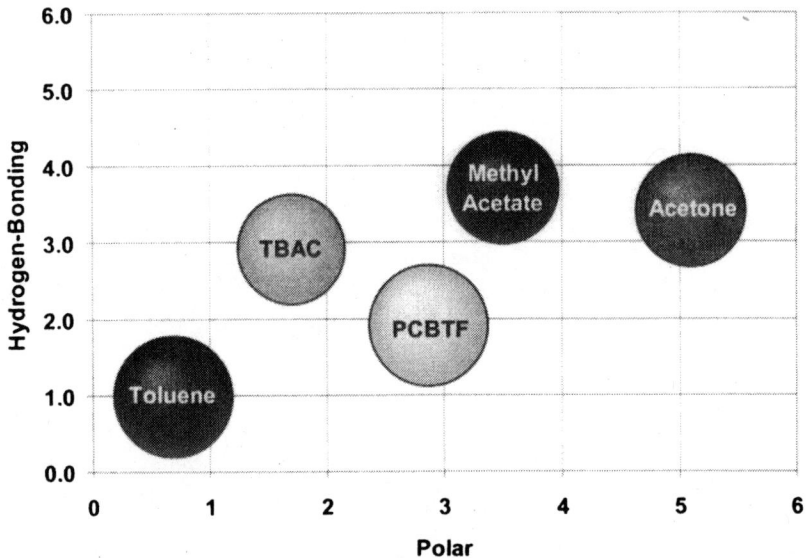

Figure 4. Calculated Hansen Solubility Parameters for Toluene and VOC-Exempt Solvents

Figure 5 compares the Brookfield viscosities of 50% solids solutions of several commercial alkyd resins in TBAC, MAK, xylene, and a blend of TBAC and MAK. TBAC alone is an effective viscosity reducer for alkyd resins, but its high evaporation rate can cause appearance problems if it is used alone in the formulation. Blending it with a slower active solvent like MAK is an excellent way to improve appearance without increasing viscosity.

Coating Formulations

Because the solvent blend plays such a critical role in all aspects of the coating operation, the coating should be formulated, applied, and tested under conditions as close to the real operation as possible. Each coating type and process has different requirements, and it is important to understand these requirements before testing begins.

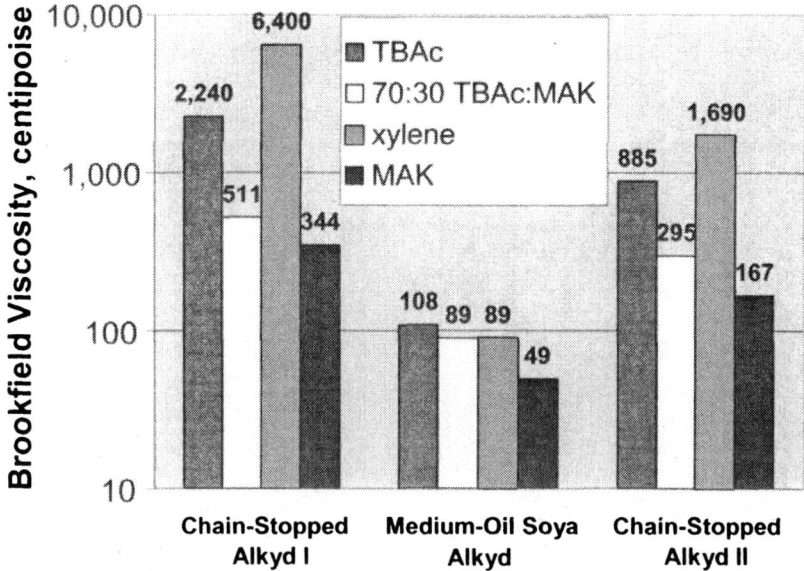

Figure 5. Viscosity of Commercial Alkyd Resin Solutions in TBAC, Xylene, MAK, and a 70:30 Blend of TBAC:MAK

In the following examples, TBAC was evaluated as a replacement for several photochemically reactive VOCs in nitrocellulose lacquers for wood, in air-drying alkyd enamels, and in two-component epoxy and urethane coatings for industrial maintenance and automotive applications. These examples illustrate the general formulation principles described in the previous paragraphs and the broad solvency and versatility of TBAC as a coating solvent.

TBAC in Nitrocellulose Lacquers

Nitrocellulose (NC) lacquers are among the oldest of paint technologies and are still used extensively in the wood furniture market because of their low cost, ease of application and repair, and unparalleled clarity and depth of image. However, conventional NC lacquers contain only 10 to 30% resin at spray viscosity and no longer comply with VOC regulations in many states.

Fortunately, NC is soluble in TBAC, which means that manufacturers of wood furniture and cabinetry will once again be allowed to use low-solids NC lacquers, but without the environmental impact of the old formulations. Table 3 shows the effect of replacing MEK, xylene, and n-BuAc with TBAC and propylene glycol methyl ether acetate (PMA) in a standard NC wood lacquer.

Table 3. Wood Lacquer Reformulation for HAP and VOC Compliance

Components	Standard Lacquer	Formulation A	Formulation B
RS 1/2 nitrocellulose	10.0	10.0	10.0
Beckosol 12-035	7.0	7.0	7.0
Diisononyl phthalate	3.0	3.0	3.0
Xylene (incl. from 12-035 resin)	35.2	4.7	4.7
MEK	10.7	0.0	0.0
N-butyl acetate	22.7	13.9	0.0
Isopropanol (from NC)	4.3	4.3	4.3
N-butanol	7.1	7.1	7.1
TBAC	0.0	41.2	50.0
PM acetate	0.0	0.0	13.9
Total lbs	100.0	100.0	100.0
Formulation Constants			
% solids	20	20	20
lbs ozone/lb solids	15.4	4.2	3.9
lbs VOC/lb solids	4.0	1.9	1.5
lbs HAP/lb solids	2.3	0.2	0.2
Lacquer Properties			
Viscosity, cps	146	185	102
Dry time, min	20	10	10
20 degree gloss	43	50	58
60 degree gloss	86	88	91
Whiteness index	65	66	67
Yellowness index	6.0	5.7	5.4

Beckosol® is a registered trademark of Reichhold Chemical.

The lacquer properties were essentially unchanged. However, the VOC and HAP^x content were significantly reduced. In this case, the ozone formation potential was reduced 75%, whereas the VOC reduction was only 63%. This is another case in which the environmental benefit exceeds the regulatory benefit.

TBAC in Two-Component Epoxy Amine Coatings

Two-component epoxy coatings are used extensively in industrial maintenance applications that require superior corrosion resistance. The epoxy resins are mixed with amine- or amido-functional curing agents just before application to give a tough, flexible, and corrosion- and chemically-resistant film.

The most common solvents used in this type of coating are toluene and xylene because they are inexpensive and do not react with the epoxy resins or amine curing agents. Unfortunately, both solvents produce large amounts of ozone when emitted into the atmosphere, especially in polluted urban environments, c.f. Figure 1, and are listed as hazardous air pollutants. Glycol ethers and ketones are also used for their polarity and excellent solubility properties. One class of solvents that is almost never used in epoxies is esters because most esters react with the amine curatives to form amides. When the amine crosslinker reacts with the solvent, it is no longer available to crosslink the epoxy, and the crosslink density of the coating decreases, affecting its cure speed, toughness, and durability.

TBAC is Stable in 2K Epoxy Systems

However, TBAC is unique because the ester group is protected by the bulky tert-butyl group (see Figure 6). It should, therefore, be less susceptible to reactions with amines and polyamides. This hypothesis was confirmed by MM2 energy calculations that estimated the activation energy of the TBAC-amine reactions to be between 4 and 7 Kcal/mole higher than the corresponding n-BuAc-amine reactions.[xi]

This suggested that the TBAC-amine reactions should be several orders of magnitude slower than with n-BuAc. This estimate was confirmed by kinetic and storage stability studies for various amines and polyamides in TBAC and other solvents.

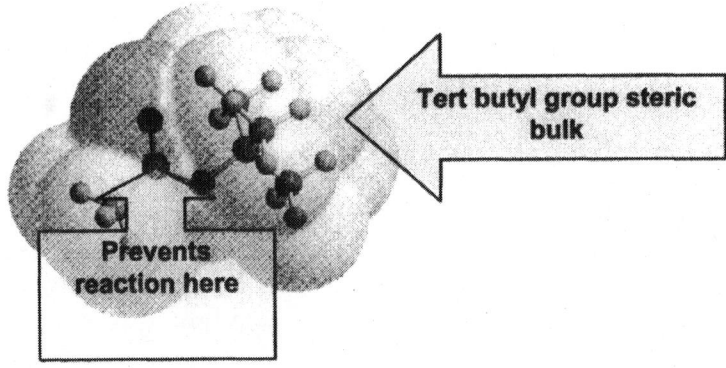

Figure 6. Space Filling Molecular Model of TBAC Showing the Steric Bulk of the Tertiary Butyl Group and Crowding Around the Ester Carbonyl

Pseudo first-order aminolysis rates were measured at several temperatures for TBAC, n-BuAc, and two epoxy curing agents. The results, shown in Figure 7, confirm that TBAC is much less reactive than n-BuAc to either the amine or, especially, amido amine curing agents. The stability of these curing agents in TBAC, n-BuAc, and toluene was also measured at 60°C for 30 days, which corresponds to one year of storage under normal conditions. The percent amine retained after 30 days in TBAC was comparable to that in xylene and much higher than that which remained in n-BuAc (see Figure 8).

2K Epoxy Reformulation

The xylene in a two-component epoxy coating formulation was replaced with a new solvent blend based on TBAC and Aromatic 100 (see Table 4). The percentage of solids and the formulation viscosities were identical. Coating properties, including flexibility, adhesion, and corrosion resistance testing, were measured and also found to be identical (see Table 5).

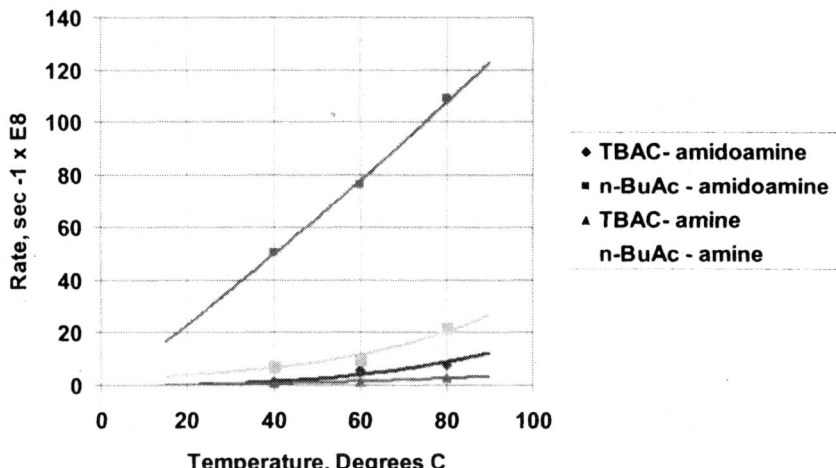

Figure 7. Effect of Temperature on the Pseudo First-Order Aminolysis Rate Constants for the Reaction of Ester Solvents and Commercial Epoxy Resin Crosslinkers

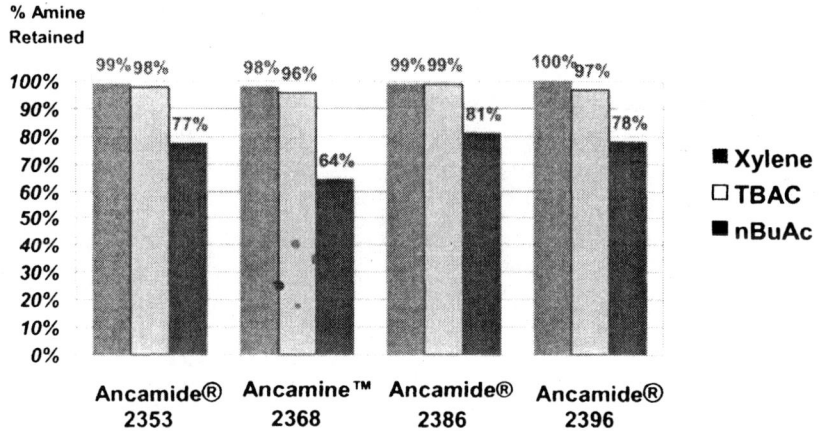

Figure 8. Percent Amine Retained in Commercial Polyamide Crosslinkers after 30 Days Storage in Solvents at 60°C. Publication pending in J. Coatings Technol. Copyright 2006 Federation of Coating Societies.)

These results confirm that TBAC can replace a large portion of xylene in two-component epoxy polyamide coating formulations with no effect on the coating performance but a large decrease in the VOC (45%) and HAP (80%) content of the coatings. The MIR reduction was 59%, another example where the environmental benefit exceeds the regulatory benefit. In our opinion, this justifies exempting negligibly reactive compounds such as TBAC. TBAC has also been evaluated as a fast solvent in two-component urethanes, alkyd coatings, and other air-dry coatings and found to be an effective viscosity reducer.

Table 4. Two-Component Epoxy Formulation Reformulated with TBAc™ Solvent

	Xylene Formulation	TBAc Formulation
Part A		
Epon® 1001X75	301	301
Pigment & fillers	260	260
Beetle® 216-8	12	12
Xylene	221	
Aromatic 100		30
ARCOSOLV® PM		76
TBAc Solvent		115
MAK		15
Part B		
Epicure® 3115	131	131
Xylene	70	
Aromatic 100		15
TBAc Solvent		55
Formulation constants		
Volume Ratio A/B	75/25	75/25
Wt % Solids mixed	62	62

Epon® and Epicure® are registered trademarks of Hexion Specialty Chemicals.
Beetle® is a registered trademark of Cytec Industries.
ARCOSOLV® is a registered trademark of BP Chemical and licensed to Lyondell Chemical Company.
TBAc™ is a registered trademark of Lyondell Chemical Company.
Ancamide® and Ancamine™ are registered trademarks of Air Products & Chemicals

TBAC in Two-Component Urethane Coatings

Two-component urethane coatings are used in a variety of applications for which durability, flexibility, and low VOCs are required. These applications include automotive refinish topcoats and maintenance coatings for bridges and concrete, as well as aircraft paints. Solvents used in these coatings must be aprotic and have low water contents so as to not react with the isocyanate crosslinker.

Table 5. Cured Coating Properties for Two-Component Epoxy Formulations

	Xylene Formulation	TBAc Formulation
Film Properties after 7 days		
Thickness, mils	1.93	1.94
DirectImpact, in.lb	160	160
Reverse Impact, in.lb	160	160
Crosshatch adhesion	100%	100%
1/8 Mandrel bend	pass	pass
800 hours Salt Spray ASTM B117	<1 mm creep 7 VF blisters	<1 mm creep 7 VF blisters

Coatings were applied to phosphated steel (Bonderite 1000) panels and allowed to cure seven days before testing.

Acetone and PCBTF are currently the primary compliance solvents for low-VOC 2K urethane coatings and equipment cleanup. The weight ratio of acetone to PCBTF in automotive refinish formulations in California is on the order of 1:2. Because of the much higher density of PCBTF (11.2 lb/gal), the volume ratio is closer to 1:1. TBAC is likely to replace both solvents because of its better performance and lower cost. If TBAC replaces this solvent blend throughout California, we estimate it will result in approximately 600,000 fewer pounds of solvent emitted and a decrease in ozone formed by approximately 200,000 lb/year.

TBAC is an active solvent for both acrylic polyols and isocyanate crosslinkers. Several suppliers of acrylic polyols are now offering their products in TBAC, and isocyanate suppliers have evaluated the stability of their most

Table 6. 2K Urethane Reformulations with TBAC

Components	Conventional	TBAC-Based	High-Solids	TBAC-Based
G-Cure 105 P70	100.0	100.0	50.0	50.0
JONCRYL 920	0.0	0.0	50.0	50.0
T-12 (1% in toluene)	1.9	1.9	0.3	0.3
FC 430 (10% in toluene)	0.3	0.3	0.3	0.3
HDI Trimer	28.3	28.3	33.9	25.4
IPDI Trimer	0.0	0.0	0.0	15.6
MAK	25.0	25.0	24.0	24.0
n-Butyl acetate	25.0	0.0	24.0	0.0
TBAC	0.0	25.0	0.0	24.0
Total lbs	205.5	205.5	206.5	213.6
% solids	46%	46%	51%	51%
lbs VOC/gal*	4.40	3.28	3.96	2.86
Properties	Conventional	TBAC-Based	High-Solids	TBAC-Based
lbs VOC/gal	4.40	3.28	3.96	2.86
Viscosity, sec #2 Zahn	21.2	21.1	20.9	20.8
Dry time, hours	3.2	3.5	7	4
20 degree gloss	88	88	90	90
60 degree gloss	95	95	95	95
DOI	90	90	90	90
Reverse impact, lbs	160	160	160	160
Direct impact, lbs	160	160	160	160
Cross hatch adhesion, %	100%	100%	100%	100%
10% acid resistance (30 min)	Pass	Pass	Pass	Pass
100 MEK double rubs	Pass	Pass	Pass	Pass

common crosslinkers in TBAC and found it to be comparable to toluene. TBAC is also being considered as a replacement for acetone/PCBTF blends in zero VOC thinners for 2K urethanes. The following formulations illustrates how TBAC can be used to replace the fast solvents in a 2K urethane to lower the VOC content of the coating without affecting the final film properties (see Table 6).

In these formulations, the fast solvent n-BuAc was replaced with TBAC. The slow solvent MAK was retained in the formulation. This substitution caused the VOC content to decrease by over 1 lb/gallon in both formulations with no effect on the final coating properties because the slow solvent is the last to leave the coating and has the greatest impact on the coating's appearance, performance, and durability.

TBAC in Alkyd Coatings

Some suppliers of alkyd resins have also begun to offer their industrial resins in TBAC or TBAC/slow solvent blends. TBAC is potentially a useful compliance tool for industrial baking and air-dry enamels based on alkyd technologies, especially short-oil and modified alkyd resins, such as phenolic- and silicone-modified alkyds, for several reasons.

First, the industrial market is more tolerant of coatings with flammable or odorous solvents than the architectural market, which relies predominantly on odorless mineral spirits. Second, industrial medium-and short-oil alkyd resins are typically more viscous and require higher solvent levels than their long-oil counterparts for the do-it-yourself or contractor-applied architectural markets. These lower solids technologies are the most impacted by VOC and HAP regulations that also limit the contents of HAP-listed toluene and xylene. Consequently, the industrial market is the most likely to benefit from TBAC as a compliance tool and to tolerate its odor and flash point.

The following alkyd formulations are based on recently published work[xii] and illustrate the effect of using TBAC as a let-down and resin solvent on the VOC content of medium-oil alkyd formulations (see Table 7). Using TBAC as a let-down solvent decreased the VOC content by almost 1 lb/gallon, but using a higher-solids alkyd resin already cut in TBAC, such as Beckosol 11-081-E2[xii], can help further reduce the VOC content.

Table 7. Medium Oil Alkyd Reformulation with TBAC and a TBAC-Based Resin

	Control		Low VOC		Lower VOC	
Formulation Components	Gal.	lbs	Gal.	lbs	Gal.	lbs
Beckosol 11-070 (50% in MS)	28.81	218.98	28.81	218.98	0	0
Beckosol 11-081-E2	0	0	0	0	18.19	156.41
Tert-butyl acetate	0	0	3.78	27.37	3.78	27.37
Mineral spirits	3.78	23.97	0	0	0	62.57
Thixotrope	0.96	13.69	0.96	13.69	0.96	13.69
Propylene carbonate	0.47	4.65	0.47	4.65	0.47	4.65
Disperse then add						
Raybo 57HS	0.58	5.47	0.58	5.47	0.58	5.47
TiO2 CR-800	7.56	260.04	7.56	260.04	7.56	260.04
Beckosol 11-070 (50% in MS)	44.3	336.68	44.3	336.68	0	0
Beckosol 11-081-E2	0	0	0	0	27.96	240.49
Tert-butyl acetate	0	0	11.34	82.12	11.34	82.12
Mineral spirits	11.34	71.9	0	0	0	96.19
Cobalt 6%	0.37	2.74	0.37	2.74	0.37	2.74
Zirconium 6%	0.49	3.56	0.49	3.56	0.49	3.56
Calcium 5%	0.88	6.84	0.88	6.84	0.88	6.84
Anti-skinning agent	0.88	6.84	0.88	6.84	0.88	6.84
Totals	100.42	955.36	100.42	968.98	73.46	968.98
VOC content, lbs/gal	3.89		2.94		2.61	

Conclusions

New environmental regulations are continually challenging the coating, adhesive, and ink industries to use less hazardous and more effective solvents. In turn, solvent suppliers are developing new solvents that perform as well or better than existing ones but have lower health and environmental impact.

TBAC is a fast-evaporating solvent that was developed specifically to replace solvents with higher ozone forming potential such as toluene, xylene, MEK, methyl isobutyl ketone (MIBK), and n-BuAc. The EPA has published a rule excluding tertiary butyl acetate from the federal definition of a VOC (40 C.F.R. § 51.100(s)(5); see also 69 FR 69304). State and local definitions may vary. TBAC is an effective viscosity reducer for several types of coatings, including two-component epoxy coatings. Ester solvents are typically not used in this type of coating because the amine curatives react with esters. However, TBAC is much more resistant to aminolysis and hydrolysis than other ester solvents and can be used in a greater number of applications due to its exceptional stability.

Because TBAC has lower health hazards and environmental impacts than many of the solvents still in use today, we anticipate it will enjoy a preferred regulatory status for years to come. As a result, we expect it will become important formulating tool in a variety of high-performance coatings, inks, adhesives and cleaners.

References

i. http://www.arb.ca gov/regact/conspro/aerocoat/aerocoat.htm
ii. http://www.epa.gov/ttn/oarpg/t1/memoranda/27601interimguidvoc.pdf
iii. http://www.epa.gov/eogapti1/module6/ozone/formation/formation.htm#tropospheric
iv. http://www.ec.gc.ca/nopp/voc/en/bkg.cfm
v. http://www.epa.gov/ozone/ods.html
vi. http://www.epa.gov/ozone/geninfo/gwps.html
vii. http://pah.cert.ucr.edu/~carter/reactdat.htm#data
viii. Propylene glycol methyl ether propionate (CAS #148462-57-1)
ix. *Hansen Solubility Parameters. A User's Handbook* by Charles Hansen, CRC Press 1999 Edition.

x. http://www.epa.gov/ttn/atw/188polls.html
xi. Figure 2 from Rodriguez, C., Cooper, C., Galick, P., Harris, S., and Pourreau, D.,"Tert-Butyl Acetate; Non-HAP Solvent for High-Solids Epoxy Formulations", J. Coat. Technol., *73,* No 922, 19-24 (2001).
xii. http://www.paint.org/thebench/current_issue.htm Volume 2, Issue 7. August 18, 2005.
xiii Beckosol 10-081 E2 is a developmental medium-oil alkyd resin from Reichhold Chemical Company.

Chapter 15

Development of UV-Curable Waterborne Polyurethane Dispersion for Soft Feel Application

Peter D. Schmitt[1], Lyuba Gindin, Sr.[2], Aaron Lockhart[3], and Phil Lunney[4]

[1]Innovation Group Leader, [2]Scientist, [3]Associate Scientist, and [4]Principal Operations Analyst, Bayer Material Science, 100 Bayer Road, Pittsburgh, PA 15205

The performance of coatings prepared from one-component waterborne polyurethane dispersions is generally inferior to one- or two-component cross-linked films in terms of chemical resistance and mechanical durability, particularly in soft-touch coatings. By incorporation of double bond functionality via reacting polyester acrylates into the backbone of the polymer we can greatly reduce cure time of these films to as little as three minutes (elevated flash off temperature, UV cure). Beyond that, the additional crosslinking effected by UV-cure imparts a new level of chemical and mechanical resistance while maintaining the desired soft feel. More so than solventborne and/or acrylic monomer containing systems, UV-Curable PUDs are environmentally friendly since they are reduced in water.

Introduction

Soft-touch coatings are used in a number of plastic applications across a wide range of markets including automotive interiors, consumer electronics, exotic packaging applications, vacuum cleaners, graphic arts, etc. Often times

© 2007 American Chemical Society

referred to as "soft-to-the-touch" or "suede-effect coatings", these systems are designed to transform the aesthetics of a plastic part from "cold and hard" to "warm and soft". However, one of the greatest limitations to the success and expansion of this technology has been its poor chemical resistance. To be more specific, soft-touch coatings with a velvet-like feel are inherently low in crosslink density, which results in susceptibility to penetration of solvents and other aggressive reagents. Martin and Shaffer reported that suntan lotion passes through the soft touch coating and attacks the plastic substrate in this manner. (1) This work showed that the choice of a more chemically resistant plastic could offset this effect, but this option is typically not pursued due to the significant increase in the total cost of finishing the interior of a car. Typically, increasing the level of crosslinking increases the coating's resistance to suntan lotion, but reduces the softness of the film, making the coating less desirable. Two component solventborne coatings, with excellent feel and robustness, were the state of the art for soft touch coatings. However, due to the strong desire in industry to reduce VOC emissions, there has been a transition toward waterborne polyurethanes for soft touch applications.

Waterborne polyurethane dispersions (PUD's) have been used in many applications because of their zero/low-VOC and physically drying characteristics. The technology is based on the preparation of high molecular weight (>100,000 grams/mole) polyurethane polymers dispersed in water. As the water evaporates from an applied film, these dispersed particles coalesce, forming a non-crosslinked finished coating. The performance of these coatings is typically inferior to most one- or two-component crosslinked films in terms of chemical resistance and mechanical durability, especially in soft-touch coatings. Hydroxy-functional PUDs can be prepared for use in conventionally curing one- and two-component crosslinked coatings. These hydroxyl-functional PUDs can be combined with nonfunctional PUDs, along with water reducible polyisocyanates to form two-component waterborne soft-touch coatings with improved overall performance. These coatings require approximately forty-five minute cure cycles (fifteen minutes ambient, thirty minutes at 82°C) before performance properties fully develop. It has been reported by Zhou and Kaltisko (2) that these two component systems provide acceptable feel with good suntan lotion resistance. The goal of the current work is to increase the resistance to suntan lotion above the level of these two component systems with a one component system by increasing the crosslink density of soft-touch coatings through UV-cure, without compromising the softness of the coating. It has been reported by Zhou and Kaltisko (2) that these two component systems provide acceptable feel with good suntan lotion resistance. The goal of the current work is to increase the resistance to suntan lotion above the level of these two component systems with a one component system by increasing the crosslink density of soft-touch coatings through UV-cure, without compromising the softness of the coating.

UV-curable waterborne PUD's have been prepared by adjusting the formulation of non-functional waterborne PUD's to incorporate polyester acrylates into the backbone for free-radical polymerization (3). The primary benefit is directly related to the incorporation of this double bond functionality, which forms a "chain link fence" type network as shown in Figure 1. A generalized structure of a UV-curable waterborne PUD is shown in Figure 2. With the additional UV crosslinking, films from this PUD have an increased level of chemical and mechanical resistance. As with most UV-curable materials, the time required to cure the finished coating has now been greatly reduced, in this case, to as little as three minutes (elevated flash-off temperature, UV cure). Equally important, is that these PUDs are high enough in molecular weight to physical dry. This eliminates the problem of insufficient UV-cure in shadow areas. Unlike solventborne/solvent-free UV coatings, which have environmentally unfriendly solvents and/or harmful monomers in the system, these coatings are reduced in water.

 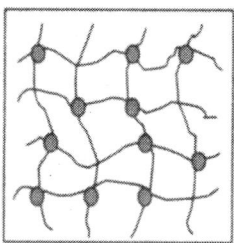

Figure 1. Crosslinked vs. non-crosslinked films
(See page 10 of color inserts.)

Experimental

A typical non-functional PUD formulation that yields films with the desired feel was chosen to be modified with acrylate functionality. (The composition of this PUD is considered proprietary by the authors.) Although this PUD provides the required softness, the films lack solvent resistance and have long dry times (10 minutes ambient flash, 30 minutes @ 82°C). Our goal was to create a new product that would develop properties quickly, have good solvent resistance (>50MEK DR immediately after UV cure) and maintain acceptable feel.

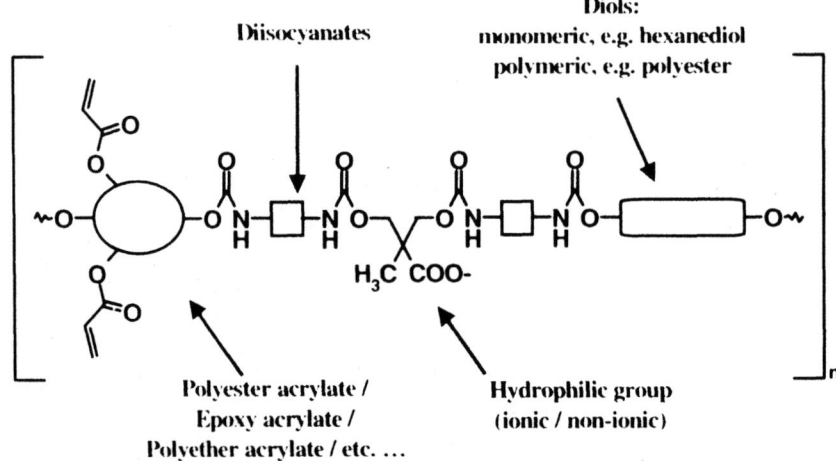

Figure 2. Generalized structure of UV-curable polyurethane dispersion (See page 10 of color inserts.)

Dispersion production process. The experimental dispersions were made using the acetonic process as outlined in Figure 3.

Screening study. A screening study followed by a statistical design study facilitated the rapid development of a candidate product. In the screening study we varied the type and the amount of ingredients that could effect the desired properties, namely hydroxy-functional polyester acrylates, short chain diols, polyether diols and isocyanate monomers.

The screening study showed that:
- Addition of the polyester acrylate reduced dry time without significantly affecting soft feel.
- Surprisingly, increasing the amount of polyester acrylate in the formulation did not affect softness of the film but improved chemical resistance.
- It was necessary to keep dimethylolproprionic acid in the formulation for better adhesion to the plastic substrate.
- Replacement of dimethylolproprionic acid with different short chain diols did not improve results.
- Addition of hydrophobic compounds with hydrocarbon nature that were claimed to improve adhesion to plastics and chemical/solvent resistance did not result in better performance.

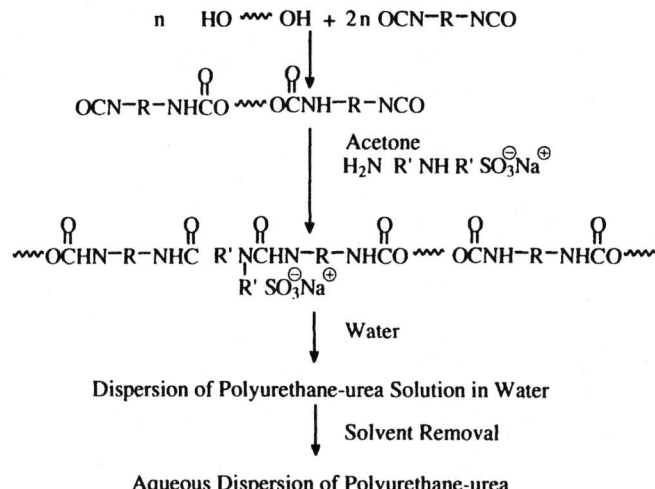

Figure 3. Diagram of acetonic dispersion production process.

Statistical design. The results of the screening study were very encouraging. The next logical step was to develop a statistical design using the findings from the screening study to identify the optimum combination and levels of components.

The factors selected for a statistical approach included the levels of Polyester Acrylate (Low and High), the percentage of chain extension (60% versus 95%), and a nominal, three-level factor representing one pure isocyanate (ISO 1) and blends of this isocyanate with two others (ISO 2 and ISO 3).

The experimental design selected for this application was a single replicate of a 2x2x3 factorial array. This design offered two advantages over alternative strategies, such as one-factor-at-a-time approaches in which only one factor is varied while the others are held constant. First, this design permitted the experimenter to observe synergistic effects or interactions among pairs of factors. Secondly, the design also allowed estimation of the three-factor interaction, which often can be used as a proxy error term for assessing the statistical significance of effects. In this case, however, the three-factor interaction was actually significant, and accounted for the failure of two formulations in the matrix to produce a viable coating. Statistical significance was thus assessed by an error term generated from readings of five replicate panels.

Preparation of UV soft-touch coating formulation. In a 250-ml beaker, 70 grams of the experimental dispersions were combined with 0.77 grams of

Byk-346 and 0.62 grams of Tego Foamex 805 under agitation using a Dispermat CV disperser at 1000 rpm. To the mixing vessel was added (under agitation at 1500 rpm) 25 grams of a matte paste (Acematt OK-412/Disperbyk 190/Water (5/1/25)). The solution was mixed for 10 minutes. In a 100 ml beaker, 9.85 grams of deionized water was combined with 6.50 grams of Butyl Carbitol. This solution was slowly added to the mixing vessel under agitation at 1500 rpm. Irgacure 819-DW (1.5 grams) was added to the mixing vessel under agitation at 500 rpm and the solution was mixed for five minutes to ensure homogeneity. The formulation was filtered into a plastic jar and left to sit overnight as needed to allow for defoaming.

Preparation and cure of soft-touch panels. The panels to be coated were cleaned by wiping with a paper towel, which was dampened with a VM & P Naphtha/Isopropanol solution (1:1). The formulated UV-curable coating was then spray-applied to the panels at approximately 4 mils (wet film thickness) using conventional spray techniques, i.e., Binks 200lss spray gun at 45 psi.

After spraying, the panels were flashed at 50°C for 10 minutes to remove any water. The coating was cured using a HP-6-High Powered Six-Inch UV Lamp System with VPS-3 Power Supply (Fusion UV Systems, Inc). The conveyor belt speed was set at 10 feet per minute and the Mercury vapor bulbs were set at 100% power. This yielded a total energy density of approximately 2700 mJ/cm^2.

Softness measurement. The resultant coatings were tested for their relative softness by hand and ranked from one to five, with one being a hard feel and five being a soft feel. This is not to be confused with testing for "most desirable feel", which was not performed because of the subjectivity of such a measurement.

MEK double rubs. The chemical resistance of each coating was tested via methyl ethyl ketone double-rubs on the same day as the initial cure, as well as after a three-day postcure. The test is conducted using a ball-peen hammer securely wrapped with several layers of cloth (8"x 8" cloth folded twice). The cloth is then saturated with MEK and the wet ball-peen is laid on coating surface, holding the hammer so that the ball-peen is at a 90-degree angle to the substrate. Without applying downward pressure, the hammer is pushed back and forth over an approximately 4" long area of the coating. One forward and back motion is counted as 1 double rub. The cloth is re-saturated with MEK after every 25 double rubs.

Chemical spot test. The coated panels were tested for resistance to isopropyl alcohol (IPA), methyl ethyl ketone (MEK), and terpene in spot tests in which a cotton ball dampened with solvent is placed on the surface of the coating. The affected area must withstand exposure to these solvents for a minimum of 2 minutes without softening, discoloration, cracking, or color transfer to the cotton ball.

Suntan lotion resistance. The coated panels were also tested for resistance to suntan lotion by three different methods. The first method (Method #1) was a room temperature spot test using "Coppertone for Kids" (SPF 30) suntan lotion. In this internal screening test, a dime-sized spot was applied to the coating and after 1hr, 2hrs, 3hrs, and 4hrs, the lotion was removed and the exposed spot was scratched with a wooden applicator stick. Method #2 followed the Ford specification WSS-M99J385-A, section 3.7.21 in which the coated surface is exposed via two separate spot tests: one at 24°C and another at 74°C. After the exposure, a visual inspection is made looking for cracking, blistering, or discoloration. The exposed spot is then rated according to the Ford visual rating scale from one to six, where one is "no effect" on the coating. Method #3 was the Motorola decorative coating test, which is similar to the Ford spot test, but called for exposing the panel, with the suntan lotion applied, to 48 hours in a controlled temperature and humidity chamber at 85°C and 85% relative humidity. Any visual and physical effects were noted.

Results and Discussion

Experimental Design. The results of the experiment were analyzed using the JMP software version 5.01 (SAS Institute of Cary, North Carolina). A brief summary of the "softness" response follows. In as much as the other two responses (initial and final MEK rubs) were somewhat correlated eith the softness response, the following summary essentially represents them all. An analysis of variance for softness is presented in Table 2.

Table 2. Analysis of variance for softness.

Effect Tests

Source	Nparm	DF	Sum of Squares	F Ratio	Prob > F
% Chain Extension	1	1	8.8888889	14.3369	0.0005
% Chain Extension*Isocyanate Blend	2	2	9.8666667	7.9570	0.0012
% Polyester Acrylate*% Chain Extension	1	1	1.8000000	2.9032	0.0962
% Polyester Acrylate*Isocyanate Blend	2	2	2.8666667	2.3118	0.1122
Isocyanate Blend	2	2	1.4000000	1.1290	0.3334
% Polyester Acrylate	1	1	0.0888889	0.1434	0.7070

The column entitled "F Ratio" in this table is analogous to a signal to noise ratio. Factor effects that exceed the inherent noise in the system have larger F Ratios than less significant factors. The last column assesses the probability that the observed F Ratio is due to chance. Thus, a very low probability is equivalent to a highly significant factor effect. Typically, a probability less than 0.05 is considered significant, although in this case, a value of 0.10 was used to compensate for a loss of sensitivity resulting from the missing two trials. Based on these considerations, it is clear that there are three significant effects, the percentage of chain extension and the two-factor interactions of percent chain extension with the isocyanate blend and with percent polyester acrylate.

The two-factor interactions in this experiment are actually the keys to understanding these results. The main effect of "% chain extension", reproduced below, suggests that an increase in the percentage of chain extension results in a corresponding increase in softness. This is typical of the results obtained from the screening study. The presence of the two-factor interactions with both "% polyester acrylate" and "isocyanate blend", however, indicates that the true effect of chain extension can only be understood by observing both pairs of factors simultaneously, as shown in Figure 4.

The two factor interaction for the highly significant "% Chain Extension" by "Isocyanate Blend" two-factor interaction is reproduced in Figure 5.

Note that the actual effect on softness of the percentage of chain extension depends completely upon the isocyanate blend under consideration. For example, with "ISO 1", there is no difference in softness between the 60% and 95% levels of chain extension. An increase in softness with increasing chain extension level is observed for the other isocyanate combinations, but note that the magnitude of the increase changes depending upon the particular isocyanate combination. Clearly, understanding of this interaction is necessary to grasp the dependency of softness upon the chain extension percentage and isocyanate blend.

A representation of the statistical inference obtained from this experiment for all responses is presented in the cube plots in Figure 6. These cube plots represent predicted values for each of the properties for particular combinations of factors and levels. For example, the combination of 95% chain extension, low level of polyester acrylate, and "ISO 1" should produce expected values of 51 initial MEK rubs, 64 final MEK rubs, and a softness equal to 3.2. Careful study of the cubes reveals the complex two-factor interactions present in this system. The shaded values represent the formulation expected to produce optimal results.

Softness measurement and MEK double rubs. Sample coating formulations incorporating the experimental dispersions were applied to test panels and cured as stated in the "Preparation and cure of soft-touch panels" section. The preferred softness as determined by a poll of Bayer Polymers personnel was a ranking of three or four on the 5 point scale. The target chemical resistance after post-cure was determined to be at least eighty-five

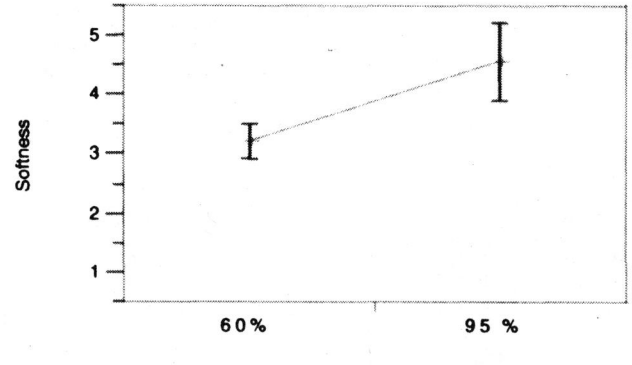

Figure 4. Effect of % chain extension on softness.

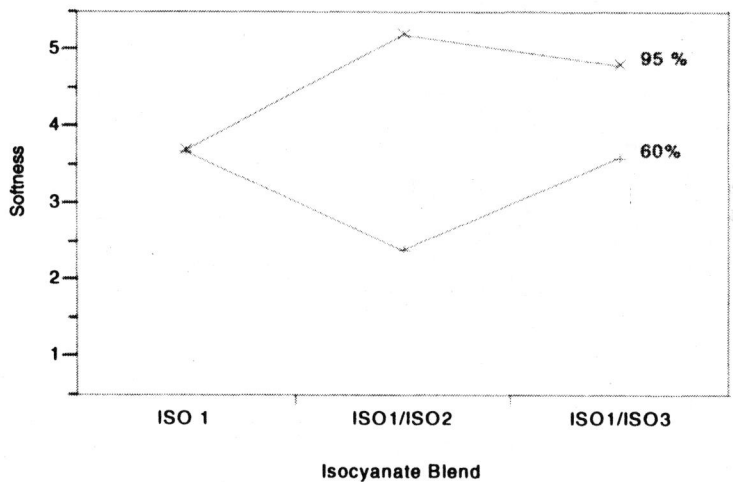

Figure 5. Combined effect of % chain extension and isocyanate content on softness (identity of ISOs is proprietary).

234

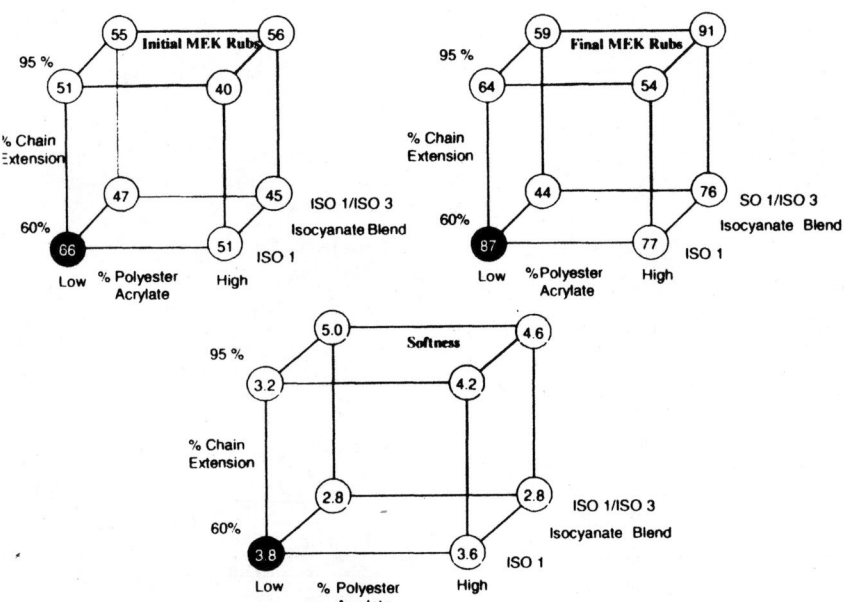

Figure 6. The predicted properties for each combination of factors.

double rubs to ensure an improved level of performance over current two-component waterborne polyurethane soft-touch with comparable feel/softness. The average of the values from five test panels for softness and number of double-rubs for Dispersion "A" were 3.8 and eighty-seven, respectively. These results can be compared to the performance of the other experimental dispersions from the statistical design in Table 3 below.

Table 3. Softness and chemical resistance of statistical design dispersions.

UV PUD	Panel 1	Panel 2	Panel 3	Panel 4	Panel 5	
A						Average
Softness	3	4	4	4	4	3.8
MEK	107	77	110	62	78	86.8
B						Average
Softness	1	2	3	3	4	2.6
MEK	99	88	82	88	86	88.6
C						Average
Softness	4	5	5	4	4	4.4
MEK	42	40	32	59	47	44
D						Average
Softness	4	3	4	4	1	3.2
MEK	74	57	65	41	82	63.8
E						Average
Softness	4	5	5	5	5	4.8
MEK	50	60	75	65	54	60.8
F						Average
Softness	5	5	5	5	5	5
MEK	55	68	50	61	60	58.8
G						Average
Softness	3	4	3	4	4	3.6
MEK	102	68	60	84	73	77.4
H						Average
Softness	3	1	3	3	1	2.2
MEK	69	80	70	80	120	83.8
I						Average
Softness	3	2	3	3	3	2.8
MEK	60	104	62	79	77	76.4
J	failed experiment					
K	failed experiment					
L						Average
Softness	3	5	5	5	5	4.6
MEK	100	100	95	80	80	91

Blending study to improve chemical resistance. Dispersion "A" was chosen as the most desirable UV-PUD because of its performance in MEK double rubs while maintaining the desired feel. This dispersion was then tested in soft touch formulations for resistance to suntan lotion via the above-mentioned internal screening test (Method #1). The desired result is minimum of four hours of exposure without removal of the coating. This level of resistance must be met in order for a coating to be considered for further suntan lotion resistance testing.

Formulations based on Dispersion "A" lasted only 2.5 hours in this test, which is the same amount of time that a two-component waterborne soft touch formulation of comparable feel resisted penetration of the suntan lotion. It was then determined that the coating formulation needed to be adjusted to increase this performance, yet maintain the desired feel. Blends were made with harder, commercially available UV waterborne PUD "M", where 10%, 25%, and 50% of the resin content was PUD "M". In each of these formulations, the chemical resistance was improved to meet the minimum requirement for this test, but the feel was dramatically reduced (softness values of 2, 1, and 1, respectively). Based on this result, it was decided that the blending resins be expanded to include urethane acrylates, which were free of reactive thinner, but were reduced in a water-soluble organic solvent. The solvent chosen was Butyl Carbitol (diethylene glycol n-butyl ether) as it is the co-solvent used in the final coating formulation.

Blends were made with Urethane Acrylates "O", "P", "Q", "R", and "S" at 14% of the total resin solids content. Urethane acrylates bearing isocyanate functional groups, namely Urethane Acrylates "T" and "U", were added to Dispersion "A" at 9% of the total resin solids content, as well as to several of the above-mentioned blends as shown below in Table 4. The initial blending study included formulations 1 through 6. This was used as an initial screening to determine which, if any, of these resins would display incompatibility problems upon blending as well as to identify the combinations with the most desirable feel. Formulations 2 through 4 yielded films with softness ratings of 2, 1, and 1, respectively. The urethane acrylates from these formulations were excluded from other blending work because of these softness ratings. It should be noted that there were no incompatibilities found in any of the blends in the study. Blends with Urethane Acrylates "R" and "S" yielded films with softness ratings of 3 and 4, respectively. Based on these results, these materials were included in further blending studies, which incorporated the NCO-bearing urethane acrylates. Each formulation was applied at the necessary wet film thicknesses to provide approximately 1.0 mil and 2.0 mil films after cure. Formulations 5 through 12 were submitted to extensive chemical resistance testing including MEK double rubs, suntan lotion resistance test Methods #1, #2, and #3, and chemical spot tests as outlined in the experimental section above.

Table 5 shows the data for Formulation #6. The performance of this formulation at a dry film thickness of 1.6-2.0 was superior to any other system tested. It should be pointed out that there were a series of systems when tested for suntan lotion resistance performed worse that Formulation #6, in spite of the fact that both formulations have comparable MEK double rub results (for example, Formulation #9). Formulation 6 at 2.0 mils dry film thickness withstood 49 MEK double rubs and gave excellent suntan lotion resistance, while Formulation 9 at 2.2 mils dry film thickness withstood 42 MEK double rubs and failed the suntan lotion test Method #1 after 8 hours of exposure.

Table 4. Urethane acrylate blends with Dispersion "A".

Formulation#	1	2	3	4	5	6
Resin	Disp. A	Disp. A/Ur. Ac. O	Disp. A/Ur. Ac. P	Disp. A/Ur. Ac. Q	Disp. A/Ur. Ac. R	Disp. A/ Ur. Ac. S
Resin Combination	Not Applicable	6/1	6/1	6/1	6/1	6/1
Softness Rating	4	2	1	1	3	4
Formulation#	7	8	9	10	11	12
Resin	Disp. A/Ur. Ac. T	Disp. A/Ur. Ac. U	Disp. A/Ur. Ac. R/Ur. Ac T	Disp. A/Ur. Ac R/Ur. Ac. U	Disp. A/Ur. Ac S/Ur. Ac. T	Disp. A/Ur. Ac. S/Ur. Ac. U
Resin Combination	10/1	10/0	10/1.4/1	10/1.4/1	10/1.4/1	10/1.4/1
Softness Rating	3	3	3	4	2/3	2

Nonetheless Dispersion "A" when blended with various urethane acrylates, can provide lacquer-drying 1K waterborne UV-curable soft touch formulations with excellent feel and suntan lotion resistance.

Table 5. Performance testing of Formulation #6

Formulation #6 Disp. "A"/Ur. Ac. "S" (6/1)	Photoinitiator TPO/819/651	DFT 1.6-2.00	Softness 4

Suntan lotion Method #1		Suntan lotion & insect repellent (Method #2)		
4 hours	No change			
6 hours	No change	Temperature	Suntan lotion	Insect Repel.
8 hours	No change	24°C	1	1
		74°C	1	1

Crosshatch	Suntan lotion Method #3
5B	The coating lifted from the panel.

MEK DR	Chemical Resistance	
Original-43	Terpene	Passed up to 15 min.
6 days-49	MEK	Failed less than 1 min.
	IPA	Passed up to 15 min.

Conclusion

Waterborne polyurethane dispersions have added greatly to the feel and performance of waterborne soft touch coatings. Their relative ease of use in one-component formulations versus two-component systems is certainly an attractive benefit, but the chemical resistance of the resultant coatings has been insufficient to pass most decorative coating specifications. Even though waterborne 2K-PUR soft touch coatings offer a wide range of "feels" and levels of performance, the amount of time required for cure is a significant obstacle, which has been overcome with the advent of UV-curable systems. The use of UV-curable polyurethane dispersions is the marriage of the best of these two worlds, combining the ease of use of a 1K system, the low VOC of waterborne coatings, and the improved performance and productivity of UV crosslinking. Dispersion "A" in particular has accomplished all of these things while maintaining a desirable feel. The ability to blend Dispersion "A" with other UV-PUDs as well as urethane acrylates provides the opportunity for varying the "feel" and performance of waterborne UV-curable soft touch formulations in the same fashion as current 2K technology.

Even though these results are promising, there remains the potential for improvement. The statistical design suggested that the general composition of the ideal dispersion would have an isocyanate content based solely on ISO 1, that the polyester acrylate content be 50% of the total polyol eqs., and that the dispersion should be chain extended to 60%. However, the results shown above suggest that using MEK double rubs to screen dispersions for chemical resistance was not a good predictor of suntan lotion resistance. To this end there are a number of dispersions that were generated through the statistical design that could be tested and optimized for improved suntan lotion resistance. For future work, the suntan lotion resistance test Method #1 will be used as the means of screening any new formulations. Beyond that, the full range of urethane acrylates and unsaturated polyesters, polyester acrylates, polyether acrylates, and polycaprolactone acrylates has yet to be explored as blending resins.

References

1. *Automotive Substrate Destruction by Suntan Lotion and Chemical Methods for Improving Resistance of Two-Component Waterborne "Soft Feel" Coatings.* Advanced Coatings Technology Conference Proceedings, Detroit, MI, Sept. 1998.K. Martin and M. Shaffer
2. *Development of Soft Feel Coatings with Waterborne Polyurethanes;* Zhou, Lichang; Koltisko, Bernard; JCT Coatings Tech; April 2005, 54-60.
3. Patent US 5,684,081 R*adiation-curable, aqueous dispersions, production and use thereof,* Wolfgang Dannhorn, et al., Wolff Walsrode AG, 11/4/97

Advanced Techniques for Measuring Coating Performance

Chapter 16

Combinatorial and High-Throughput Development of Polymer Sensor Coatings for Resonant and Optical Sensors

Radislav A. Potyrailo

Chemical and Biological Sensing Laboratory, Materials Analysis and Chemical Sciences, General Electric Company, Global Research Center, Niskayuna, NY 12309

Polymeric sensor materials are attractive in a variety of sensors including resonant and optical devices. Polymeric sensor materials play a critical role in the effort to advance selectivity, response speed, and sensitivity of chemical and biological determinations in gases and liquids. Desirable capabilities of polymeric sensor materials originate from their numerous functional parameters, which can be tailored to meet specific needs. However, when the functional complexity of polymeric sensor materials increases, the ability to rationally design preferred materials properties becomes increasingly limited. In this study, we demonstrate new opportunities in polymer sensor materials research for resonant and optical sensors that were opened up with the use of combinatorial and high-throughput experimentation methodologies.

Sensor materials are one of the cornerstones of the sensor development process that includes discovery, optimization, and commercialization (*1*). Application of polymeric sensor coating materials brings desired diversity in transduction principles for chemical and biological detection in gases and liquids (*2-8*). While one cannot skip the specific development phases of sensor materials shown in Figure 1, applying new research methodologies can easily reduce the duration of certain material development phases. The multidimensional nature of the interactions between the function and the composition, preparation method, and end-use conditions of sensor materials often makes difficult their rational design for real-world applications (*9-14*). These practical challenges in rational sensor material design provide tremendous prospects for combinatorial and high-throughput research, which is the applied use of technologies and automation for the rapid synthesis and performance screening of relatively large numbers of compounds (*15-18*).

In this study, we critically analyze new opportunities in polymer sensor materials research for resonant and optical sensors that were opened up with the use of combinatorial and high-throughput methodologies. We demonstrate that by applying new advanced analytical instruments and data mining tools, it is possible to accelerate discovery and optimization of sensor materials.

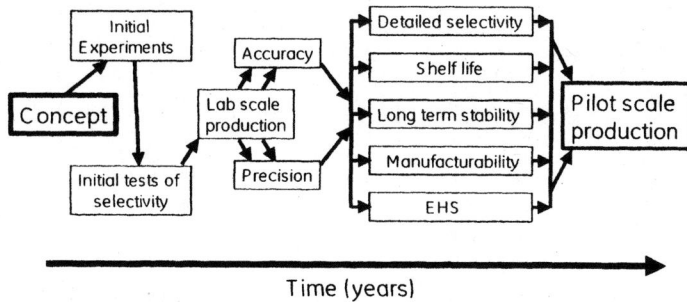

Figure 1. Sensor coatings development at GE from concept to pilot scale.

Combinatorial Screening Process of Sensor Materials

A combinatorial screening process flow to design and optimize sensor materials involves several key process steps. These include (*19*) planning, materials synthesis, characterization of materials response to analytes and interferences, incorporation of results into a database, data analysis, data mining, and scale-up of combinatorial leads (Figure 2). Certain aspects of the process will depend on the throughput of materials fabrication and capabilities of performance testing. As we have shown, with screening of 50 – 100 coating formulations per day, certain aspects of the process can be manual (*20,21*).

New knowledge generated from combinatorial experimentation can significantly assist in a more rational sensor materials design. On the other hand, to be applicable for practical applications, predictive models should also take into the account material development aspects that are less straightforward to adapt for combinatorial screening, such as shelf-life and long-term stability of response.

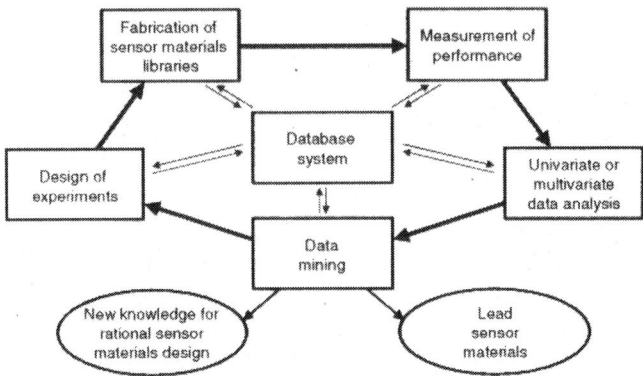

Figure 2. Typical combinatorial cycle for sensor materials development. Reprinted with permission from reference 19. Copyright 2006, Wiley-VCH.

Polymer Coatings for Resonant Sensors

In this direction of development of polymeric sensor materials, homoor copolymers are applied for sensing without any additional polymer-modification steps. In gas sensing, several models (*22-26*) can be applied to calculate polymer responses based on the knowledge of polymer – analyte interaction mechanisms that include dispersion, dipole induction, dipole orientation, and hydrogen bonding interactions (*7,27*). These models can be further applied to construct sensors with transducers based on electrical, radiant, mechanical, and thermal energy (*1,28-42*).

One of the widely accepted models is based on the linear solvation energy relationships (LSER) (*22,23*). This approach permits design of polymer-based sensors from understanding how particular interactions contribute to the overall sorption process. The LSER method systematically explores the role of solubility properties and fundamental interactions in selectivity and diversity of sensor polymers. These polymer solubility studies were initially performed with the goal of understanding of development principles of more effective stationary phases for gas chromatography by studying of nearly 2000 compounds (*23,43*), and were further expanded into sensor applications (*22,23*). The LSER method calculates polymer/gas partition coefficients, the thermodynamic parameter which is defined as the ratio of the analyte concentration in the polymer sensor film, CE, to the analyte concentration outside the film, C_o (*44,45*):

$$K = CF/C_o. \qquad (1)$$

These partition coefficients are calculated as a linear combination of terms that represent several molecular types of interactions (*22,23*):

$$\log K = c + rR_2 + s\pi_2^H + a\sum \alpha_2^H + b\sum \beta_2^H + l\log L^{16} \qquad (2)$$

where R_2, π_2^H, α_2^H, β_2^H, and $\log L^{16}$ are solvation parameters that characterize the solubility properties of the vapor in a sorbent polymer, coefficients r, s, a, b, and l are the corresponding sorbent polymer parameters, and c is the regression constant. The LSER method has proven to be very effective at correlating the polymer-vapor sorption properties with R > 0.95 correlation between the predicted and experimentally obtained partition coefficients for single vapors (46,47).

The LSER method was applied as a guide to select a combination of polymers to construct a resonant acoustic-wave sensor array based on the thickness-shear mode (TSM) devices for determination of chlorinated and other organic solvent vapors including dichloroethylene isomers, in the headspace above groundwater (48). Figure 3 shows our LSER results of response of four polymers to three isomers of dichloroethylene (chlorinated organic solvent vapors of industrial and environmental interest), in the presence of water vapor as an interferent (19). We tested the sensor system in the field (see Figure 4) (49). With the available polymers (see Figure 3), developed sensors demonstrated detection limits of several ppm. However, to meet the requirements for detection of trace contaminants in groundwater at ppb levels, the need was to develop sensor materials with higher sensitivity.

We have found that silicone block polyimide polymer had at least 100 times more sensitive response for detection of chlorinated organic solvent vapors than other known polymers (50,51). Assuming only the mass-loading response when deposited onto the TSM devices, silicone block polyimide polymers have partition coefficients of over 200,000 to part-per-billion concentrations of trichloroethylene. However, the base silicone block polyimide polymer did not provide the required discrimination in detection of different analytes and interfering species. Thus, new families of materials related to silicone block polyimide polymer were developed for selective part-per-billion detection of chlorinated solvent vapors in presence of interferences.

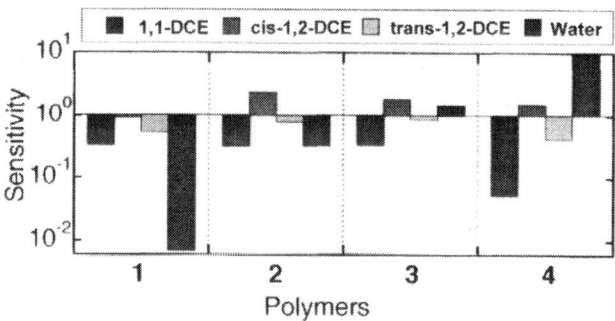

Figure 3. Computer modeled response pattern of polymers 1 – 4 to three isomers of dichloroethylene and water vapor. Polymers 1 – 4 are poly(isobutylene), poly(epichlorohydrin), poly(vinyl propionate), and poly(ethylenimine), respectively. Reprinted with permission from reference 19. Copyright 2006, Wiley-VCH.

Figure 4. Testing of operation of GE resonant sensor system in the field.

For high-throughput screening of these developed polymer sensor materials, we have designed and built a 24-channel sensor system shown in Figure 5. This sensor system is based on TSM resonators and can measure the mass, viscosity, and density changes in polymers (*41,42*). These operation modes are well known and were used by others in numerous practical applications with individual sensors and small sensor arrays (*39,40*).

For practical uses of sensor materials, we developed a materials screening strategy that included three levels of screening (*41,42*). The primary screen was the discovery screen where materials were exposed to a single analyte concentration. The secondary screen was the focused evaluation where the best subset of these materials was exposed to analytes and interferences. Finally, the tertiary screen involved evaluation of material performance under conditions mimicking the long-term application.

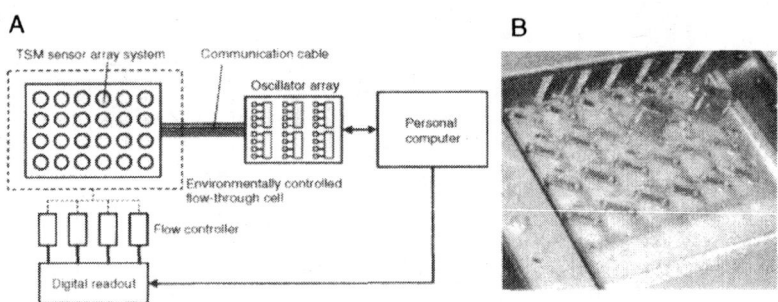

Figure 5. 24-channel TSM sensor array for high-throughput materials characterization. (A) Setup schematic for gas-sorption evaluations of polymeric coatings. (B) Photo of 24 sensor crystals (including two reference sealed crystals) in a gas flow cell. (A) Reprinted with permission from reference 42. Copyright 2004, American Institute of Physics.

Typical results of the primary screen of gas-sorption properties of the 24 channel sensor array are presented in Figure 6. Performance of polymeric materials was evaluated with respect to the differences in partition coefficients for analytes perchloroethylene (PCE), trichloroethylene (TCE), and cis-dichloroethylene (cis-DCE). Gases were introduced into the gas cell with a four-channel computer-controlled gas flow system. Three vapors were introduced in sequence (cis-DCE, PCE, TCE) with no blank gas (nitrogen) in between. Typical exposure time per vapor was 80 min. Blank gas was introduced between three replicates of three-gas exposures.

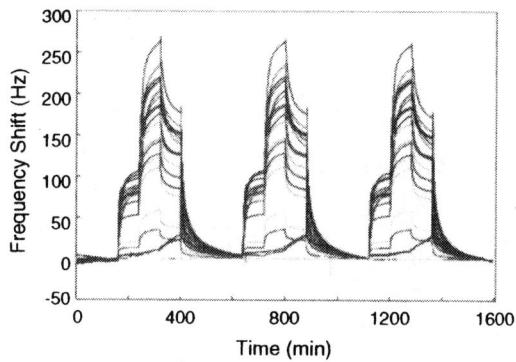

Figure 6. Primary screen of gas-sorption properties of polymer sensor coatings. Triplicate exposures of coatings to analvtes (cis-D CE, PCE, and TCE) with exposures to nitrogen gas in between.

Results of the secondary screening of several families of polymers with respect to the differences in partition coefficients for analytes (PCE, TCE, and cis-DCE) and interferences (carbon tetrachloride, toluene, and chloroform) are summarized in Figure 7. Normalization of all partition coefficients for different vapors to the partition coefficient for TCE showed a noticeable difference in response pattern (see Figure 7A). For the detailed evaluation of diversity of the fabricated materials, the principal components analysis (PCA) tools (52) were applied as shown in Figure 7B and 7C. PCA is a multivariate data analysis tool that projects the data set onto a subspace of lower dimensionality with removed collinearity. PCA achieves this objective by explaining the variance of the data matrix X in terms of the weighted sums of the original variables with no significant loss of information. These weighted sums of the original variables are called principal components (PCs). Upon applying the PCA, the data matrix X is expressed as a linear combination of orthogonal vectors along the directions of the principal components (53):

$$X = t_1 p^T_1 + t_2 p^T_2 + ... + t_A p^T_K + E \qquad (3)$$

where t_i and p_i are, respectively, the score and loading vectors, K is the number of principal components, **E** is a residual matrix that represents random error, and T is the transpose of the matrix.

The capability for discriminating of six vapors using eight polymers was evaluated using a scores plot (Figure 7B). It demonstrated that these six vapors are well separated in the PCA space when measurements are done with these eight polymers. To understand what materials induce the most diversity in the response, a loadings plot was constructed (Figure 7C). The bigger the distance between the films of the different types, the better the differences between these films. The loadings plot also demonstrates the reproducibility of the response of replicate films of the same materials. Such information served as an additional input into the materials selection for the tertiary screen.

It is critical to have the knowledge about the long-term stability of polymeric materials employed in practical sensors. Aging of base polymers and copolymers is difficult or impossible to model (54). To address this aspect of sensor materials research, our tertiary screen involved evaluation of materials performance under conditions mimicking the long-term application (41,42). In these experiments, polymeric materials were coated onto the sensors of the 24-channel sensor array and were aged in the flow-through cell. Figure 8 demonstrates replicate responses of two polymers in the array to a nonpolar analyte vapor (TCE), a polar interference vapor (water vapor), and their mixture before and after the aging study. Several important findings followed from this preliminary study. First, the magnitude of the response to a nonpolar analyte vapor decreased (Figure 8B) and even slightly reversed (Figure 8D) upon aging. Second, the magnitude of the response to a polar interference vapor increased upon aging. Third, because of these changes, the response patterns of these two polymers were altered. In combination with the tertiary screen, our high-throughput screening approach permitted rapid examination of polymers with high partition coefficients and their implementation in field deployable sensors. Several identified sensor coatings were field-tested (55) as shown in Figure 9.

Detailed determination of analyte-polymer interactions using combinatorial experimentation provides the foundation for better understanding materials parameters that affect sensor selectivity. Our on-going efforts in high-throughput screening of sensing polymers for the long-term stability promise to supply additional experimental data applicable for the development of more practical sensor models.

Formulated Polymer Coatings for Optical Sensors

The functionality of polymeric sensor coatings is dramatically expanded by formulating base polymers with a variety of functional additives. A majority of formulated sensor materials is designed to expand the range of types of species detectable by sensors. Often, these sensor materials are multicomponent

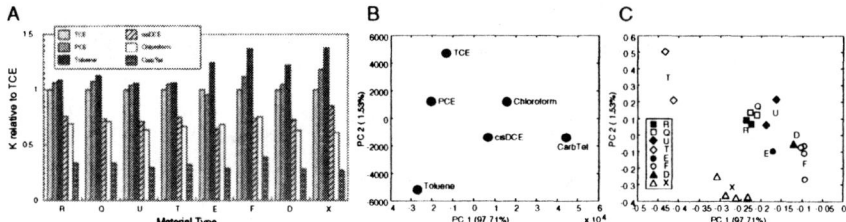

Figure 7. Secondary screen of gas-sorption properties of eight polymer sensor coatings. (A) Partition coefficients for analytes (PCE, TCE, and cis-DCE) and interferences (carbon tetrachloride, toluene, and chloroform) normalized to the partition coefficient for TCE. (B) PCA scores plot that demonstrates the discrimination ability of six vapors with eight types of polymers. (C) PCA loadings plot that demonstrates the differences in response of eight polymer coatings to six vapors. Reprinted with permission from reference 42. Copyright 2004, American Institute of Physics.

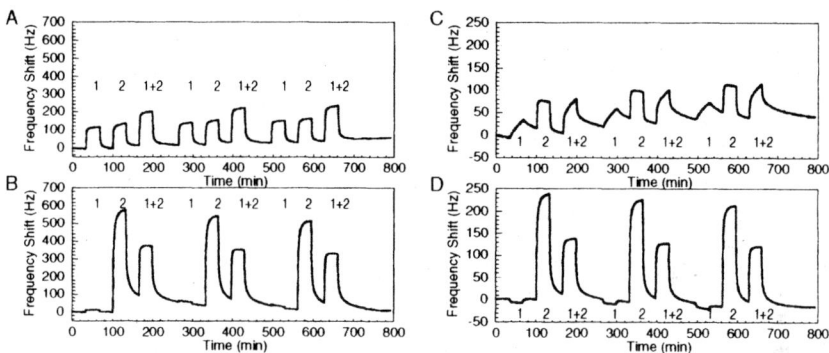

Figure 8. Tertiary screen for evaluation of long-term stability of two polymer sensor coatings. Replicate responses to TCE (1) and water (2) vapors and their mixture (1+2) are performed before and after aging. First coating, (A) before and (B) after aging. Second coating, (C) before and (D) after aging.

Figure 9. Field testing s polymer coatings developed using high-throughput screening approach.

formulations that require a careful attention to the concentration of each formulation component and the ratio of components. Table 1 shows typical complexity of formulation compositions of polymeric coatings for ion selective electrodes, optical sensors, sensors based on composite conductive polymers, conjugated doped polymers, and polymer-immobilized biomolecules (*56-60*). Depending on the analyte-reagent response mechanism, sensors with formulated polymeric coatings are built based on transduction principles that involve radiant, electrical, mechanical, and thermal energy (*5,61-68*). Besides analyte-complexing reagents, incorporation of other reagents into polymers can provide plasticization, modification of electrical conductivity, and improvement of dispersion of additives (*11,12;58,62,64,69-72*). These changes of polymeric properties serve to tailor dynamic range, selectivity, accuracy, sensitivity, long-term stability, and response time of formulated materials (*56,73,74*). An extensive optimization is required to identify sensor formulations that yield largest sensor sensitivity, smallest response to interferences, shortest response time, enhanced stability, and other desired parameters (*73-76*).

Optimization of formulated sensor materials is a cumbersome process because it typically involves evaluation of numerous polymeric matrices or multiple additives at multiple concentrations. Theoretical predictions often are limited by practical issues, such as poor solubility and compatibility of formulation components (*11,77-79*). These practical issues represent significant knowledge gaps that prevent a more efficient application of rational approaches of development of formulated sensor materials. Thus, combinatorial methodologies have been demonstrated for the development of multicomponent formulated sensor materials for gaseous (*11,80-82*) and ionic (*12,14,83,84*) species. By combining manual and automated steps in preparation of discrete sensor film arrays, it was possible to reduce the time to screen sensor materials by at least 1000 (*11*).

In our high-throughput screening, we employ discrete and gradient sensor coatings arrays as shown in Figure 10. Discrete sensor regions are straightforward to produce using existing liquid-dispensing robotic equipment. Depending on the application, an ink jet printing (*85,86*), liquid dispensing robots (*11,84*), and microarrayers (*87*) can be applied. Jnk jet printing typically produces dispensed volumes of around several picoliters while the liquid dispensing robots can dispense several nanoliters and up to several microliters (*88*).

Spatial gradients in polymers can be generated by varying chemical nature of monomers, molecular constitution of polymers and their supramolecular structure or morphology (*89*). In polymeric sensor materials, additional parameters of gradients are also required that can include concentrations of formulation additives, thickness, extend of crosslinking, and some others. Gradient sensor materials can be produced using solvent-assisted polymerization, (*10*) fiber drawing (*90-92*), or draw coating (*14,93,94*). Other potentially useful techniques can include ink jet printing (*85,86*) and electrospinning (*95,96*). If material requires an additional curing or polymerization, it can be further performed, for example, using gradient temperature heaters (*97*) or gradient UV curing setups (*94*).

Table 1. Typical formulated polymeric sensor materials.

Sensor type	Formulation components	Function	Ref.
Ion selective electrode	Combination of polyurethane and terpolymer mixture of poly(vinyl chloride), poly(vinyl acetate) and poly(hydroxypropyl acrylate)	Polymer matrix	(56)
	Combination of dioctyladipate and octyldiphenylphosphate	Plasticizer mix	
	Methylmonensin	sodium ionophore	
	Potassium tetrakis-(4-chlorophenyl) borate	Lipophilic anionic additive to facilitate phase transfer	
	Thioether derivatized calix[4]arene (TDC)	Free ligand	
	$Ag(TDC)NO_3$	Silver-ligand complex for stability improvement	
	Tetrahydrofuran	Common solvent	
Optochemical sensor for ions	Copolymer of vinyl chloride, vinyl acetate and vinyl alcohol	Polymer matrix	(57)
	Bis(2-ethylhexyl)sebacate and 2-nitrophenyl octyl ether	Plasticizer mix	
	Valinomycin	potassium ionophore	
	Potassium tetrakis-(4-chlorophenyl) borate	Lipophilic anionic additive to facilitate base transfer	
	Chromoionophore VI	Fluorescent pH indicator	
	Nile red	Reference fluorophore	
	Tetrahydrofuran	Common solvent	
Composite resistor polymeric film for vapors	poly(vinyl acetate)	Polymer matrix	(58)
	carbon black	Electrically conductive additive	
	Hypermer surfactants	Improve dispersion of carbon black articles	
	Tetrahydrofuran	Common solvent	
Conducting conjugated polymer biosensor	Polyaniline (electrochemically polymerized)	conducting polymer matrix	(59)
	Glucose oxidase enzyme (electrochemically polymerized)	Sensor moiety	
Polymeric film for biosensing	Combination of poly(lactic acid) and poly(glycolic acid)	Polymer matrix	(60)
	recombinant human KGF antibodies	Sensor moiety	
	Sodium bis(ethylhexyl)sulfosuccinate	Stabilizer of proteins in organic solvents	
	Methylene chloride, chloroform, *n*-heptane, *n*-octane, *iso*-octane	Solvent mixture	

 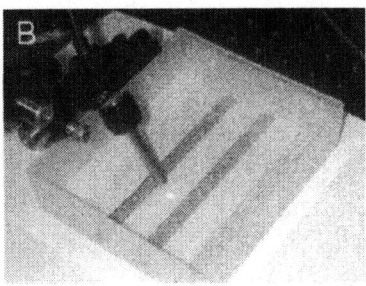

Figure 10. Examples of optical sensor coatings arrays: (A) Discrete sensor coating array and (B) Gradient sensor coating array. Reprinted with permission from reference 102. Copyright 2005, American Institute of Physics.

Optimization of concentration of formulation components can require significant effort due to nonlinear relationships (*10,73,74,76,98-100*). A detailed optimization of formulated sensor materials can be performed using our recently developed a preparation and analysis approach of gradient sensor materials libraries (*101*). To produce composition gradients, individual coatings were flow-coated onto a flat substrate and allowed to merge when still containing solvents. This method combines fabrication of gradients of materials composition with recording the materials response before and after analyte exposure and taking the ratio or difference of responses. Thus, if reagent concentration is too low in the coating, such signal ratio or difference will be respectively low. However, if reagent concentration is too high in the coating, the signal ratio or difference will be also diminished compared to the optimum reagent concentration. The number of measured regions n_L to resolve the property change along the length of the gradient coating array is given by (*102*):

$$n_L = (1/k)\, n_p\, (\Delta P/\Delta L), \tag{4}$$

where t_r is sensor material response time, n_s is the number of scanning probes that simultaneously acquire data from different sensor elements of the array, n_d is the number of collected data points from each sensor element in the array, r is the integration time of data acquisition, and t_s is the time required for a scanning probe to travel between neighbor sensor elements in the array. In general, dynamic experiments not only provide an acceleration in the screening rate (*103*), but also considerably reduce or even eliminate variation sources, which randomly affect sensors response during the one-at-a-time analysis. As seen from eq. (5), depending on the time constant of the sensor response and the capabilities of the scanning system, it may be advantageous to employ dynamic imaging to increase the number of elements in the screened array.

Imaging of a discrete array of sensor materials through several narrow bandpass optical filters was implemented that permitted quantitative determination of rapid optical dynamic responses selectively at desired wavelengths (84). Dynamic imaging was performed in a background-corrected absorbance mode for quantitative analysis of imaging data. For deposition of discrete sensor regions, we employed a liquid dispensing robot initially used for deposition of organic coatings of high viscosity with a good reproducibility of deposition (104).

Acquisition of kinetic data was performed from an array of sensor coatings robotically formulated using a common hydrogel polymer and several preselected colorimetric dyes. Chlorine was chosen as a model sensor system to demonstrate the applicability of the approach. In a typical experiment, a sequence of images was collected with a rate of 0.25 – 1 Hz depending on the expected kinetic response of the sensor coatings and the overall planned exposure time. Figure 12A shows an absorbance image of the array at the advanced stage of the experiment where eight coating formulations were deposited with several replicates. Both positive and negative absorbance values were recorded upon exposure of the whole sensor array to 2 ppm of chlorine when referenced to an unexposed sensor coating array. The data from the individual images was further quantitatively processed to obtain kinetic information about performance of individual sensor coatings upon exposure to 2 ppm of chlorine. Three data points from representative sensor coating formulations a, b, and c and from a coating-free region d (all circled in Figure 12A) were averaged and three replicates of these regions were selected for plotting dynamic data. Figure 12B demonstrates kinetic responses of representative sensor coating formulations a, b, and c and a coating-free region d as determined by quantitative reflected light dynamic imaging. These differences in dynamic response were not intuitive yet reproducible across all replicates. If observed during conventional one-at-a-time analysis, these differences could be considered to be experimental artifacts possibly due to a

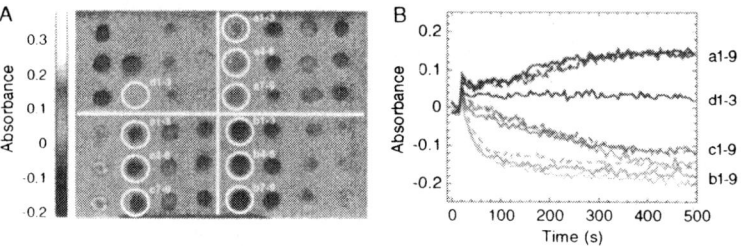

Figure 12. Screening of dynamic responses of formulated optical ionic sensor coatings. (a) Quantitative dynamic imaging performed in absorbance mode and referenced to an unexposed sensor coating array. (b) Kinetic responses of representative sensor coating formulations a, b, and c, and a coating-free region d. Numbers are replicate CCD pixels employed for quantitation. Reprinted with permission from reference 102. Copyright 2005, American Institute of Physics.

variation of experimental conditions. However, while eventually being reproduced, this analysis would require much more time to accomplish.

Formulated sensing polymeric materials can also be incorporated into structures of other materials for combinatorial evaluation of properties of those materials. Coatings with different levels of barrier properties to an analyte gas (e.g. O_2 and H_2O) are required for a variety of applications ranging from food and medical packaging to protection of electronic components. Traditional methods of evaluation of barrier properties of materials require a large coating area (*105*). A sensor probe that exhibits a change in its optical property upon exposure to an analyte gas can be incorporated in a polymer layer material to which an array of baffler coatings is applied (see Figure 13A). Thus, gas-transport properties of the barrier coating arrays can be determined by the time-dependent change in the optical signature of this polymer-immobilized sensor probe under the baffler coating. Such technique would allow mapping of spatial variations of permeability across the array of barrier coatings (*80,106*).

Such testing approach of barrier coatings utilizing sensors was developed (*107*) and as a result of these developments, a high-throughput screening method has been implemented that quickly screened barrier coating arrays (*108-110*). For the identification of the most suitable sensor film formulation for the determination of water vapor, a variety of polymer-immobilized reagents were screened. The parameters of interest were sensor material sensitivity, response reproducibility, response time, simple (preferably two-component) sensor formulation, and adequate photostability. Reagents malachite green, methylene blue, crystal violet, prodan, rhodamine 6G, DCM, nile red, and 7-hydroxycoumarmn were incorporated into Nafion polymer. These reagents were solvatochromic, acid-base, leuco, and others. This relatively small number of reagents was selected from our large database of reagents reported for moisture-sensing applications (*5*). Reagents were incorporated in Nafion polymer, known as one of the best candidates for moisture sensitive where n_P is the number of resolution regions, ΔP is a *desired* resolution of a property of interest in the gradient coating array, ΔL is an *available* spatial resolution of measurements of a property of interest in the gradient coating array, and k is the coefficient related to instrument detection sensitivity.

We applied our method for optimization of sensor material formulations for analysis of ionic and gaseous species (*101*). Figure 11 shows optimization of concentration of Pt octaethylporphyrin in a polystyrene film for detection of oxygen by fluorescence quenching (*19*). This data demonstrates the simplicity yet tremendous value of such determinations for the rapid assessment of sensor film formulations. It shows if the optimal concentration has been reached or exceeded depending on the nonlinearity and decrease of the sensor response at the highest tested additive concentration. Unlike traditional concentration optimization approaches (*74,99*) the new method provides a more dense evaluation mesh, and opens opportunities for time-affordable optimization of concentrations of multiple formulations components with tertiary and higher gradients (*101*).

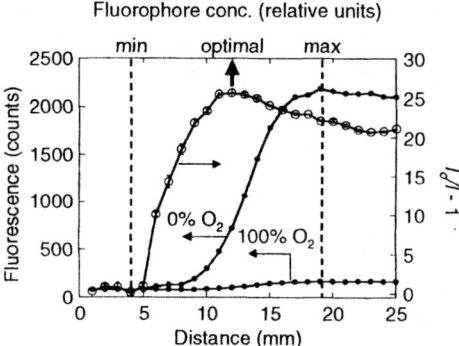

Figure 11. Optimization of formulated sensor materials using gradient sensor films. Variation of fluorescence intensity I at different oxygen concentrations and sensitivity $I_o/I-1$ of an oxygen-sensitive film of polystyrene polymer with incorporated Pt octaethylporphyrin fluorophore as a function of distance along the sensor film with a gradient change in fluorophore concentration. Reprinted with permission from reference 19. Copyright 2006, Wiley-VCH.

Capabilities of combinatorial screening of sensor materials are also superior when the need is to evaluate dynamic response of new sensor formulations. Dynamic spectroscopic scanning or imaging of the whole sensor materials arrays can be easily performed. When monitoring a dynamic process (e.g. a response time) of sensor materials arranged in an array with a scanning system, a maximum number N of elements in sensor library is given by (*102*):

$$N = t_r n_s / [n_d(\tau + t_s)], \qquad (5)$$

optical sensor materials (*111-113*). Sensor coatings arrays (8 reagents x 6 replicates) were positioned into the gas flow cell and were analyzed before and after exposure to dry and moist (80% relative humidity) air using colorimetric and fluorescence analysis. Optical spectra collected in dry and moist air were subtracted in order to obtain a relative ranking of sensor coating sensitivity toward moisture. Sensitivities of different coatings toward moisture presented as spectral absorption differences are shown in Figure 13B. The largest sensitivity toward moisture was demonstrated by sensor formulations 7, 5, and 3.

As a result of the screening experiments, a formulation containing crystal violet reagent (*112,114*) was selected. For a demonstration of a "combinatorial factory" screening process of moisture baffler coatings, poly(vinylidene chloride) latex-based baffler polymer films with different thickness (~15 - 70 μm) were deposited onto a 50-μm PET film. This structure was adhered to a 1.5-μm thick sensor film cast onto a glass support. Figure 13C demonstrates typical results of moisture barrier screening of an array of poly(vinylidene chloride) polymer films with increasing thickness. In this experiment, the laminated arrays were formed in the absence of moisture and were further placed into 90% relative humidity atmosphere at 40 °C to monitor the increase of absorbance at 630 nm. Overall, this polymer-screening approach demonstrated accurate determination of moisture transport down to 0.9 g/m² per day, a good correlation with conventional detection methods (*109*) and provided an opportunity for the development of several promising coating formulations.

The key well-recognized feature of formulated sensor materials is in a suite of opportunities to tailor materials responses to a specific application. Several aspects of formulated sensor materials development that still require primary attention are the understanding of ageing and drift mechanisms. Our tools developed for the combinatorial screening of long-term stability of polymeric coatings, can also be applicable for studies of these phenomena in formulated sensor coatings. Future discoveries of synergistic effects in formulated sensor materials promise to provide additional capabilities, as has been demonstrated in other types of formulated materials (*115,116*).

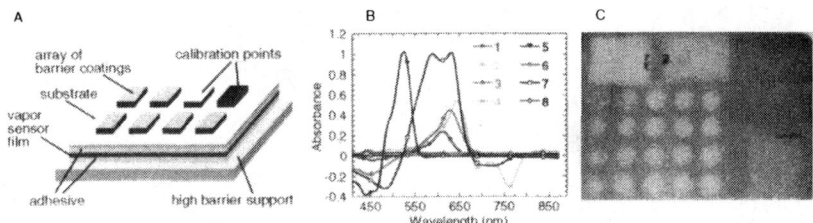

Figure 13. Application of moisture-sensitive sensor films for high-throughput screening of libraries of moisture-barrier coatings. (A) Laminated structure with a library of moisture-barrier coatings on top of a moisture-sensitive sensor film. (B) Moisture sensitivities of optical sensor coatings 1 – 8. (C) Photograph of typical screening results of transport of moisture through an array of moisture-barrier coatings and associated color change of moisture-sensitive sensor film from yellow (no moisture) to green (moisture present) after 72h of exposure at 40 °C and 90% relative humidity. (C) Adapted from (109).

Outlook

Since the first reports on polymeric coatings for sensor applications that go back to 1960-s (*117-121*), new knowledge of molecular interactions has been created and micro/nano-scale fabrication and analytical characterization tools have been built. These advances brought innovative concepts in coatings development for chemical and biological sensing. Considering the predominant mechanisms of sample/sensor interactions, we can identify four broad research directions on polymer-based sensor materials as shown in Figure 14. In the first direction, homo- or co-polymers are used for sensing without any additional polymer-modification steps. In the second direction, sensor capabilities are expanded by formulating polymers with additives. In the third direction, sensor capabilities are expanded by engineering polymeric structures with desired morphology. Finally, in the fourth direction, sensor capabilities are expanded by providing a shape-induced improvement in sensor selectivity. Individual sensing modalities can be further combined to achieve additional sensor response features. All these directions of polymeric sensor materials will continue to benefit from the combinatorial screening technologies (*117-121*).

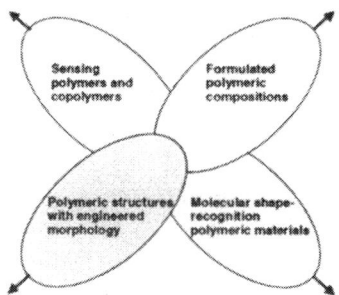

Figure 14. Development trends of polymeric sensor materials.

Acknowledgements

We gratefully acknowledge NIST, US DOE, and GE components for support of our combinatorial sensor research.

References

1. Hagleitner, C.; Hierlemann, A.; Brand, O.; Baltes, H., CMOS Single Chip Gas Detection Systems - Part I, In *Sensors Update, Vol. 11*; H. Baltes; W. Göpel and J. Hesse, Eds.; VCH: Weinheim, 2002; 101-155.
2. *Fiber Optic Chemical Sensors and Biosensors*; Wolfbeis, O. S., Ed.; CRC Press: Boca Raton, FL, 1991.
3. Harsanyi, G. *Polymer Films in Sensor Applications*; Technomic: Lancaster, PA, 1995.
4. *Polymers in Sensors. Theory and Practice*; Akmal, N.; Usmani, A. M., Eds.; American Chemical Society: Washington, DC, 1998; Vol. 690.
5. Potyrailo, R. A.; Hobbs, S. E.; Hieftje, G. M., Fresenius' *J. Anal. Chem.* **1998**, *362*, 349-373.
6. Albert, K. J.; Lewis, N. S.; Schauer, C. L.; Sotzing, G. A.; Stitzel, S. E.; Vaid, T. P.; Walt, D. R., *Chem. Rev.* **2000**, *100*, 2595-2626.
7. Grate, J. W., *Chem. Rev.* **2000**, *100*, 2627-2648.

8. Adhikari, B.; Majumdar, S., *Prog. Polym. Sci.* **2004**, *29*, 699-766.
9. Janata, J.; Josowicz, M., *Nature Materials* **2002**, *2*, 19-24.
10. Dickinson, T. A.; Walt, D. R.; White, J.; Kauer, J. S., *Anal. Chem.* **1997**, *69*, 3413-3418.
11. Apostolidis, A.; Klimant, I.; Andrzejewski, D.; Wolfbeis, O. S., *J. Comb. Chem.* **2004**, *6*, 325-331.
12. Chojnacki, P.; Werner, T.; Wolfbeis, O. S., *Microchim. Acta* **2004**, *147*, 87-92.
13. Potyrailo, R. A., *Macromol. Rapid Comm.* **2004**, *25*, 77-94.
14. Potyrailo, R. A., *Polymeric Materials Science and Engineering. Polymer Preprints* **2004**, *90*, 797-798.
15. Jandeleit, B.; Schaefer, D. J.; Powers, T. S.; Turner, H. W.; Weinberg, W. H., *Angew. Chem. Int. Ed.* **1999**, *38*, 2494-2532.
16. *Combinatorial and Artificial Intelligence Methods in Materials Science;* Takeuchi, I.; Newsam, J. M.; Wille, L. T.; Koinuma, H.; Amis, E. J., Eds.; Materials Research Society: Warrendale, PA, 2002; Vol. 700.
17. *High Throughput Analysis: A Tool for Combinatorial Materials Science;* Potyrailo, R. A.; Amis, E. J., Eds.; Kluwer Academic/Plenum Publishers: New York, NY, 2003.
18. *Combinatorial and Artificial Intelligence Methods in Materials Science II;* Potyrailo, R. A.; Karim, A.; Wang, Q.; Chikyow, T., Eds.; Materials Research Society: Warrendale, PA, 2004; Vol. 804.
19. Potyrailo, R. A., *Angew. Chem. Int. Ed.* **2006**, *45*, 702-723.
20. Potyrailo, R. A.; Chisholm, B. J.; Olson, D. R.; Brennan, M. J.; Molaison, C. A., *Anal. Chem.* **2002**, *74*, 5105-5111.
21. Potyrailo, R. A.; Chisholm, B. J.; Morris, W. G.; Cawse, J. N.; Flanagan, W. P.; Hassib, L.; Molaison, C. A.; Ezbiansky, K.; Medford, G.; Reitz, H., *J. Comb. Chem.* **2003**, *5*, 472-478.
22. Grate, J. W.; Abraham, M. H., *Sens. Actuators B* **1991**, *3*, 85-111.
23. Abraham, M. H., *Chem. Soc. Rev.* **1993**, *22*, 73-83.
24. Maranas, C. D., *Ind. Eng. Chem. Res.* **1996**, *35*, 3403-3414.
25. Wise, B. M.; Gallagher, N. B.; Grate, J. W., *J. Chemometrics* **2003**, *17*, 463-469.
26. Belmares, M.; Blanco, M.; Goddard III, W. A.; Ross, R. B.; Caldwell, G.; Chou, S.-H.; Pham, J.; Olofson, P. M.; Thomas, C., *J. Comput. Chem.* **2004**, *25*, 1814-1826.
27. Grate, J. W.; Abraham, H.; McGill, R. A., Sorbent polymer materials for chemical sensors and arrays, In *Handbook of Biosensors and Electronic Noses. Medicine, Food, and the Environment;* E. Kress-Rogers, Ed.; CRC Press: Boca Raton, FL, 1997; 593-612.
28. Grate, J. W.; Kaganove, S. N.; Patrash, S. J.; Craig, R.; Bliss, M., *Chem. Mater.* **1997**, *9*, 1201-1207.
29. Brecht, A.; Gauglitz, G., *Sens. Actuators B* **1997**, *38-39*, 1-7.
30. Potyrailo, R. A.; Hobbs, S. E.; Hieftje, G. M., *Anal. Chem.* **1998**, *70*, 1639-1645.

31. Phillips, C.; Jakusch, M.; Steiner, H.; Mizaikoff, B.; Fedorov, A. G., *Anal. Chem.* **2003**, *75*, 1106-1115.
32. Rathgeb, F.; Gauglitz, G., *Anal. Chim. Acta* **1998**, *372*, 333-340.
33. Hierlemann, A.; Lange, D.; Hagleitner, C.; Kerness, N.; Koll, A.; Brand, O.; Baltes, H., *Sens. Actuators B* **2000**, *70*, 2-11.
34. Hagleitner, C.; Hierlemann, A.; Lange, D.; Kummer, A.; Kerness, N.; Brand, O.; Baltes, H., *Nature* **2001**, *414*, 293-296.
35. Hagleitner, C.; Hierlemann, A.; Brand, O.; Baltes, H., CMOS Single Chip Gas Detection Systems - Part II, In *Sensors Update,* Vol. 12; H. Baltes; W. Göpel and J. Hesse, Eds.; VCH: Weinheim, 2003; 51-120.
36. Sepaniak, M. J.; Datskos, P. G.; Lavrik, N. V.; Tipple, C., *Anal. Chem.* **2002**, *74*, 568A-575A.
37. Lavrik, N. V.; Sepaniak, M. J.; Datskos, P. G., *Rev. Sci. Instrum.* **2004**, *75*, 2229-2253.
38. Ward, M.; Buttry, D. A., *Science* **1990**, *249*, 1000-1007.
39. Ballantine, D. S., Jr.; White, R. M.; Martin, S. J.; Ricco, A. J.; Frye, G. C.; Zellers, E. T.; Wohltjen, H. *Acoustic Wave Sensors: Theory, Design, and Physico-Chemical Applications;* Academic Press: San Diego, CA, 1997, pp 436.
40. Thompson, M.; Stone, D. C. *Surface-Launched Acoustic Wave Sensors: Chemical Sensing and Thin-Film Characterization;* Wiley: New York, NY, 1997, pp 196.
41. Potyrailo, R. A.; Morris, W. G.; Wroczynski, R. J., Acoustic-Wave Sensors For High-throughput Screening of Materials, In *High Throughput Analysis: A Tool for Combinatorial Materials Science;* R. A. Potyrailo and E. J. Amis, Eds.; Kluwer Academic/Plenum Publishers: New York, NY, 2003; ch. 11.
42. Potyrailo, R. A.; Morris, W. G.; Wroczynski, R. J., *Rev. Sci. Instrum.* **2004**, *75*, 2177-2186.
43. Abraham, M. H.; Andonian-Haftvan, J.; Whiting, G. S.; Leo, A.; Taft, R. S., *J. Chem. Soc., Perkin Trans.* **1994**, *2*, 1777-1791.
44. Grate, J. W.; Snow, A.; Ballantine, D. S., Jr.; Wohltjen, H.; Abraham, M. H.; McGill, R. A.; Sasson, P., *Anal. Chem.* **1988**, *60*, 869-875.
45. Grate, J. W.; Patrash, S. J.; Abraham, M. H., *Anal. Chem.* **1995**, *67*, 2162-2169.
46. Grate, J. W.; Patrash, S. J.; Abraham, M. H.; Du, C. M., *Anal. Chem.* **1996**, *68*, 913-917.
47. Hierlemann, A.; Zellers, E. T.; Ricco, A. J., *Anal. Chem.* **2001**, *73*, 3458-3466.
48. Potyrailo, R. A.; May, R. J.; Sivavec, T. M., *Sensor Lett.* **2004**, *2*, 31-36.
49. Potyrailo, R. A.; Sivavec, T. M.; Bracco, A. A., Proc. SPIE-Int. Soc. Opt. Eng., *3856* 140-147 (1999).
50. Sivavec, T. M.; Potyrailo, R. A. *Polymer coatings for chemical sensors;* US Patent 6,357,278 B1: 2002.
51. Potyrailo, R. A.; Sivavec, T. M., *Anal. Chem.* **2004**, *76*, 7023-7027.
52. Beebe, K. R.; Pell, R. J.; Seasholtz, M. B. *Chemometrics: A Practical Guide;* Wiley: New York, NY, 1998.
53. Wise, B. M.; Gallagher, N. B. *PLS_Toolbox Version 2.1 for Use with MATLAB;* Eigenvector Research, Inc.: Manson, WA, 2000.

54. Ulmer II, C. W.; Smith, D. A.; Sumpter, B. G.; Noid, D. I., *Comput. Theor. Polym. Sci.* **1998**, *8*, 311-321.
55. Shaffer, R. E.; Potyrailo, R. A.; Salvo, J. J.; Sivavec, T. M.; Salsman, L. *GE/Nomadics In-Well Monitoring System for Vertical Profiling of DNAPL Contaminants. Final Technical Report of Work Performed Under Contract DE-AC26-01NT41188, OSTI ID: 834346,* http.//www.osti.gov/servlets/purl/834346-tEuKN5/native/; US Dept. of Energy Info. Bridge: 2003.
56. Liu, D.; Meruva, R. K.; Brown, R. B.; Meyerhoff, M. E., *Anal. Chim. Acta* **1996**, *321*, 173-183.
57. Koronczi, I.; Reichert, J.; Heinzmann, G.; Ache, H. J., *Sens. Actuators B* **1998**, *51*, 188-195.
58. Arshak, K.; Moore, E.; Cavanagh, L.; Harris, J.; McConigly, B.; Cunniffe, C.; Lyons, G.; Clifford, S., *Composites A: Appl. Sci. Manufact.* **2004**,
59. Sangodkar, H.; Sukeerthi, S.; Srinivasa, R. S.; Lab R.; Contractor, A. Q., *Anal. Chem.* **1996**, *68*, 779-783.
60. Cho, E. J.; Tao, Z.; Tang, Y.; Tehan, E. C.; Bright, F. V.; Hicks, W. L., Jr.; Gardella, J. A., Jr.; Hard, R., *Appl. Spectrosc.* **2002**, *56*, 1385-1389.
61. Wolfbeis, O. S., *Anal. Chem.* **2004**, *76*, 3269-3284.
62. Bakker, E.; Bühlmann, P.; Pretsch, E., *Chem. Rev.* **1997**, *97*, 3083-3132.
63. Janata, J., *Crit. Rev. Anal. Chem.* **2002**, *32*, 109-120.
64. Bakker, E., *Anal. Chem.* **2004**, *76*, 3285-3298.
65. Gunter, R. L.; Delinger, W. G.; Manygoats, K.; Kooser, A.; Porter, T. L., *Sens. Actuators A* **2003**, *107*, 219-224.
66. Kooser, A.; Gunter, R. L.; Delinger, W. D.; Porter, T. L.; Eastman, M. P., *Sens. Actuators B* **2004**, *99*, 474-479.
67. Kröger, S.; Danielsson, B., Calorimetric biosensors, In *Handbook of Biosensors and Electronic Noses. Medicine, Food, and the Environment*; E. Kress-Rogers, Ed.; CRC Press: Boca Raton, FL, 1997; 279-298.
68. Rick, J.; Chou, T.-C., *Biosens. Bioelectron.* **2005**, *20*, 1878-1883.
69. Freud, M. S.; Lewis, N. S., *Proc. Natl. Acad. Sci. USA* **1995**, *92*, 2652-2656.
70. Kolytcheva, N. V.; Müller, H.; Marstalerz, J., *Sens. Actuators B* **1999**, *58*, 456-463.
71. Wang, E.; Chen, H. B.; Patel, H.; Sadaragani, I.; Romero, C., *Anal. Chim. Acta* **1999**, *397*, 287-294.
72. Penco, M.; Sartore, L.; Bignotti, F.; Sciucca, S. D.; Ferrari, V.; Crescini, P.; D'Antone, S., *J. Appl. Polymer Sci.* **2004**, *91*, 1816-1821.
73. Collaudin, A. B.; Blum, L. J., *Sens. Actuators B* **1997**, *38-39*, 189-194.
74. Basu, B. J.; Thirumurugan, A.; Dinesh, A. R.; Anandan, C.; Rajam, K. S., *Sens. Actuators B* **2005**, *104*, 15-22.
75. Walt, D. R.; Dickinson, T.; White, J.; Kauer, J.; Johnson, S.; Engelhardt, H.; Sutter, J.; Jurs, P., *Biosens. Bioelectron.* **1998**, *13*, 697-699.
76. Eaton, K., *Sens. Actuators B* **2002**, *85*, 42-51.
77. Mills, A.; Lepre, A.; Wild, L., *Anal. Chim. Acta* **1998**, *362*, 193-202.

78. Bedlek-Anslow, J. M.; Hubner, J. P.; Carroll, B. F.; Schanze, K. S., *Langmuir* **2000**, *16*, 9137-9141.
79. Wang, J.; Musameh, M.; Lin, Y., *J. Am. Chem. Soc.* **2003**, *125*, 2408-2409.
80. Potyrailo, R. A. *Devices and methods for simultaneous measurement of transmission of vapors through a plurality of sheet materials*; US Patent 6,567,753 B2: 2003.
81. Potyrailo, R. A.; Brennan, M. J. *Method and apparatus for characterizing the barrier properties of members of combinatorial libraries*; US Patent 6,684,683(B2): 2004.
82. Amis, E. J., *Nature Materials* **2004**, *3*, 83-85.
83. Potyrailo, R. A. *Combinatorial development and accelerated performance testing of polymers;* Second Dutch Polymer Institute workshop 'High Throughput Experimentation / Combinatorial Material Research', May 15-16 2003, Eindhoven University of Technology: Eindhoven, the Netherlands, 2003.
84. Hassib, L.; Potyrailo, R. A., *Polymer Preprints* **2004**, *45*, 211-212.
85. Lemmo, A. V.; Fisher, J. T.; Geysen, H. M.; Rose, D. J., *Anal. Chem.* **1997**, *69*, 543-551.
86. de Gans, B.-J.; Kazancioglu, E.; Meyer, W.; Schubert, U. S., *Macromol. Rapid Commun.* **2004**, *25*, 292-296.
87. Schena, M.; Shalon, D.; Davis, R. W.; Brown, P. O., *Science* **1995**, *270*, 467-470.
88. Schena, M. *Microarray Analysis*; Wiley: Hoboken, NJ, 2003.
89. Shen, M.; Bever, M. B., *J. Mater. Sci.* **1972**, *7*, 741-746.
90. Potyrailo, R. A.; Wroczynski, R. J.; Pickett, J. E.; Rubhisztajn, M., *Macromol. Rapid Comm.* **2003**, *24*, 123-130.
91. Potyrailo, R. A.; Szumlas, A. W.; Danielson, T. L.; Johnson, M.; Hieftje, G. M., *Meas. Sci. Technol.* **2005**, *16*, 235-241.
92. Potyrailo, R. A.; Wroczynski, R. J., *Rev. Sci. Instrum.* **2005**, *76*, 062222.
93. Meredith, J. C.; Karim, A.; Amis, E. J., *Macromolecules* **2000**, *33*, 5760-5762.
94. Potyrailo, R. A.; Olson, D. R.; Brennan, M. J.; Akhave, J. R.; Licon, M. A.; Mehrabi, A. R.; Saunders, D. L.; Chisholm, B. J. *Systems and methods for the deposition and curing of coating compositions*; US Patent 6,544,334 B1: 2003.
95. Huang, Z.-M.; Zhang, Y.-Z.; Kotaki, M.; Ramakrishna, S., *Composites Sci. Technol.* **2003**, *63*, 2223-2253.
96. Kameoka, J.; Czaplewski, D.; Liu, H.; Craighead, H. G., *J. Mater. Chem.* **2004**, *14*, 1503-1505.
97. Potyrailo, R. A.; Olson, D. R.; Medford, G. F.; Brennan, M. J., *Anal. Chem.* **2002**, *74*, 5676-5680.
98. Mills, A., *Sens. Actuators B* **1998**, *51*, 60-68.
99. Papkovsky, D. B.; Ponomarev, G. V.; Trettnak, W.; O'Leary, P., *Anal. Chem.* **1995**, *67*, 4112-4117.
100. Levitsky, I.; Krivoshlykov, S. G.; Grate, J. W., *Anal. Chem.* **2001**, *73*, 3441-3448.

101. Potyrailo, R. A.; Hassib, L., In *MACRO 2004 - World Polymer Congress, the 40th IUPAC International Symposium on Macromolecules, July 4 - 9* IUPAC: Paris, France, 2004.
102. Potyrailo, R. A.; Hassib, L., *Rev. Sci. Instrum.* **2005**, *76*, 062225.
103. Potyrailo, R. A., *Trends Anal. Chem.* **2003**, *22*, 374-384.
104. Chisholm, B. J.; Potyrailo, R. A.; Cawse, J. N.; Shaffer, R. E.; Brennan, M. J.; Moison, C.; Whisenhunt, D. W.; Flanagan, W. P.; Olson, D. R.; Akhave, J. R.; Saunders, D. L.; Mehrabi, A.; Licon, M., *Prog. Org. Coat.* **2002**, *45*, 313-321.
105. Mayer, W. N. *Apparatus and method for online or offline measurement of vapor transmission through sheet materials*; US Patent 6,009,743: 2000.
106. Potyrailo, R. A. *Devices and methods for measurements of barrier properties of coating arrays*; US Patent 6,383,815 B 1: 2002.
107. Combinatorial Methodology for Coatings Development, NIST ATP Award 1111/1999 - 10/31/2002, In http://iazz.nist.gov/atpcf/prjbriefs/prjbrief.cfm?ProjectNumber=99-01-6069: 2003; accessed Sept. 30, 2003.
108. Saunders, D. L.; Potyrailo, R. A. *High throughput screening for moisture barrier characteristics of materials*; US Patent Application 20010034063: 2001.
109. Grunlan, J. C.; Mehrabi, A. R.; Chavira, A. T.; Nugent, A. B.; Saunders, D. L., *J. Comb. Chem.* **2003**, *5*, 362-368.
110. Saunders, D.; Grunlan, J.; Akhave, J.; Licon, M.; Murga, M.; Chavira, A.; Mehrabi, A. R., Combinatorial Study and High Throughput Screening of Transparent Barrier Films Using Chemical Sensors, In *High Throughput Analysis: A Tool for Combinatorial Materials Science*; R. A. Potyrailo and E. J. Amis, Eds.; Kluwer Academic/Plenum Publishers: New York, NY, 2003; Ch. 14.
111. Zhu, C.; Bright, F. V.; Wyatt, W. A.; Hieftje, G. M., *J. Electrochem. Soc.* **1989**, *136*, 567-570.
112. Sadaoka, Y.; Matsuguchi, M.; Sakai, Y.; Murata, Y.-U., *Sens. Actuators B* **1992**, *7*, 443-446.
113. Zinger, B.; Shier, P., *Sens. Actuators B* **1999**, *56*, 206-214.
114. Sadaoka, Y.; Matsuguchi, M.; Sakai, Y.; Murata, Y.-U., *J. Mater. Sci.* **1992**, *27*, 5095-5100.
115. Whisenhunt Jr., D. W.; Carter, R.; Shaffer, R.; Bulsiewicz, W.; Flanagan, W., High Throughput Screening of the Thermal Stability of Colorants in a Polycarbonate Matrix, In *Combinatorial and Artificial Intelligence Methods in Materials Science II MRS Symposium Proceedings*; R. A. Potyrailo; A. Karim; Q. Wang and T. Chikyow, Eds.; Materials Research Society: Warrendale, PA, 2004; Vol. 804; 137-142.
116. Chambers, B. D.; Taylor, S. R.; Kendig, M. W., *Corrosion* **2005**, *61*, 480-489.
117. King, W. H., Jr., *Anal. Chem.* **1964**, *36*, 1735-1739.
118. Shaw, G., *Trans. Faraday Soc.* **1967**, *63*, 2181-2189.
119. Bergman, I., *Nature* **1968**, *218*, 396.
120. Hormats, E. I.; Unterleitner, F. C., *J. Phys. Chem.* **1965**, *69*, 3677-3681.
121. Shatkay, A., *Anal. Chem.* **1967**, *39*, 1056-1065.

Chapter 17

Using Quartz Crystal Microbalance–Heat Conduction Calorimetry to Monitor the Drying and Curing of an Alkyd Spray Enamel

Allan L. Smith

Masscal Corporation, 96 A Leonard Way, Chatham, MA 02633

Chemical processes occurring in a thin film can be characterized using quartz crystal microbalance/heat conduction calorimetry, (QCM/HCC), a new technique that simultaneously measures small mass changes, heats of reaction, and viscoelastic changes in films. Published viscoelastic data obtained from quartz crystal microbalances are shown to be consistent with the WLF formalism of polymer viscoelasticity. Another data set from the literature is used to show how the loss compliance $J''(t)$ of the film can be calculated from the motional resistance of the film which, in turn, can be measured directly by instruments such as the Masscal G1. Measurement principles of QCM/HCC are presented, and the technique is illustrated with a study of the drying and curing of a commercial alkyd spray enamel. By monitoring the process in alternating atmospheres of nitrogen and oxygen, it is possible to measure separately the drying of the film and its oxidative curing.

Drying and Curing of Coatings

When an organic coating is applied to a surface, a complex set of chemical and physical changes occurs before that coating can be useful in its intended environment. Further, the effectiveness of the final product can depend strongly on the properties determined by the competition and interaction of these reactions. It is helpful to categorize the two main types of processes as *drying* and *curing,* even though the two usually occur at the same time. Sliva (*1*) has summarized test methods used to determine drying, curing, and film formation of organic coatings. For the purposes of this article, we define *drying* as the physical evaporation of organic solvents and other volatile components present during the initial application of the film to a surface. Theoretical models have been developed to predict the drying behavior of polymer films on inert substrates (*2*) and these models contain no reaction chemistry. Miranda (*3*) has reviewed the process and the measurement of curing, and he defines *curing* as the cross-linking of polymeric systems. In all organic coatings except lacquers, cross-linking of the polymers introduce desirable properties such as increased hardness, durability and resistance to solvents. Cross-linking can be produced by oxidative, reactive, or catalytic means (*3*).

Heat and mass are two of the basic variables measured in experimental thermodynamics. From a thermodynamic point of view it should be straightforward to develop a method of distinguishing between the drying and the curing of a finish at constant temperature. In the drying process, the fluid coating (resin + solvents + pigments + propellants + catalysts) loses mass by evaporation, an endothermic process requiring heat to be absorbed by the coating. The curing process, on the other hand, may involve either gain or loss of mass, for example as oxygen diffuses into the coating to generate the radicals inducing cross-linking, but the formation of chemical bonds in cross-linking is always exothermic. As the coating dries and cures it changes from a viscous fluid to a viscoelastic solid, and the film's mechanical properties, as measured by its modulus, change dramatically. Kaye and coworkers have defined many terms related to the "nonultimate mechanical properties of polymers" (*4*), including the loss compliance in simple shear deformation. Dynamic mechanical analysis (DMA) is the principal technique used to determine viscoelastic properties of polymer systems. DMA has been used to characterize the rates of cross-linking in thermoset polymers (*5*).

Quartz Crystal Microbalance/Heat Conduction Calorimetry (QCM/HCC)

QCM/HCC is a new measurement technique uniquely positioned to measure the critical properties of films described above. QCM/HCC has been demonstrated to simultaneously measure the small mass changes, heats of reaction, and viscoelastic changes in films held in an isothermal environment and subjected to changes in gas composition (6-8). The Masscal G1™ Nanobalance/Microcalorimeter incorporates a quartz crystal resonator in intimate thermal contact with a sensitive heat flow sensor. The resonator is coated with a thin film sample (0.001-10 μm), and three quantities are monitored as the film is exposed to a programmed controlled gas mixture: (a) the mass m(t) (to ± 2 ng), (b) the thermal power P(t) (to ± 500 nW), and (c) the motional resistance R_{mot}(t). We show below that in the thin film limit, the difference in motional resistance ΔR_{mot} of the coated and uncoated resonator is proportional to $\rho^2 h^3 J''$, where ρ is the density, h is the thickness, and J'' is the shear loss compliance of the coating. Thus, the Masscal G1 provides simultaneous real-time gravimetric, calorimetric, and dynamic mechanical monitoring of a single thin film sample such as a coating.

The first part of this section is an abbreviated version of the theory presented in Smith and Shirazi (8), a reference that also contains diagrams of the apparatus. A flat quartz disc with electrodes on both surfaces can be forced to oscillate in a transverse acoustic mode (motion parallel to the surface) by an RF voltage applied at the acoustical resonance frequency of the plate. This device is called a *transverse shear mode (TSM) quartz plate resonator*. TSM quartz plate resonators have been used as sensitive microbalances for thin adherent films since the late 1950's, following the pioneering work of Sauerbrey (9). A TSM resonator whose frequency is continuously monitored when sample is deposited on its surface is known as a quartz crystal microbalance (QCM). The resonant frequency of a quartz TSM resonator of thickness h_q is

$$f_0 (\mu_q/\rho_q)^{1/2}/2h_q \tag{1}$$

where μ_q and ρ_q are the shear modulus and density of quartz. The shift in frequency due to deposition of a thin, stiff film is proportional to the deposited mass per unit area of the film, $\Delta m/A$:

$$\Delta f = -(2f_0^2/(\mu_q\rho_q)^{1/2})\Delta m/A = -(2f_0^2/(\mu_q\rho_q)^{1/2}) h_f\rho_f = -C\, h_f\rho_f \tag{2}$$

where h_f and ρ_f are the thickness and density of the film. For a 5 MHz crystal, C = 56.6 Hz/(μg/cm^2).

The width of the resonance for an uncoated 5 MHz resonator is 10-20 Hz, and the mechanical damping within the quartz that gives rise to this broadening can be determined by measuring the *motional resistance* R of the resonator, typically ~ 10 ohms. When thin, stiff films are deposited on the QCM surface the increase in R is small, but softer, thicker films (i.e., rubbery polymers 5-10 microns thick) can increase R by hundreds of ohms. The impedance of a TSM resonator damped by a finite viscoelastic film can be described as the sum of the impedance of the quartz and the impedance of the film. Theoretical predictions for the mechanical response of a model quartz crystal microbalance with a thin sample layer in a fluid bath have been given by White and Schrag (*10*). The film impedance is a function of four film parameters: the thickness h_f, the density ρ_f, the shear storage modulus G_f', and the shear loss modulus G_f''. Shear moduli must be determined at the TSM resonant frequency (5 MHz). In the theory of viscoelastic solids, the *compliance* can be used instead of the *modulus* to quantify storage and loss behavior. The shear storage compliance is defined as

$$J' = \frac{G'}{(G'^2 + G''^2)} \tag{3}$$

and the shear loss compliance is defined as

$$J'' = \frac{G''}{(G'^2 + G''^2)} \tag{4}$$

An important step in any analysis of motional resistance data for polymeric coatings is to recognize how G' and G" are related to each other and to independent variables such as temperature, measurement frequency, and film composition. The basis for understanding these relationships is the Williams-Landel-Ferry theory of polymer viscoelasticity (*11,12*) and the time-temperature superposition principle. Key concepts in this formalism are as follows:

(1) Master curves of G' and G" versus reduced measurement frequency f_{red} can be constructed for any polymer, and these curves are similar in shape from one polymer to another.

(2) The reduced frequency is proportional to the actual frequency f at which the moduli are measured,

$$f_{red} = f \, a_T, \tag{5}$$

where a_T is called the shift factor. The shift factor depends on the free volume of the polymer.

(3) The temperature dependence of the shift factor is given by

$$\log a_T = -c_1^g (T - T_g)/(c_2^g + T - T_g), \tag{6}$$

where T_g is the glass transition temperature, and the "universal" constants c_1^g and c_2^g have the values $c_1^g = 17.44$ and $c_2^g = 51.60$. Equation (6) is often referred to as the WLF equation.

(4) Since free volume depends both on temperature and on the mass

fraction of absorbed vapor, changes in both G' and G" can thus be related to these independent variables.

(5) For a glassy polymer (i.e., measurement temperature less than the glass transition temperature, T_g), the loss modulus G" is much less than the storage modulus G'.
(6) At T_g the shear and loss modulus become comparable, and the loss tangent $\delta = \tan^{-1}(G"/G')$ is a maximum.
(7) Above T_6 both G' and G" decrease with temperature.

Two convenient measurements of the film properties obtainable by QCM/HCC are (a) the difference in resonant frequency Δf = f(crystal + film) – f(crystal), and (b) the difference in motional resistance ΔR = R(crystal + film) – R(film). Voinova et al (13) present equations for both quantities in a power series expansion in the thickness h_f. Utilizing equations (3) and (4), the result for Δf and ΔR, to third order in h_f, is

$$\frac{\Delta f}{f_0} = \frac{\rho_f h_f}{\rho_q h_q}\left\{1 + \frac{4h_f^2 \pi f_0^2 \rho_f J'}{3}\right\} \quad (7)$$

and

$$\Delta R = \frac{2L_q}{3\pi Z_q}\omega^4 \rho_f^2 h_f^3 J" \quad (8)$$

Thus, the motional resistance change is proportional to the <u>square</u> of the film density, the <u>cube</u> of the film thickness, and the <u>loss compliance</u> of the film. For a 5 MHz QCM, typical values for L_q and Z_q are 0.0402 Henry and 8.84×10^6 Pa s/m, respectively. We demonstrate the correctness of Eq. (8) with the data of Lee, Hinsberg, and Kanazawa (14). They prepared a series of poly (n-butyl acrylate) films of thickness from 6 to 4000 nm and measured the motional resistance of the film in both air and water. Using the data from their Figure 5(a) for motional resistance in air, we plotted ΔR versus the cube of the film thickness, as suggested by Eq. (8). The resulting plot is linear with an R^2 of 0.9996. The slope is 6.1×10^9 ohm/nm³ ± 5%. Using a value 1.08 g/cm³ for the density of poly (n-butyl acrylate), Eq. (8) gives a value for the loss compliance J" of 5.0×10^{-9} Pa⁻¹. Lee et al (14) report values based on the Kanazawa relations for the shear storage modulus G' and shear viscosity η of 3×10^7 Pa and 0.13 Pa s, respectively. With the relationship G" = 2μfη, Eq. (4) yields a loss compliance J" = 4.5×10^{-9} Pa⁻¹ from these data, in good agreement with the value we derive, 5.0×10^{-9} Pa⁻¹, from Eq. (8).

Two sets of measurements of the shear storage modulus and the shear loss modulus of a thin film with a TSM resonator exist in the literature. Lucklum et at (*15,16*) measured G' and G" for a polyisobutylene film as a function of temperature, determined at 15 MHz using coating thicknesses of from 0.2 to 1.0 µm. Their data are shown in Figure 1. Katz and Ward (*17*) measured G' and G" at 5 MHz for a 9.25 µm polystyrene film as a function of the mass fraction of adsorbed 2-chlorotoluene vapor at 25°C. Their data are shown in Figure 2. Polystyrene is a glassy polymer at room temperature; Ferry (*11*) gives T_g = 97°C. Although polyisobutylene is classified as a rubbery material at low frequencies (T_g = -70°C), the Lucklum data show that at a measurement frequency of 15 MHz the glass transition temperature (i.e. the temperature at which G' and G" are equal) is ~50°C. This is consistent with the WLF theory, which states that the glass transition temperature increases with measurement frequency.

The data of Katz and Ward (Figure 2) show that as the mass fraction of absorbed solvent in polystyrene increases, G' decreases substantially by over two orders of magnitude while G" first increases and then decreases. The point where G' ~ G" occurs at a mass fraction of 4%. Katz and Ward interpret this change to be due to rotational relaxation of the 2-chlorotoluene solvent. Another interpretation of the data of Katz and Ward is possible. It is well known that the addition of a diluent of low molecular weight to an undiluted polymer decreases the glass transition temperature T_g. Indeed, this is the basis for adding plasticizers to stiff polymers to make them more pliable. Ferry (*11*) describes a theoretical approach to predicting the shift in T_g by combining his shift factor a_T with a linear dependence of T_g on the weight fraction of the diluent:

$$T_g = T_{g2} - kw_1. \tag{9}$$

Here 1 is the diluent or solvent and 2 is the polymer, w_1 is the weight fraction of solvent in the polymer, and k ranges from 200°C to 500°C for various solvents in polystyrene. Martin, Frye, and Senturia (*18*) extend these ideas of Ferry to derive a new equation for the dependence of the shift factor a_T on both temperature and vapor absorption. They say of this relationship, "This free volume treatment indicates the interchangeability of temperature and vapor absorption in determining film elastic properties."

It is not a coincidence that the polyisobutylene data of Figure 1 look similar in shape to the polystyrene data of Figure 2, even though the independent variable differs for the two plots. Either an increase in temperature or an increase in solvent volume fraction will decrease the shift factor a_T and thus move the frequency regime from high frequencies (glassy polymers) to lower frequencies (rubbery polymers). We used the data of Katz and Ward to compute

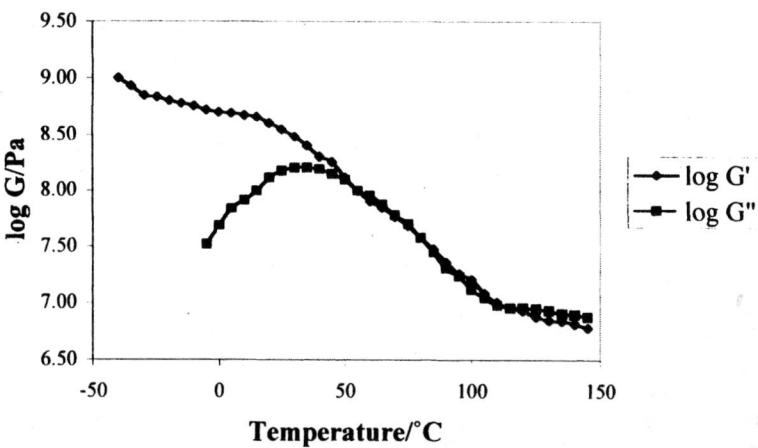

Figure 1. Shear storage modulus G' and shear loss modulus G" for polyisobutylene. Data are from Ref. 15.

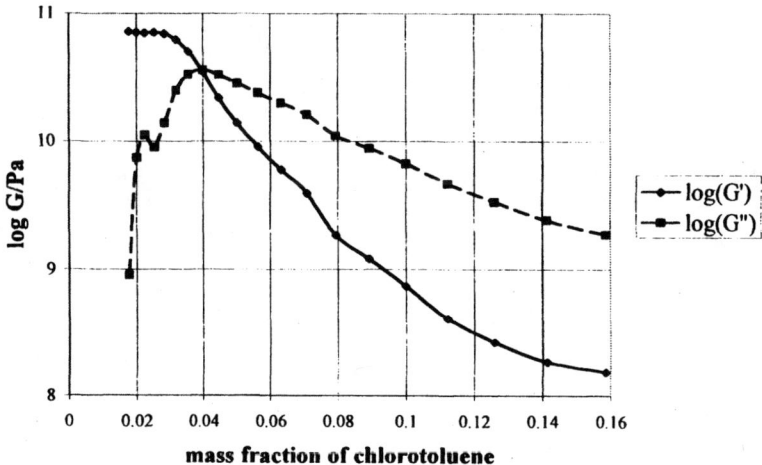

Figure 2. Shear storage modulus G' and shear loss modulus G" for polystyrene plasticized with 2-chlorotoluene. Data in this figure are from Ref. 17.

the dependence of the shear storage compliance and the shear loss compliance on the mass fraction of 2-chlorotoluene absorbed in polystyrene. We find that J' remains constant at 1.4×10^{-11} Pa^{-1} for 0.02<mass fraction <0.08, increasing to 4.4×10^{-11} Pa^{-1} at a mass fraction of 0.16. The results for J" are shown in Figure 3. Notice that as the 2-chlorotoluene mass fraction increases J" increases by three orders of magnitude. We compute from Eq. (7) that the error in using the Sauerbrey equation is only 0.13%. From Eq. (8) we calculate that the change motional resistance varies from 0.1 ohms to 447 ohms.

QCM/HCC Measurements on an alkyd spray enamel

We present here an example of the application of the QCM/HCC technique to the drying and curing of Decrolon, a commercial aerosol spray enamel made by Sherwin-Williams. The resin in Decrolon is a vinyl toluene alkyd, and the volatile organic solvents are acetone (20%), toluene (13%) and light aliphatic naphtha (8%), plus propellants.

Short-term drying of enamel

An uncoated QCM crystal was characterized by weighing on a 5-place balance and then measuring its resonant frequency and motional resistance in a Masscal G1 nanobalance/calorimeter. The larger gold electrode of this crystal was then sprayed with a thin coat of Decrolon. Within three minutes the crystal was installed in the Masscal G1 sample chamber at 30.0 °C and purged briefly with N_2 so as to arrest any air-induced curing. Ten minutes into the experiment, a 5 std cm^3/sec flow of dry N_2 was established; the resulting heat flow, mass change and motional resistance are shown in Figure 4. Evaporation of the solvents causes the large negative thermal power peak, the loss in mass of the film, and the decrease in motional resistance or loss compliance. At the end of all experiments on this film, the QCM crystal was reweighed. The mass of the film was 250 ±30 µg. From the difference in frequency of the crystal before and after coating and the area of the coating (1.6 cm^2), the film mass was calculated to be 252 ±5 µg using the Sauerbrey equation. Assuming a film density of 1.5 g/cm^3, the film thickness is 1.08 µm. The loss in mass per unit area shown in Figure 4 is 40 µg/cm^2, and the area of the sprayed film was 1.6 cm^2. Thus the evaporative mass loss of the solvents was 64 µg, or 26% of the original film mass.

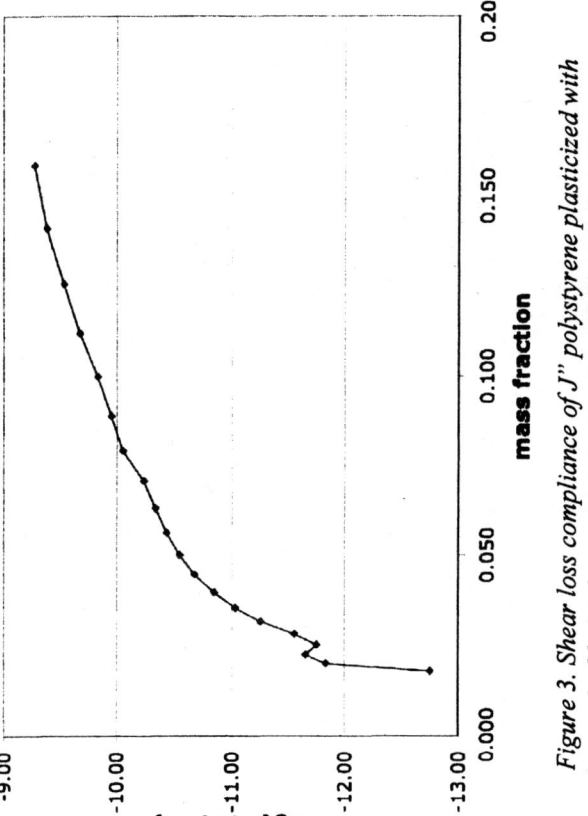

Figure 3. Shear loss compliance of J" polystyrene plasticized with 2-chlorotoluene. Data in this figure are from Ref. 17.

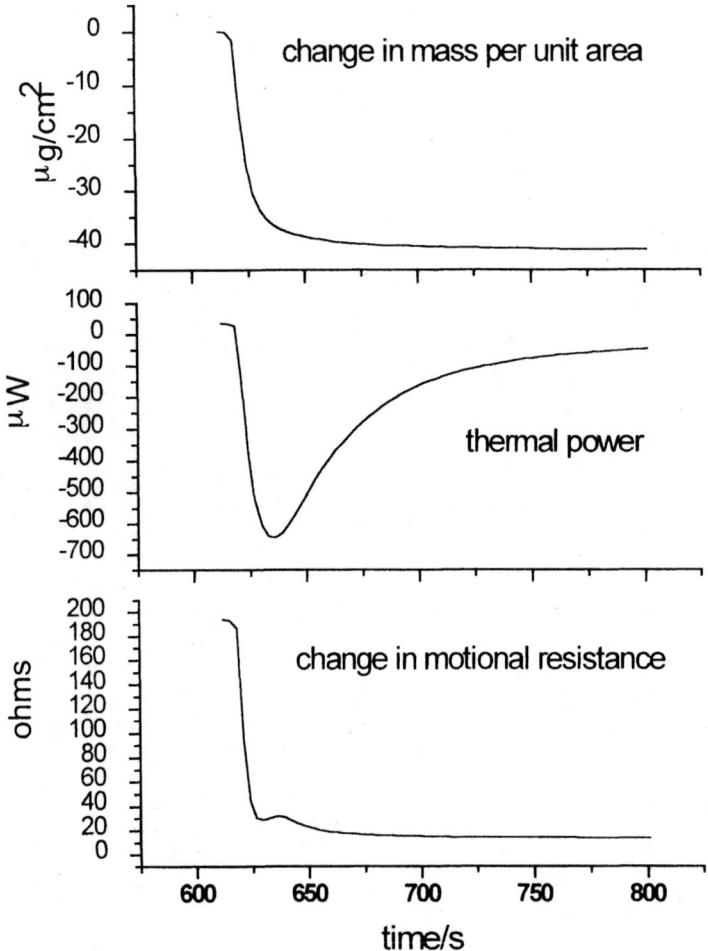

Figure 4. Short-term drying of uncured Decrolon film.

Toluene uptake and loss in uncured and air-cured enamel

As long as the alkyd enamel film remains under the nitrogen flow, no oxidative cross-linking can occur. We refer to a film in this state as *dried* but *uncured*. To examine the uptake and release of toluene vapor from such a film, two mass flow controllers were used to control the flow of a variable mixture of pure N_2 and of N_2 saturated with toluene vapor into the sample chamber of the Masscal G1. The toluene bubbler chamber temperature was 28.9 °C. The system was programmed to provide five 900 second steps of increasing toluene vapor pressure from 0 to 3600 Pa followed by five steps of equivalent decreases. A complete cycle thus took 9000 seconds, and a single run included two cycles. The data collected are shown in Figure 5. The top panel shows the partial pressure of toluene produced by the flow program, and the remaining three panels show the variables measured as the Decrolon film was exposed to the toluene vapor. All thermal power measurements have been corrected for the heat dissipated by the oscillating crystal, a function of the motional resistance of the crystal. Details of this correction will be presented elsewhere.

After the run was completed, the QCM crystal was removed and allowed to cure in air for 3 days. It was reweighed on the 5-place balance and then was then reinserted in the G1, and the same flow program protocol was followed. The raw data were similar to those of Figure 5 but with a smaller mass of toluene absorbed. Analysis of the QCM/HCC data, construction of sorption isotherms and determinations of sorption enthalpies follows the methodology introduced in the study of water and ethanol sorption in an aliphatic polyurethane (*7*). The prompt changes in mass and motional resistance and the peaks in thermal power produced whenever the partial pressure of toluene changes indicates that this film of Decrolon quickly absorbs and desorbs toluene vapor and that the rate of sorption is not significantly limited by the rate of diffusion of toluene in the resin. The sorption isotherms of toluene in Decrolon calculated from the data of Figure 5 and from the data on the air-cured Decrolon film are shown in Figure 6. Notice that the air-cured film absorbs less toluene than the dried but uncured film.

To determine the sorption enthalpy of toluene in Decrolon, the integral $\int Pdt$ of the thermal power over one step of constant toluene activity yields the integrated heat, Q, associated with that sorption or desorption process. The change in mass of the film for that step is equated to the mass of toluene vapor adsorbed or desorbed, $m_{toluene}$. Finally, the enthalpy of sorption is calculated for each step by the relationship $\Delta H_{sorption} = Q/(m_{toluene}/MM_{toluene})$, where MM is the molar mass of toluene. The average of all steps shown in Figure 5 give $\Delta H_{sorption}$ = -21.2 ± 2.8 kJ/mol. Since the molar enthalpy of condensation of toluene is -38.0 kJ/mol at 25°C (*19*), we can conclude that the interaction of toluene with the alkyd resin is weaker than the interaction of toluene with itself in the pure liquid state.

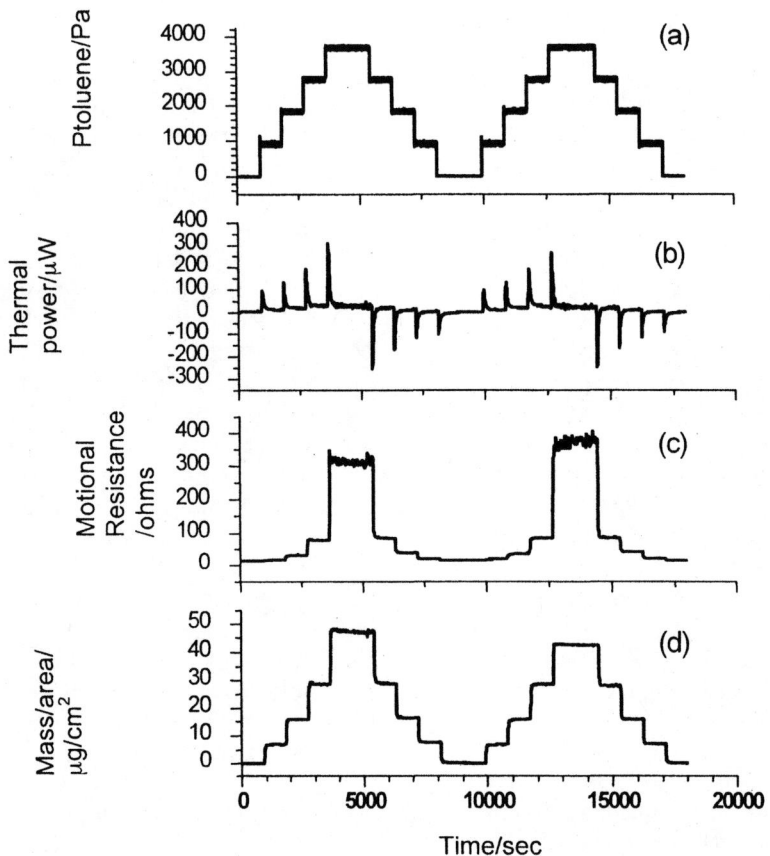

Figure 5. Sorption of toluene vapor in an uncured Decrolon film. (a) toluene vapor partial pressure; (b) net thermal power (Eq. 11); (c) motional resistance; (d) change in mass per unit area.

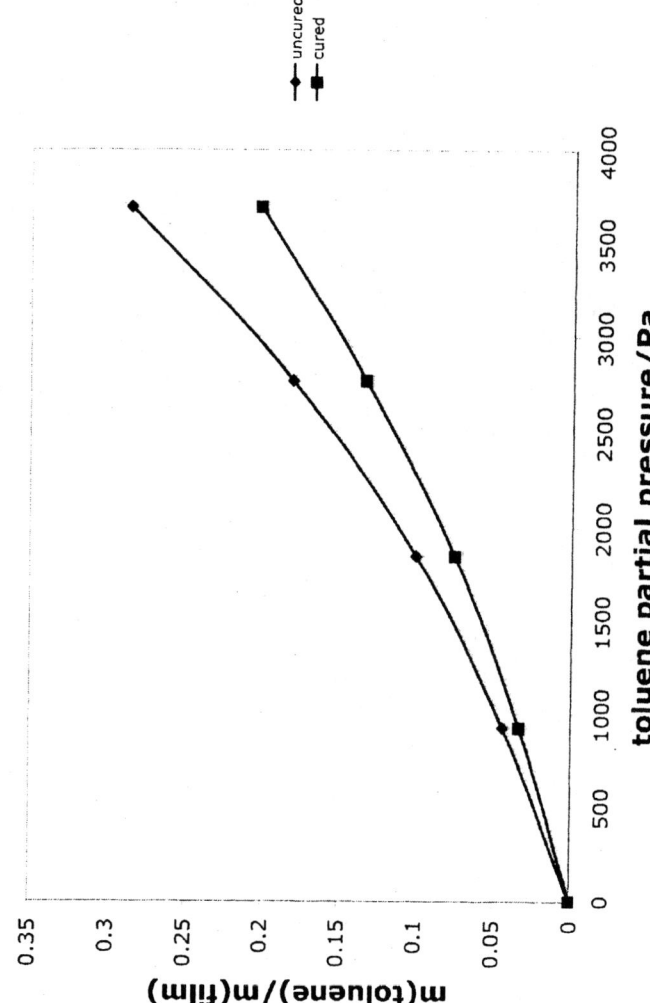

Figure 6. Toluene sorption isotherms for uncured and cured Decrolon.

The motional resistance data for the uncured (Figure 5) and cured Decrolon film are converted to loss compliance using Eq. (8) and plotted in Figure 7 as a function of the mass of toluene per gram of film. Notice that the loss compliance of the cured film is lower than that of the uncured film for all values of the toluene content of the films. Since lower loss compliance is generally correlated with greater stiffness of a polymer film we can conclude that these data are consistent with the fact that curing increases the stiffness of the resin coating.

Monitoring the O_2-induced curing of enamel with the QCM/HCC.

A second, thicker Decrolon film was prepared on another QCM crystal and installed within a few minutes of its deposition in the sample chamber of the Masscal G1. The mass of this film was 1.33±0.11 mg and its thickness was calculated to be 4.9 μm. The film was dried for four hours within the Masscal G1 under a 5 cm^3/min flow of nitrogen. Two mass flow controllers were used to provide a variable mixture of nitrogen and oxygen to the sample chamber. The flow controller software was programmed to alternate between 5 cm^3/min of N_2 and 5 cm^3/min of O_2 at 10 minute intervals, and a 12-hour experiment was recorded. We call this mode of operation "modulated environment chemistry", because it is only during the presence of oxygen that air-curing of the alkyd enamel can occur. If drying was not complete in the first four hours, the mass and motional resistance might still decrease slowly under N_2 flow as the film dries, but O_2-initiated cross-linking cannot occur. The resulting data are shown in Figure 8.

There is much detailed information of interest in the data shown in this Figure, and a complete analysis, particularly of the rapid short-term changes, must wait for another publication. But the long-term trends in mass, motional resistance, and thermal power show clearly the onset of a slow chemical process at about 6 hours into the run, well beyond the expected drying period of the thin film. A more exothermic process begins in the thermal power curve at 6 hours and reaches its maximum amplitude at 8.5 hours; this is consistent with the exothermic process expected in cross-linking. By 10 hours, the mass of the film has increased by 0.3700 and the loss compliance has decreased by 50%. As can be seen from Figures 2 and 3, a decrease in loss compliance of a film below its glass transition temperature is consistent with the stiffening of a glassy film due to cross-linking. An extension of the run shown in Figure 8 for the next several days produced a film whose loss compliance was only 20% of its value at the beginning of exposure to O_2. We interpret the changes in all three signals starting at 6 hours to be the onset of the curing of the alkyd resin induced by the O_2. These long-term trends clearly show that the QCM/HCC can separate the measurement of the drying process from that of the curing process.

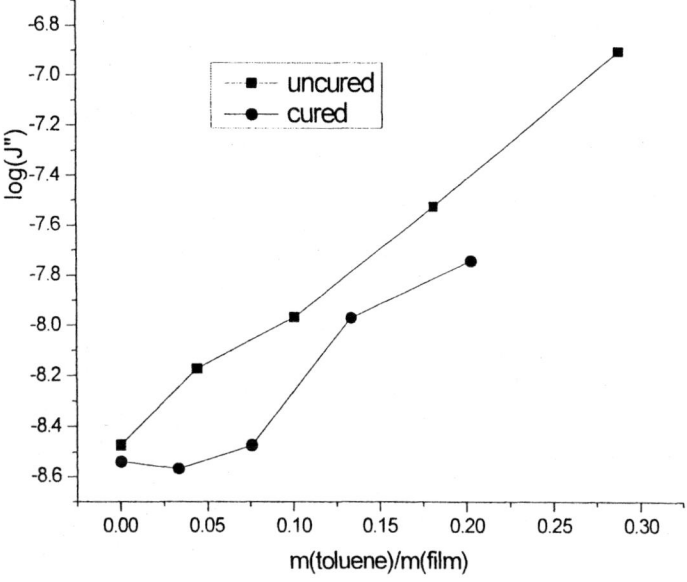

Figure 7. Loss compliance of uncured and cured Decrolon as a function of toluene content.

Conclusions

In this chapter we have described briefly how a new measurement technology, quartz crystal microbalance/heat conduction calorimetry, has the capability to simultaneously measure three critical properties for understanding thin film reactions: thermal power, mass, and motional resistance. We have explained the connection of the measured motional resistance to the loss compliance of the film and we demonstrate that viscoelastic properties of polymers measured at 5 MHz with the Masscal G1 nanobalance/calorimeter can be understood in terms of the WLF theory of polymer viscoelasticity. As an illustration of the measurements, we present data on the drying and curing of an alkyd enamel spray finish. The data demonstrates the usefulness of making these measurements in real time under controlled changes in environment, not only to determine reaction properties under different gases but also to be able to separate the changes related to drying and curing. The measurements made were highly sensitive to changes in the film and we believe the information provided by this technology will enable coatings researchers both to monitor and to control the drying and curing process. Watching paint dry is not boring if you have the right perspective.

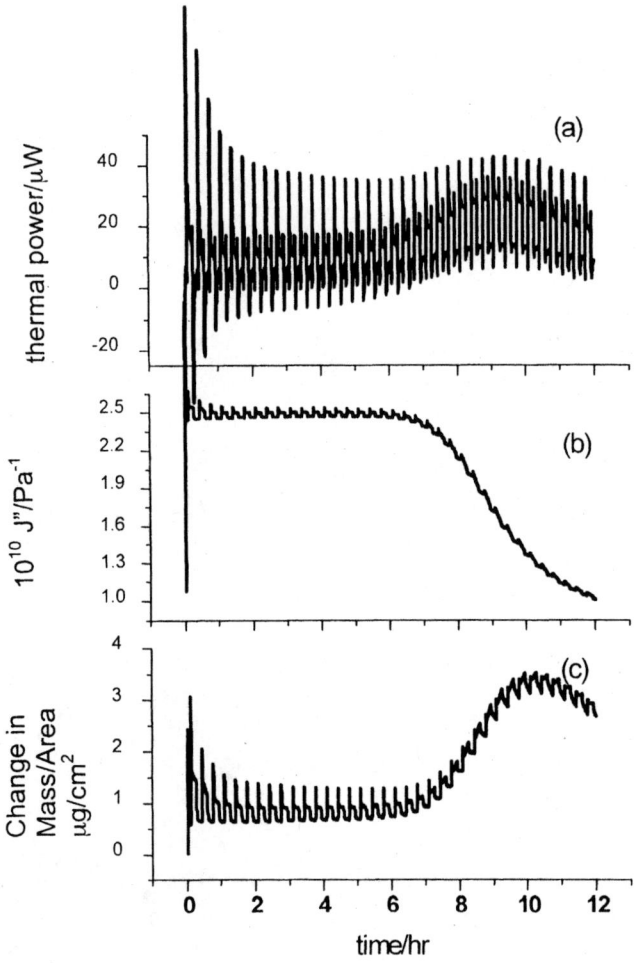

Figure 8. Change in mass per unit area, loss compliance, and thermal power in a Decrolon film when exposed to alternate flows of N_2 and O_2 in 10 minute intervals.

References

1. Sliva, T. J. Drying Time. In *Paint and Coating Testing Manual;* 14th Edition of the Gardner-Sward Handbook ed.; Koleske, J. V., Ed.; ASTM: Philadelphia, 1995; Vol. MNL 17; pp 439.
2. Alsoy, S.; Duda, J. L. *AIChE* **1999**, *45*, 896.
3. Miranda, T. J. Curing: The Process and Its Measurement. In *Paint and Coating Testing Manual;* Koleske, J. V., Ed.; ASTM: Philadelphia, 1995; Vol. MNL 17; pp 407.
4. Kaye, A.; Stepto, R. F. T.; Work, W. J.; Aleman, J. V.; Malkin, A. Y. *Pure and Appl. Chem.* **1998**, *70*, 701.
5. Nichols, M. E.; Gerlock, J. L.; Smith, C. A. *Polymer Degradation and Stability* **1997**, *56*, 81.
6. Smith, A. L.; Shirazi, H. M. *J. of Thermal Analysis and Calorimetry* **2000**, *59*, 171.
7. Smith, A. L.; Mulligan, S. R.; Shirazi, H. M. *J. Polymer Sci. Part B Polymer Physics* **2004**, *42*, 3893.
8. Smith, A. L.; Shirazi, H. M. *Thermochimica Acta* **2005**, *432*, 202.
9. Sauerbrey, G. *Z Physik* **1959**, *155*, 206.
10. White, C. C.; Schrag, J. L. *J. Chem. Phys.* **1999**, *111*, 11192.
11. Ferry, J. D. *Viscoelastic Properties of Polymers,* 3rd ed.; Wiley: New York, 1980.
12. Weissman, P. E.; Chartoff, R. P. Extrapolating Viscoelastic Data in the Temperature-Frequency Domain. In *Sound and Vibration Damping with Polymers;* American Chemical Society, 1990; pp 111.
13. Voinova, M. V.; Jonson, M.; Kasemo, B. *Biosensors and Bioelectronics* **2002**, *17*, 835.
14. Lee, S.-W.; Hinsberg, W. D.; Kanazawa, K. K. *Anal. Chem.* **2002**, *74*, 125.
15. Lucklum, R.; Hauptmann, P. *Faraday Discussion* **1997**, *107*, 123.
16. Lucklum, R.; Hauptmann, P. *Electrochimica Acta* **2000**, *45*, 3907.
17. Katz, A.; Ward, M. D. *J. Appl. Phys.* **1996**, *80*, 4153.
18. Martin, S. J.; Frye, G. C.; Senturia, S. D. *Anal. Chem.* **1994**, *66*, 2201.
19. *CRC Handbook of Chemistry and Physics,* 80th ed.; CRC Press: Boca Raton, 1999.

Chapter 18

Predicting Service Life Performance: Our Analytical Toolbox

Karlis Adamsons

Performance Coatings, Marshall R&D Laboratory, DuPont Company, Philadelphia, PA 19146

Abstract

Chemical characterization of multi-layered automotive coating systems for predicting service life requires a strong analytical toolbox. This report provides a summary of sampling techniques and measurement technologies that allow principle investigators routine access to locus specific chemical information or the more detailed chemical depth profiles. A variety of measurement technologies have been developed that would (for the most part) be commonly available or readily affordable in many academic and industrial research laboratories. This overview describes an analytical toolbox that helps serve our needs in design and service life testing of coating systems in the arena of high performance automotive finishes.

Introduction

The need for a measurement technology toolbox for chemical depth profiling of automotive coating systems has been rapidly increasing during the course of the last decade. Specifically, the success of service life predictions for high performance automotive coating systems is (to a significant extent) a function of appropriate measurement technologies to monitor chemical composition distribution and changes.

Over the last decade a variety of powerful tools have been identified or developed to allow surface, near-surface and interface specific measurements, as well as the detailed depth profiling of single- or multi-layer coating systems [1-7]. Bulk analysis measurement technologies such as IR (in transmission mode), UV-VIS [7], and DSC [8,9] have been successfully applied in characterizing the general composition of a coating layer. Near-surface analysis measurement technologies such as IR (in attenuated total reflectance [ATR], diffuse reflectance [DRIFT], or photoacoustic [PAS] modes) [7], Raman [6] and K+IDS MS [11] have been used frequently in characterizing the uppermost ~2-10 μm. Surface specific measurement technologies such as time-of-flight secondary ion mass spectrometry [ToF-SIMS] [1,6] or X-ray photoelectron spectroscopy [XPS or ESCA] [6,10] have proven very useful in obtaining an intimate look at the top ~10-100 Angstroms. The bulk and near-surface approaches allowed characterization at a given, albeit relatively broad, locus. The surface specific approaches were traditionally not applied for purposes of chemical depth profiling, but rather focused on the uppermost regions. A combination of sampling techniques and these aforementioned measurement technologies offer an efficient way to obtain the chemical depth profiles. Direct analysis of slab microtomed sections (cut coplanar to the coating system surface) has been done by IR [3,4,6,7,14,15] DSC [8,9] ESCA [10], ToF-SIMS [1,3,6] and K+IDS MS [11] techniques. Acid digests (with and without microwave digestion) of these slab microtomed sections have been studied by AA and ICP techniques. Also, indirect analysis of these sections (following solvent extraction) has been done by UV-VIS [3,4,6,7,14,15], HPLC [7] and GC[1] techniques.

The original research driver for this work was the ability to monitor UVA concentration as a function of depth. This enabled coating system developers a means of determining UVA permanence and effectiveness in the field [7]. UVA migration, photo-stability, thermal-stability and solubility/segregation could now be studied. Knowledge of the UVA distribution and overall content has shown promise in predicting probable success (or failure) of a coating system, and thus, can provide more effective and timely feedback during new product development.

A subsequent research driver for continuing work was to identify or develop tools that would permit tracking the chemical impact of climate and other environmental factors on automotive finishes. Factors of primary concern are solar radiation (primarily UV wave-lengths) [12] heat (surface temperature; thermal expansion/contraction), moisture (dew: rain: humidity), pollutants (acid rain; ozone, aerosols), and biological (plant, insect or animal residues). Chemical degradation by photooxidative and hydrolytic mechanisms [16] can now be studied as a function of depth.

A final research driver for these efforts was to identify or develop tools that would allow investigators a way to determine the distribution of coating layer or system components [13-15]. This information would better define the uniformity or non-uniformity of components resulting from the original application and cure stages. Wet-on-wet application and cure of CC/BC systems is common in the automotive industry (for the purpose of saving time and energy), but may be prone to mixing of components across interface boundaries. More recently efforts in wet-on-wet-on-wet applications and cure of CC/BC/PR systems are underway, including chemical depth profiling studies. These tools can document the extent of component mixing and perhaps offer insight into the effect on product service life. Also, additional concerns may exist over factors such as pigment de-mixing in a given colorant layer, segregation of components at interfaces or into domains, and trapping of slower solvents. The chemical depth profiling tools described herein should offer many opportunities for researchers to monitor composition and associated changes/kinetics on the micron scale [3-9,13-19]. A well-equipped analytical toolbox may be critical in the reduction of future product design/development/testing cycles.

Experimental

Standard microtomy:

A standard microtome was used in obtaining ~5 μm thick normal angle cross-sections of the automotive coating systems. A small amount of Fluorolube® was used to wet the cross-sections and (effectively) anchor them onto a quartz slide substrate.

Sliding (Slab) microtomy of automotive finishes:

A Leica Polycut model SM-2500E Slab Microtome was used in sectioning the automotive coating systems co-planar to the surface. The thickness of the slab sections was typically cut in the ~5-10 μm range. Once cut, they were stored between sheets of weighing paper.

Direct analysis of the slab microtomed sections:

The slab sections were extracted by tetrahydrofuran (THE) or methylene chloride (Aldrich, high purity spectroscopy grades) at room temperature for a period of ~1 day. Unless otherwise indicated, methylene chloride was the preferred solvent during extractions due to its ability to swell the 3D network. Since this solvent at swelling and extraction of these thin sections, it was used on both the 1K acrylo-silane-melamine and 2K acrylo-urethane coating systems.

Acid and microwave digest of the microtomed sections:

Slab microtomed sections were carefully weighed and put into aqueous HCl, H_2SO_4 or HNO_3 solutions and placed into a microwave digest unit to create a particle-free solution. These solutions were routinely used in AA and ICP type analyses.

Through-film (T-Mode) UV-VIS Spectroscopic Analysis:

Absorbance measurements were obtained on ~5µm thick layers. Each layer was placed between two 25 mm diameter sodium chloride salt plates (~4 mm thick), which had been wetted with the minimally absorbing Fluorolube® (International Crystal Labs) to minimize scattering. Absorbance spectra were obtained on a Hewlett Packard diode array UV-VIS spectrophotometer. Absorbance of the NaCl salt plates plus Fluorolube was subtracted from the total spectrum in each case to obtain the absorbance profiles of each layer.

Infrared (IR) Attenuated Total Reflectance (ATR) analyses:

IR ATR-mode analyses were done on Nicolet Nexus 470 FT-IR ESP Spectrophotometer equipped with a Nicolet Smart Dura Sampl/R ATR module (Thunderdome).

Ultraviolet- Visible (UV- VIS) analyses:

UV-VIS analyses were done on a Hewlett Packard Model 8452A Diode Array Spectrophotometer using a standard 1 cm^2 x 5 mL capped quartz cell.

Differential Scanning Calorimetry (DSC) Analyses:

DSC analyses were done on a DuPont 912 Dual Sample DSC interfaced with the DuPont 2100 Thermal Analyst System.

Gas Chromatography (GC) Analyses:

GC analyses were done on a Hewlett Packard Model 6890 Series GC System equipped with a flame ionization detector (FID).

High Performance Liquid Chromatography (HPLC) Analyses:

HPLC analyses were done on a Hewlett Packard Model 1100 Series HPLC System equipped with a diode array detector.

Time-of-Flight Secondary-Ion MS (ToF-SIMS) Analyses:

ToF-SIMS analyses were done on a PhI-7200 ToF-SIMS Spectrometer equipped with a focused 25 KeV Gallium (Ga) ion beam (gun) which permitted an effective spot size ~2 µm diameter and X-Y translating stage.

Electron Spectroscopy for Chemical Analysis (ESCA):

ESCA (also known as X-Ray Photoelectron Spectroscopy or XPS) analyses were done on an Ulvac PhI Quantera 5X7 ESCA Spectrometer.

Atomic Absorption (AA) Analyses:

AA analyses were done on a Perkin Elmer (PE) AAnalyst 300 spectrometer. Sample digestion was done using two approaches. Open acid digestion was done on a hot plate if the element of interest was not considered volatile. Closed acid ("bomb" type) was done if the element of interest was considered volatile or associated with a volatile species.

Potassium Ionization of Desorbed Species (K^+IDS) MS Analyses:

K^+IDS MS analyses were done on a Finnigan Thermo Quest TSQ Mass Spectrometer.

Results & Discussion

Depth profiling of the chemical composition has proven very useful in better understanding factors such as component distribution, segregation of species, interlayer mixing (or strike-in), UVA/HALS permanence or effectiveness, solvent trapping, pigment de-mixing, catalyst location, uniformity of network crosslinking (or lack thereof) and tracking species/network degradation.

During the last decade we have successfully combined the sampling technique of sliding (slab) microtomy with numerous measurement technologies to obtain the depth profiles required to better define these factors. Sectioning co-planar to a coating's surface down to a scale of 5-10 µm/section for a sample area of ~30x30 mm^2 offered the opportunity to apply numerous measurement technologies. A variety of IR-based (Transmission-, ATR-, DRIFT-, and PASmode), Raman microprobe, laser desorption mass spectrometric [LDMS], ToFSIMS, ESCA and thermal [DSC; TGA] analyses were possible directly with the thin sections produced. A second group of analyses were made possible following solvent extraction of the sections, including UV-VIS (to monitor UVA or degradation species), high performance liquid chromatography [HPLC] (to monitor UVA, HALS, or degradation species) and potassium ionization of desorbed species mass spectrometric [KID$^+$S MS] (to monitor non-volatile oligomeric or degradation species) analysis. A third group of analyses were also made possible following an acid/microwave digest step, including atomic absorption [AA] and inductively coupled plasma [ICP] analysis. Table I provides a group/technique summary.

UV-VIS Analyses for Depth Profiling:

Observations and constraints for UVA depth profiling of CC cross-sections monitored in transmission-mode by UV-VIS microscopy [20] are as follows:

(1) Cross-sections must be relatively thin in order to allow sufficient transmission to obtain a quality spectrum, typically ~5-7 µm thick. (2) UVA migration (or decomposition) is significant even after 1 year outdoor exposure in Hialeah, Florida. (3) UVA loss rate is a function of depth. A gradient becomes apparent on longer Florida or Xe boro/boro exposures. Rate of loss is fastest in uppermost section, and slowest in deepest sections. (4) Unexposed CC was stored in a cabinet at STP conditions of ~26°C, ~50% RH, and in the dark. (5) There are other UV absorbing species in the CC that are apparent in the composite UV-VIS spectra. Figure (1) shows a typical UVA (benzotriazole type) depth profile for an unexposed CC layer.

Observation and constraints for UVA depth profiling of CC cross-sections monitored in transmission-mode after methylene chloride extraction of the slab sections:

(1) Solvent extraction efficiency of newly prepared coating systems is 95-98% for Ciba's Tinuvin 900. (2) UVA migration (or decomposition) is significant even after ~1 year outdoor exposure in Hialeah, Florida. (3) UVA loss rate is found to be a function of depth. A gradient becomes apparent on longer Florida or Xe boro/boro exposures. Rate of loss is fastest in the uppermost section, and slowest in deepest sections. (4) Unexposed CC was stored in a cabinet at STP conditions of ~26°C, ~50% RH, and in the dark. (5) CC samples exposed longer were found to be more brittle, and (as a result) slab sections would break/fracture more readily. Figure (2) shows a typical UVA depth profile for an outdoor exposed CC/BC bilayer coating system.

The original strong UVA absorption in the uppermost slab sections is significantly reduced due to factors such as migration and structure degradation.

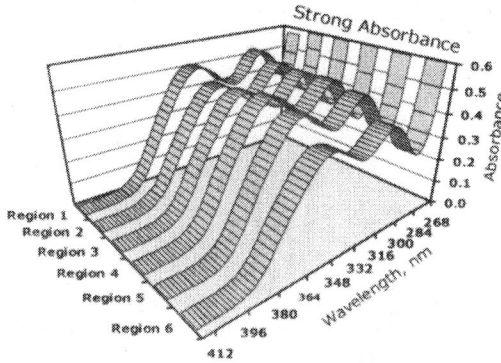

Figure 1. UV-VIS UVA (benzotriazole type) CC layer depth profile obtained by analysis of a ~5 μm thick cross-section of the unexposed multilayered coating system. Region 1 is uppermost (@ surface). Region 6 is last CC section prior to BC. This system is unexposed. The UV-VIS microscope with X-Y translating stage permitted viewing of ~6 consecutive areas across the CC layer. Fluorolube® was used to anchor the cross-section to the stage. The XYZ plot orientation was selected to highlight uniformity throughout the clearcoat.

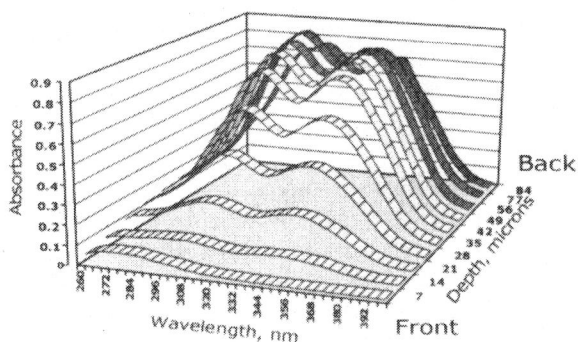

Figure 2. UV-VIS UVA (benzotriazole type) CC layer depth profile of an outdoor exposed coating system obtained by analysis of slab microtomed sections that were extracted by methylene chloride. Section 1 is uppermost (@ surface). Section 6 is last CC section prior to BC. Each section was ~7μm thick. The UVA content was normalized to the weight of each section obtained prior to solvent extraction. The XYZ plot orientation was selected to highlight the lower UVA content in the uppermost (surface) layers.

IR Analysis • POI as a Function of Depth

Photo-oxidation index (POI) monitoring as a function of exposure time and conditions has been done routinely by Ford and its CC suppliers to help determine probable service life. Several kinds of IR analyses have been used to obtain mid-IR range spectra, including transmission-, ATR-, DRIFT- and PAS-mode experiments [20-22]. The most recent development in this area has been to combine sampling by slab microtomy with ATR-mode analysis for convenient and rapid analysis. This adds another dimension to the characterization of the automotive paint/coating systems. Figure (3) shows POI depth profiles from a study comparing two identical CC/BC systems, except for UV-absorber (UVA) fortification. In system #1 there was no fortification (UVA or HALS). In case #2 there was fortification, 1.0% UVA and 1.0% HALS.

Exposure Time] x [Exposure Conditions] x [Coating System Depth]

Experiments such as this give the investigator information as to the extent of change at a given (depth) locus. Now it is possible to determine if chemical changes are mostly surface/near-surface specific, CC/BC interface specific, or relatively uniform throughout a given paint/coating system layer. Knowledge of the changes and the associated kinetics will help product designers create a more robust (i.e., durable) system. Studies involving acrylic/melamine, acrylic/silane, or acrylic/urethane CC systems have typically shown fastest changes at/near to the surface. The primary mechanisms involved in these changes are photo-oxidation and hydrolysis. In cases such as this it may be possible to select UV-absorbers that stratify in the CC providing most of the fortification at/near the surface.

DSC Analysis • Network Tg(s) as a Function of Depth

Slab microtomy sections cut at 10 μm's thick were analyzed by DSC to obtain a Tg depth profile. The study of a commercial GEN III® CC system involved a comparison of an unexposed panel and one exposed to 3 years in Hialeah, FL. The unexposed CC showed a uniform Tg throughout, while the exposed CC showed a significant (~20%) elevation in Tg for the uppermost section. Of particular note was the observation that the Tg for the bulk of the CC was effectively the same, regardless of the outdoor exposure. The same CC system was also exposed in Europe (Wuppertal, Germany) for 4 years, Mexico (Mexico City) for 3 years, and Asia (Okinawa, Japan) for 1 year. In all cases the Tg was relatively uniform below the uppermost (10 μm) section. In all cases the Tg of the uppermost section was somewhat elevated over deeper sections (~18% in Europe, ~8% in Mexico, and ~12% in Asia). Table II provides an overview.

Table I. Chemical Depth Profiling using Slab Microtomed Sections

- Direct Analysis of Sections:
  - UV-VIS (Transmission-mode)
  - IR (Transmission-mode)
  - IR (ATR-mode)
  - IR (PAS-mode)
  - ToF-SIMS
  - ESCA (XPS)
  - DSC
- Solvent Extraction of Sections:
  - GC
  - HPLC
  - UV-VIS
  - K^+IDS MS
- Acid/Microwave Digest of Sections:
  - AA
  - ICP

Table II. DSC Tg Depth Profiles for Automotive Clearcoat

Exposure Location	Exposure Time Years	Tg Elevation Top 10 Microns
Virgin CC	0	Uniform
Hialeah, Florida	3	~20% Elevation
Wuppertal, Germany	4	~18% Elevation
Mexico City, Mexico	3	~8% Elevation
Okinawa, Japan	1	~12% Elevation

NOTE: % Change in Degrees Centigrade

HPLC Analysis • Interlayer Mixing as a Function of Depth

Slab microtomy sections Cut at ~5 μm's thick were analyzed by using methylene chloride to extract (~95-97% in acrylo-melamine CC/BC systems) the UVA, followed by HPLC analysis to monitor the Tinuvin 900 content. Methylene chloride was used in this case because it tends to be very efficient for extraction of free-add UVA compounds from acrylic/melamine type binder systems. The extraction efficiency in this study was found to be ~96%-98% of expected loading levels. The BC was not fortified with Tinuvin 900 (or any other UVA species). The HPLC UVA analysis results were normalized to the weight of each slab section. Sections used were ~0.7"x0.7" in area and cut at room temperature (permitting cutting to 5 μm's for these unexposed samples).

Figure (4) shows the UVA depth profile for the CC/BC system described above, comparing the wet-on-wet with the wet-on-dry applications. There appears to be significant mixing of UVA from the CC into the BC in the case of wet-on-wet application, and considerably less in the case of wet-on-dry application. This is not unexpected. However, this study helps document the extent of mixing for such materials under these application conditions. UVA mixing is confirmed to ~10-15 μm's deep into the BC in the wet-on-wet case, while only ~5 μm's deep in the wet-on-dry case. Slab microtomy for these materials is limited to ~4-5 μm's per section at room temperature. This follows optimization of the cutting blade (carbide) angle and speed.

The UVA migration kinetics appears to be relatively slow compared to the rate of mixing during the application/cure timeframe. This was concluded as due to the UVA distributions observed in the two cases after ~2 weeks storage. Table III shows the very similar UVA distribution before and after this storage period. The belief, however, is that migration does occur, but is a function of the network chemistry, crosslink density, storage conditions and any exposure conditions.

GC Analysis • Solvent Trapping as a Function of Depth

Depth profiles for trapped solvents were determined using a combination of slab microtome sectioning, methylene chloride extraction, and GC analysis to monitor the trapped solvent content. Experimental automotive coating systems were prepared with co-solvent blends. Mixtures of methyl amyl ketone (MAK) and methyl ethyl ketone (MEK) were used in most of these studies. When freshly prepared CC/BC automotive systems were prepared followed by slab microtomy co-planar to the surface at ~10 μm thick sections and then followed

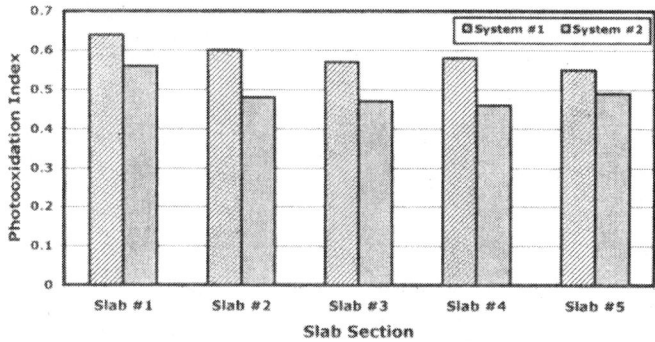

Figure 3. IR photooxidation index (POI) analysis of two automotive coating systems with identical UV fortification in the CC layer, without UV fortification in the System #1 BC, and with UV fortification in the System #2 BC. Each slab microtomed section is 10 μm thick. IR ATR-mode analysis was used on the topside of each section. Note slab #1 is the uppermost 10 μm section. The substrate is TPO in both cases. System #1 has CC-UVA ~3%, CC-HALS ~1.5%, BC-UVA 0%, and BC-HALS 0%. System #2 has CC-UVA ~3%, CC-HALS 1.5%, BC-UVA 1%, and BC-HALS 1%. The UVA is Tinuvin 400 (Ciba) and HALS is CGL-052 (Ciba).

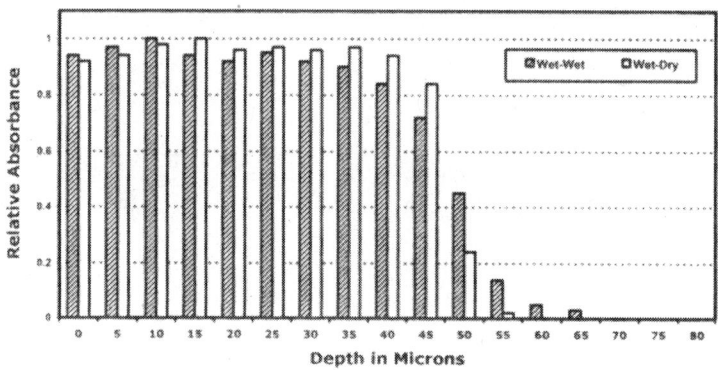

Figure 4. HPLC UVA depth profiles of Tinuvin 900 in both unexposed wet-on-wet and wet-on-dry application for the solventborne acrylo-melamine CC/BC system. The CC/BC interface is ~50 μm below the CC surface in both cases. The panels were stored at ~25°C, ~50 RH, and in the dark for two weeks prior to slab microtomy, methylene chloride extraction and HPLC analysis.

Table III. HPLC Tinuvin 900 UVA Monitoring CC/BC System

Depth Locus Microns	R.A. @ Initial Wet-on-Dry	R.A. @ 2 Weeks Wet-on-Dry
0 to 5	94	94
5 to 10	97	96
10 to 15	98	98
15 to 20	100	98
20 to 25	96	100
25 to 30	97	96
30 to 35	96	96
35 to 40	97	95
40 to 45	95	92
45 to 50	85	86
50 to 55	25	28
55 to 60	3	5
60 to 65	0	0
65 to 70	0	0

System: Solventborne Acrylo-Melamine CC/BC
Storage: ~25C, ~50 RH, Dark, Methylene Chloride Extraction

immediately by solvent extraction, both MAK and MEK could be observed. When these same systems were stored for a period of ~1 month (~25°C, 50% RH, dark), no MEK was observed by the same GC analysis, but MAK was found. This is probably a function of MEK being a more volatile (faster) solvent and a smaller (more compact) molecule than MAK, allowing the MEK molecules to migrate through the network and to evaporate more efficiently.

During ~1 month of storage a slightly sigmoid-like MAK depth profile developed. Note than the ~1"xl" area that was sectioned required almost immediate introduction of each section into a 5 mL aliquot of methylene chloride and then quickly capped. The concern was that the 10 μm thick sections would allow the remaining solvent to quickly evaporate. The weight used for normalizing the GC signal for MAK was determined after the extraction step. The extracted slab sections were dried in an oven for 1 hour at 100°C. Additional studies are in progress to explore these phenomena in greater detail and to compare to commercial coating systems.

ToF-SIMS Analysis • Pigment Distribution as a Function of Depth

Cross-section analysis by ToF-SIMS has been previously reported as quite successful in characterization of photo-oxidation and chlorine content as a function of depth in a multilayered automotive paint/coating system [3,6] The gallium (Ga) ion gun was used and allowed a sampling spot size ~2 μm's diameter. This permitted a resolution of ~2 μm's in full step mode and ~1 μm's in half step mode using the X-Y translating stage. Depth profiles were obtained using line scans or by rastering (to get area maps).

A variation on this approach was explored using slab sectioning of a full automotive paint/coatings system co-planar to the surface. The resolution in this work is a function of the section thickness. For practical handling purposes the limitation is ~4-5 μm's per section for unexposed sections and ~5-7 μm's per section for typical exposed sections. Although the study monitored chlorine content per se, it was correlated to the chlorinated copper phthalocyanine pigment content. However, one must be careful in doing quantitation in this manner. The efficiency of the ToE-SIMS experiment is a function of the binder/network surrounding the pigment particles in the BC layer. Also, one must be cautious in comparing chlorine content between coating layers, since the efficiency is now due to different sources of chlorine and different binder/network composition.

ESCA Analysis • C'atalyst Distribution as a Function of Depth

Slab microtomy sections were analyzed directly by ESCA at the topside of each layer. Since ESCA analysis is highly surface specific, the section thickness (usually ranging from ~5-10 µm) is not an issue. This technique is valuable for the study of surface layers to a depth of ~100 Angstroms. The technique allows identification of elements (except hydrogen) and their associated covalent binding or complexation. As a result it is often possible to determine the types of chemical functionality inherent at that locus. Note ESCA is also commonly identified as X-ray photoelectron spectroscopy (or XPS).

Figure (5) shows a tin catalyst (dibutyl tin dilaurate) depth profile in a model CC/BC system. The initial catalyst loading in the CC formulation was ~1% w/w. The CC thickness used was ~55 µm, with BC below. All ESCA analyses were performed from the topside of each ~5µm thick section. Note, the ESCA sampling depth herein was somewhat <100 Angstroms and (thus) very surface specific to the top of each section. The reported study shows a significant elevation of tin catalyst for the top most section and evidence for mixing of CC and BC layers during wet-on-wet application prior to cure and effectively no mixing after wet-on-dry application.

AA Elemental Analysis:

Atomic Absorption (AA) and Inductively Coupled Plasma (ICP) analyses have been successfully conducted to obtain depth profiles based on one (usually) or more elements. Common applications in the study of automotive coating systems have been to monitor silicon (Si) when used as a binder crosslinking species, tin (Sn) when organo-silicon catalysts are used in clearcoats, and copper (Cu) when pigments such as copper phthalocyanine are used in basecoats (as a colorant). Certainly, numerous other elements could be monitored as a function of depth. The key in success of this approach was to obtain accurate weights of each slab microtomed section and to fully digest sections prior to AA or ICP analyses.

Open hot plate acid digestion of slab microtomed sections was found to be problematic if the element of interest (eg. organo-tin and organo-silicon species) is volatile or associated with a volatile complex. Microwave digestion using a closed "bomb" type sample container was determined to be a much better choice to avoid loss through volatility.

K$^+$IDS MS Analysis • Degradation Products as a Function of Depth

Sampling has been done by combining slab microtomy sectioning with solvent (tetrahydrofuran, methylene chloride, or xylene) extraction. Commonly,

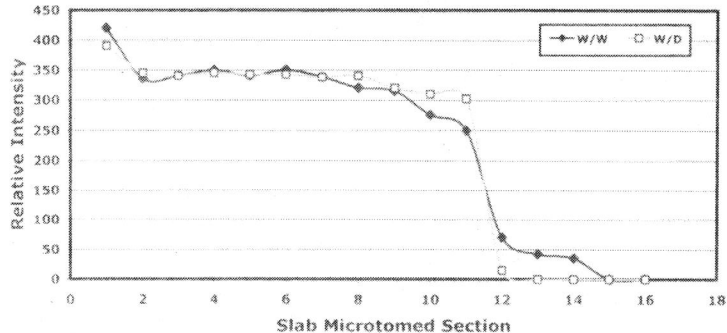

Figure 5. ESCA determined elemental depth profiles of two CC/BC systems containing a tin catalyst, dibutyl tin dilaurate. One CC/BC system was applied wet-on-dry and the other wet-on-wet. Each slab microtomed section was ~5 μm thick and cut at ~25°C. All analyses were done from the topside of each section.

sections used ranged from ~0.7" x 0.7" x 8 μm thick to ~1" x 1" x 10 μm thick. Each section was extracted into 5 mL organic solvent over a period of at least 24 hours. Then each extract solution was evaporated down to ~0.5 mL prior to use in the K^+IDS MS analysis.

Preliminary results indicate that some of the degradation species are related to the melamine type cross-linkers used in the acrylic/melamine formulations. The core of the melamine structures is a triazine ring. However, a variety of masses are observed that correspond to a triazine ring with different kinds of substructures attached to the triazine core. Further studies of coating systems subjected to varying periods of outdoor (Hialeah, FL) and accelerated (Xenon boro/boro; QUV 340) conditions are in progress.

Conclusions

A toolbox of sampling techniques and measurement technologies has been described to allow locus specific chemical analysis or chemical depth profiling of high performance multi-layer automotive coating systems. The toolbox can be divided into one basic depth sectioning technique and three measurement technology categories. A sliding (slab) microtome is used to cut a coating layer or system into sections in the 5-10 micron thickness range. The first category of testing allows for direct analysis of each section's chemistry. IR, Raman, LDMS, ToF-SIMS, ESCA, and thermal (DSC/TGA) analyses have proven useful in locus specific bulk studies. The second category of testing involves a prior solvent extraction (and concentration) step to isolate the species of interest.

UV-VIS, HPLC and KIDS MS analyses have proven useful in depth profiling of solvent extractable species (eg. UVAs, trapped solvents, degradation species). The third category of testing involves a prior acid and/or microwave digest step to reduce the sample section into its component elements and ions. AA and ICP analyses have proven useful in elemental depth profiling of species.

Various combinations of these tools have been used for the purpose of service life prediction. UVA permanence and effectiveness can be monitored directly (through solvent extraction and UVA species analysis) or indirectly (by tracking chemical degradation of the coating system network). If these studies are done as a function of depth and exposure type/time, then the product investigator will have a reasonable level of confidence on the expected service life.

Acknowledgements

Credit for the success of this work belongs to many of my DuPont and non-DuPont collaborators. My principle DuPont collaborators include Gregory Blackman (CR&D), Bruce Henderson (DPC), Li Lin (DPC), Robert Matheson (DPC), Kathy Lloyd (CR&D), William Simonsick Jr. (DPC), Kate Stika (CR&D), Dennis Swartzfager (CR&D), Dennis Walls (CR&D) and Phil Yaneff (DPC). My principle non-DuPont collaborators include David Bauers (Ford), Roscoe Carter (Ford), Nancy Cliff (Ciba), John Gerlock (Ford), Gottfried Haacke (Cytec), Mouhcine Kanouni, Mark Nichols (Ford), and Cindy Peters (Ford).

References

1. Andrawes, F.; Valcarcel, T.; Haacke, G; Brinen, J.; Depth distribution of light stabilizers in coatings analyzed by supercritical fluid extraction gas – chromatography and time-of-flight secondary ion mass spectrometry; *Anal. Chem.*, **1998**, *70(18)*, 3762-3765.
2. Haacke, G; Depth analysis of automotive coatings; PMSE-195, Book of Abstracts, 215-th ACS National Meeting, Dallas, TX., March 29-April 2, 1998.
3. Adamsons, K.; Chemical depth profiling of multi-layer automotive coating systems. *Prog. Org. Coat*, **2002**, *45(2-3)*, 69-81.
4. Adamsons, K.; Chemical depth profiling of automotive coating systems using slab microtome sectioning with IR/UV-VIS spectroscopy and optical microscopy. *J. Coat. Tech.*, **2002**, *74(924)*, 47-54.
5. Urban, M.W.; Allison, C.L.; Johnson, G.L.; Di Stefano, F.; Stratification of butyl acrylate/polyurethane (BA/PUR) latexes: ATR and step-scan photoacoustic studies, *Appl. Spec.*, **1999**, *53(12)*, 1520-1527.

6. Adamsons, K.; Litty, L.; Lloyd, K.; Stika, K.; Swartzfager, D.; Walls, D.; Wood, B.; Depth profiling of automotive coating systems on the micrometer scale, ACS Symposium Series, **1999**, *722*, 257-287.
7. Adamsons, K.; Chemical depth profiling of automotive coating systems using IR, UV-VIS and HPLC methods. ACS Symposium Series, **2002**, *805*, 185-211.
8. Arai, T.; Tomimasu, H.,; Measuring method of chemical distribution in coated layers by ESCA, *Kami Pa Gikyoshi*, **1989**, *43(2)*, 213-225.
9. English, AD.; Spinelli, H.J.; Degradation chemistry of primary crosslinks in high-solids enamel finishes. Solar-assisted hydrolysis.; ACS Symposium Series, **1984**, *243*, 257-269.
10. Haacke, G.; Observation of glass transition temperature and crosslink density gradients in acrylic/melamine coatings, relaxation phenomenon in polymers; Proceedings for Relaxation Phenomenon in Polymers Symposium, Dearborn, MI., Nov. 8, **1994**,167-173.
11. Haacke, G.; Andrawes, F.F.; Campbell, B.H.; Migration of light stabilizers in acrylic/melamine clearcoats. *J. Coat. Tech.*, **1996**, *68(855)*, 57-62.
12. Haacke, G.; Brinen, J.S.; Depth dependence of the glass transition temperature and crosslink density in two-layer acrylic/melamine coatings. FATIPEC Congress, 22-nd, **1994**, Vol.3, 182-191.
13. Haacke, G.; Brinen, J.S.; Larkin, P.J.; Depth profiling of acrylic/melamine formaldehyde coatings. *J. Coat. Tech.*, **1995**, *67(843)*, 29-34.
14. Vyoerykkae, J.; Paaso, J.; Tenhunen, M.; Tenhunen, J.; Litti, H.; Vuorinen, T.; Stenius, P.; Analysis of depth profiling data obtaining by confocal Raman microspectroscopy. *Appl. Spec.*, **2003**, *57(9)*, 1123-1128.
15. Scherzer, T.; Depth profiling of the degree of cure during the photopolymerization of acrylates studied by real-time FT-IR attenuated total reflection spectroscopy. *Appl. Spec.*, **2002**, *56(11)*, 1403-1412.
16. Scherzer, T.; Depth profiling of the conversion during the photopolymerization of acrylates using real-time FTIR-ATR spectroscopy. *Vibr. Spec.*, **2002**, *29(1-2)*, 139-145.
17. Cliff, N.; Kanouni, M.; Adamsons, K.; Yaneff, P.V.; Migration of reactable UVAs and HALS in automotive plastic coatings and their impact on durability. Proceedings of the International Waterborne, High-Solids, and Powder Coatings Symposium, 30-th, **2003**, 29-46.
18. Cliff, N.; Kanouni, M.; Yaneff, P.V.; Adamsons, K.; Migration behavior of light stabilizers in automotive coatings for plastic substrates. *Polym. Matrs. Sci. &Eng.*, **2003**, *89*, 173.
19. Simonsick, Jr., W.J.; Organic coatings characterization by potassium ionization of desorbed species. *Prog. Org. Coat.*; **1992**, *20(3-4)*, 411-423.
20. Gerlock, J.L.; Smith, C.A.; Cooper, V.A.; Dusbiber, T.G.; Weber, W.H.; On the use of Fourier transform infrared spectroscopy and ultraviolet spectroscopy to assess the weathering performance of isolated clearcoats from different chemical families; Polym. Degrad. Stab. **1998**; *62(2)*; 225.

21. Gerlock, J.L.; Smith, C.A.; Carter III R.O.; Dearth, M.A.; Korniski, T.J.; Kaberline, T.J.; DeVries, J.E.; Cooper, V.A.; DuPuie, J.L.; Dusbiber, T.G.; Paint weathering tests: transition from art to science; Surf. Coat. Aust. **1997**, *34(7)*; 14-16.
22. Bauer, D.R.; Paputa Peck M.C.; Carter III R.O.; Evaluation of accelerated weathering tests for a polyester-urethane coating using photoacoustic infrared spectroscopy; J. Coat. Technol.; **1987**; *59(755)*; 103-109.

Chapter 19

Evaluation of the Protective Properties of Novel Chromate-Free Polymer Coatings Using Electrochemical Impedance Spectroscopy

E. Kus[1], M. Grunlan[2], W. P. Weber[2], N. Anderson[3], C. Webber[3], J. D. Stenger-Smith[3], P. Zarras[3], and F. Mansfeld[1,*]

[1]Corrosion and Environmental Effects Laboratory (CEEL), The Mork Family Department of Chemical Engineering and Materials Science, Department of Materials Science and Engineering, University of Southern California, Los Angeles, CA 90089–0241
[2]Department of Chemistry, University of Southern California, Los Angeles, CA 90089–1661
[3]Polymer Science and Engineering Branch (Code 498200D), Naval Air Warfare Center Weapons Division, Department of the Navy, 1900 North Knox Road (Stop 6303), China Lake, CA 93555–6106

The protective properties of different types of novel chromate-free polymer coatings have been evaluated by means of electrochemical impedance spectroscopy (EIS). The first set of samples included steel panels coated with pentasiloxanes where different crosslinking/cure chemistries were compared for their effect on the protective properties of the coatings. The second set contained Al 2024-T3 coated with a poly(2,5-bis(N-methyl-N-hexylamino)phenylene vinylene (BAM-PPV) based powder coating. Samples were exposed to 0.5 N NaCl and EIS data were recorded as a function of exposure time. The EIS data suggest that the nonfluorinated pentosiloxane provided better corrosion protection than the two fluorinated pentasiloxanes and that room temperature UV cure with a photo-acid catalyst provided better corrosion protection than phenol-catalyzed thermal crosslinking. The sample with the BAMPPV coating showed very good corrosion resistance during exposure for 50 days. Delamination properties were examined by means of scribing the sample after a certain

© 2007 American Chemical Society

period of exposure and evaluating the resulting changes in the impedance spectra. The delaminated area was observed with SEM at the end of the test.

Introduction

EIS has been used in the evaluation of protective properties of different chromate-free polymer coatings in this study. The first set of samples was developed as non-toxic, environmentally friendly polymer coatings that minimize biofouling and protect steel from corrosion. Since siloxanes and fluorosilicones have been shown to be effective non-toxic foul release materials, siloxanes that contain pendant perfluoroalkyl groups have been prepared and the corrosion protection properties of films that were UV cured with photo-acid catalyst have been studied using EIS (*1*). Similar fluorinated siloxanes have also been converted to solid films by thermal cure with α,ω-diaminoalkanes (*2*).

The second set of samples, BAM-PPV based powder coatings, were developed as an alternative to high volatile organic compounds (VOCs) based spray applications. BAM-PPV has been demonstrated to provide corrosion prevention when applied onto Al 2024-T3 substrates using xylene solutions. These BAM-PPV coatings were shown in neutral salt fog experiments to inhibit corrosion for 336 hours. A minimum of 336 hours is a military requirement for alternatives pretreatment coatings. The BAM-PPV pretreatment coating is a potential alternative to the current military chromate conversion coatings (CCCs) (*3*). Since hexavalent chromium has been identified as a health threat and is highly regulated due to its toxicity, these polymers were considered to reduce these hazardous materials, eliminate chromium (VI) from coating formulations and facilitate compliance with new environmental VOC regulations.

The aim of this study is to help assess the long-term effectiveness of the new novel chromate-free and zero-VOC polymer coatings and to provide an improved understanding of the mechanisms of corrosion protection. Other properties of these new types of coatings such as resistance to biofouling are addressed in other parts of this project. EIS is a non-destructive technique that allows monitoring of changes in the coating properties and corrosion rates of the substrate at locations where the coating has delaminated. The impedance spectra were analyzed using appropriate equivalent circuits (EC) (*4 – 7*). The resulting fit parameters can be related to the properties of the coating and the metal surface and their changes with exposure time to a corrosive solution.

Experimental Approach and Data Analysis

The corrosion behavior of the coated samples was evaluated in 0.5 N NaCl solution (open to air) by EIS. A Gamry PCI4/300 potentiostat and Gamry EIS300 Software were used for the impedance measurements. The spectra were recorded in the frequency range of 100 kHz – 5mHz at the corrosion potential E_{corr} or an applied potential. The samples were exposed to the test solution for several weeks and impedance spectra were evaluated as a function of exposure time. The impedance spectra have been plotted as Bode plots, where the logarithm of the impedance modulus, |Z|, and the phase angle Φ are shown as functions of the logarithm of the frequency f of the applied signal.

The first set of samples (pentasiloxanes) was supplied by the USC Department of Chemistry. Two fluorinated pentasiloxanes (A, where $R_f = CH_2CH_2CF_3$ and B, where $R_f = CH_2CH_2(CF_2)_5CF_3$) and a nonfluorinated pentasiloxane (C, where $R = CH_3$) have been prepared as previously reported (1, 2). Depending on the terminal epoxy group type, two different crosslinking\cure chemistries were utilized. The first type of crosslinking reaction involved room temperature UV cure with a photo-acid catalyst ("A", "B", "C") (1). The second crosslinking reaction was the phenol-catalyzed thermal crosslinking of A or C with α,ω-diaminoalkanes [NH_2-$(CH_2)_n$-NH_2, where n = 6 ("A6" or "C6"), 8 ("A8" or "C8"), or 12 ("A12" or "C12")] (2).

Impedance spectra for the pentasiloxanes coated samples were obtained at -0.6 V since a stable E_{corr}, could not be determined due to the high ohmic resistance of the more or less pore-free coatings. The amplitude of the applied ac signal was 20 mV. The spectra were analyzed using the COATFIT module of the ANALEIS software (5) or the open boundary finite length diffusion (OFLD) model (6).

The second set of samples (BAM-PPV) was supplied by the Naval Air Warfare Center Weapons Division. The BAM-PPV (1 wt %) was dispersed into a commercial triglycidyl isocyanurate (TGIC) polyester resin using an acoustic blending technique with an average particle size ranging from 45-57 microns. The resins were coated onto Al 2024-T3 panels (0.032" x 3"x 6", Q Panel Lab Products) via electrostatic coating (e-coat). The spectra for the BAMPPV coated Al 2024 samples were obtained at -0.8 V vs. SCE and an ac signal amplitude of 20 mV was applied. The samples were scribed after a certain exposure time for testing of coating delamination from the defects. Scribed surfaces were examined by means of scanning electron microscopy (SEM) at the end of the exposure period.

Results and Discussion

Set I. Pentasiloxanes Coatings on Steel

Samples with UV cured coatings A and B were exposed for 70 days and the sample with coating C were exposed for 35 days. Figure 1 shows the impedance spectra for these samples after 7, 21 and 35 days exposure. In the spectra for sample A the frequency dependence of the phase angle at the lowest frequencies indicates that the spectra agree with the OFLD model (Figure 2) that describes processes in which diffusion occurs in a porous layer (*6*). In the EC in Figure 2 C_c is the coating capacitance, R_p is the polarization resistance and OFLD is the term that models the diffusion impedance Z_{OFLD} that in Boukamp's notation is given by:

$$Z_{OFLD} = \{\tanh B\, (j\omega)^{1/2}\} / Y_0(j\omega)^{1/2} \qquad (1)$$

$$\text{where } B = d/(D)^{1/2} \text{ and } Y_0 = (\sigma^{1/2})^{-1}$$

In Eq. 1, B is the diffusion parameter with diffusion length d and diffusion coefficient D and σ is the Warburg coefficient. For polymer coatings d is assumed to be equal to the coating thickness (*6*).

The impedance spectra for the fluorinated samples A and B and for the nonfluorinated sample C showed different changes with exposure time. The spectra for sample A agreed with the OFLD model and there was a significant decrease of the impedance values at low frequencies with time. The spectra for sample B agreed with the one-time-constant (OTC) model and showed only small changes during exposure which can be seen more clearly in the frequency dependence of the phase angle (Figure 1). The spectra for sample C showed capacitive behavior and did not change much with time for 35 days. The results of the analysis of the impedance spectra based on the discussed models are given elsewhere (*8,9*). The Warburg coefficient σ for sample A decreased with time and the fit parameter B increased with time as the delaminated area increased (6) (Figure 3).

Figure 4 and Figure 5 show the impedance spectra for samples A, A6, A8, A12 (Figure 4) as well as C, C6, C8, C12 (Figure 5) for 7, 21 and 35 days of exposure to 0.5 N NaCl. The spectra for these samples agree with the coating model (4 – 7) (Figure 6) that describes the properties of the polymer coating layer and the corrosion reactions occurring on the steel surface at delaminated

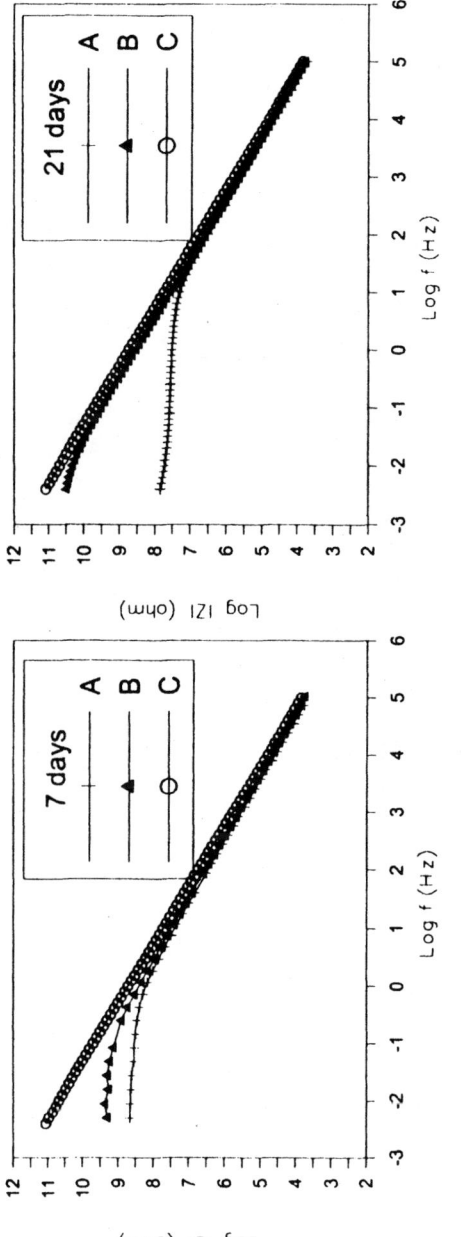

Figure 1. Impedance spectra of samples A, B, C for 7, 21 and 35 days exposure. (Reproduced with permission from reference 8. Copyright 2005 The Electrochemical Society.) Continued on next page.

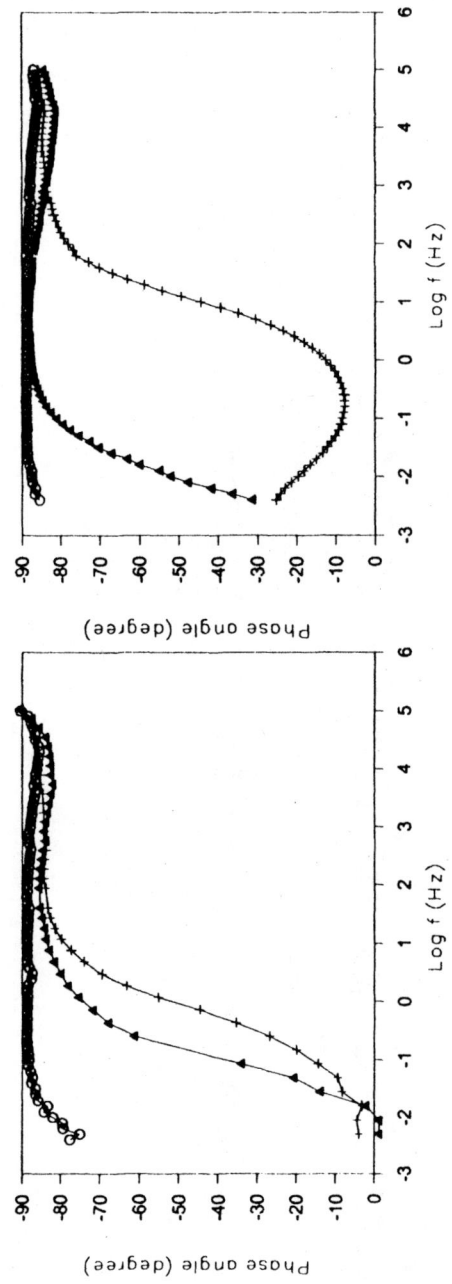

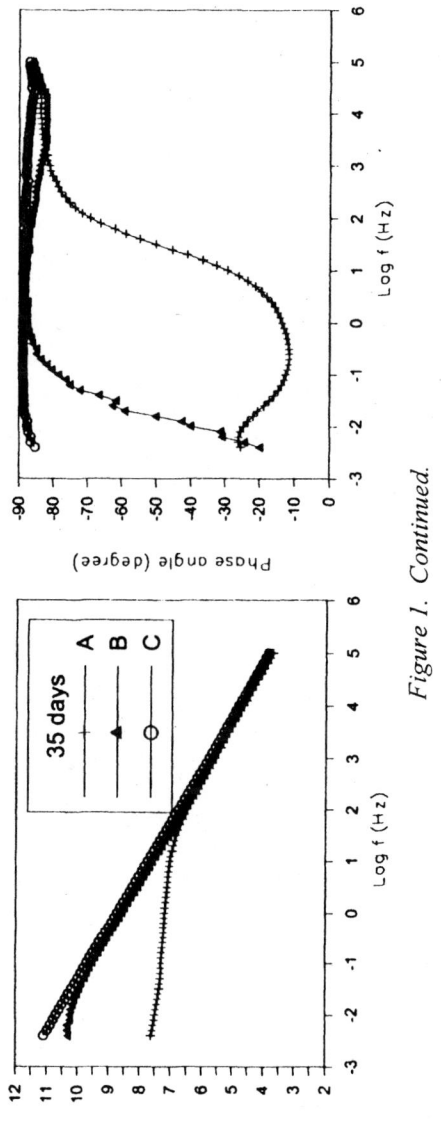

Figure 1. Continued.

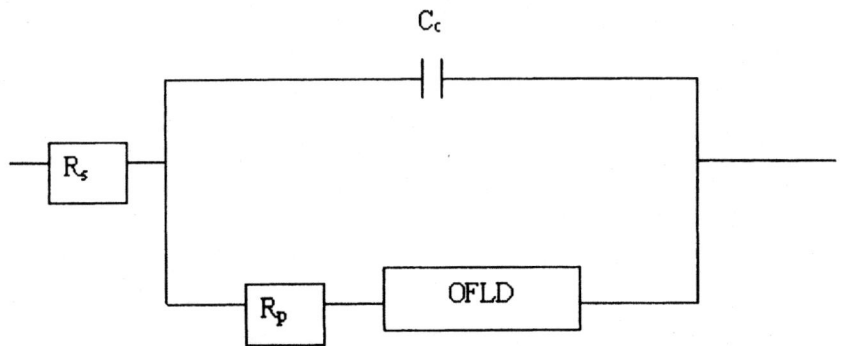

Figure 2. OFLD model (Reproduced with permission from reference 8. Copyright 2005 The Electrochemical Society.)

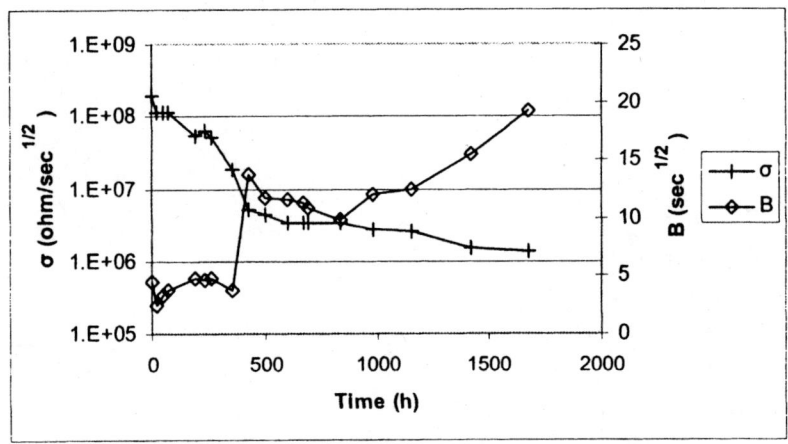

Figure 3. Time dependence of warburg coefficient σ and fit parameter B for sample A. (Reproduced with permission from reference 8. Copyright 2005 The Electrochemical Society.)

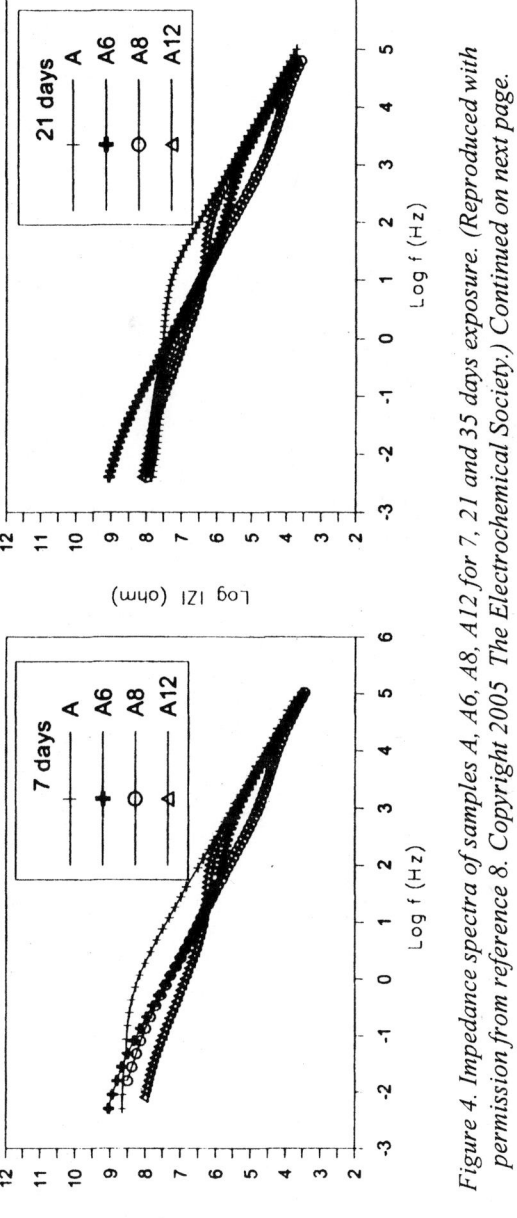

Figure 4. Impedance spectra of samples A, A6, A8, A12 for 7, 21 and 35 days exposure. (Reproduced with permission from reference 8. Copyright 2005 The Electrochemical Society.) Continued on next page.

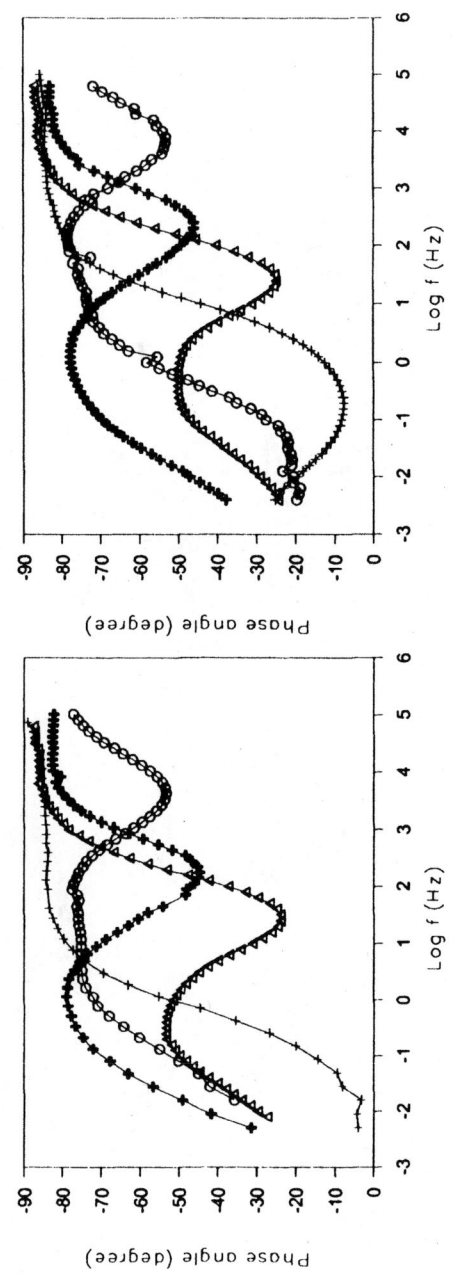

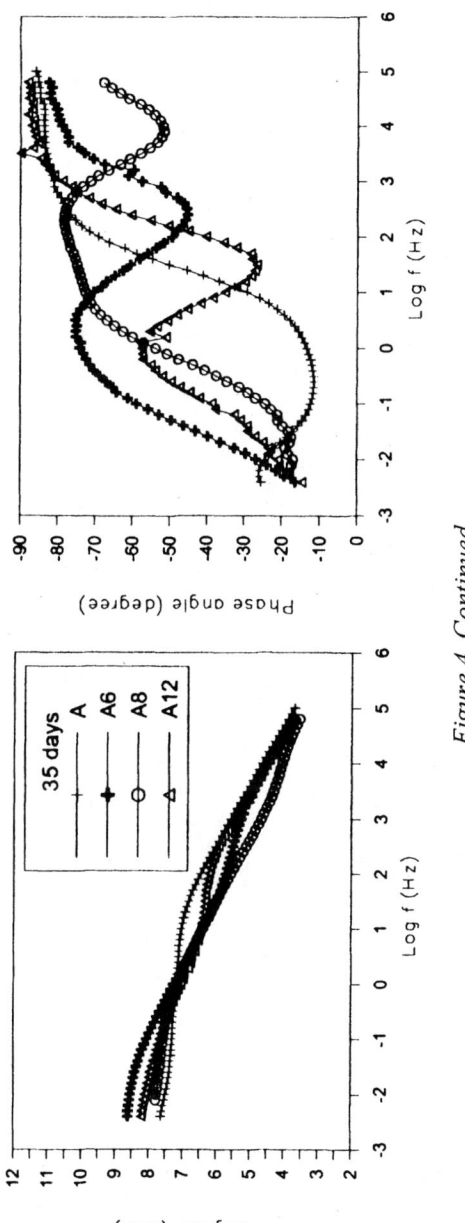

Figure 4. Continued.

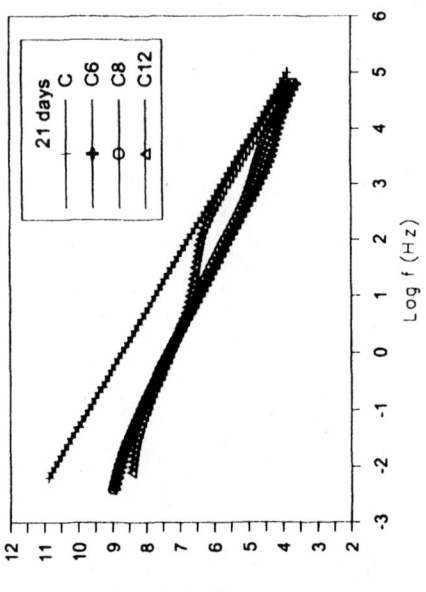
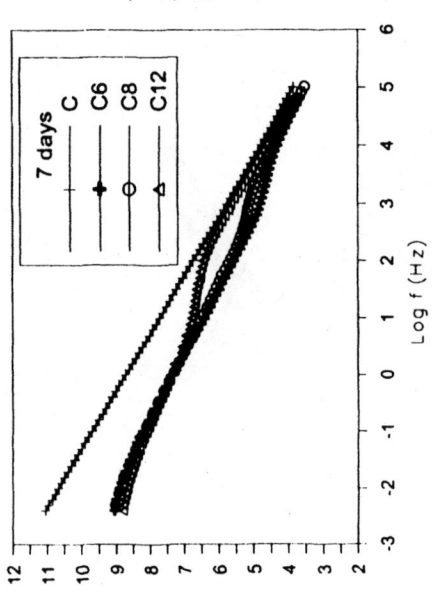

309

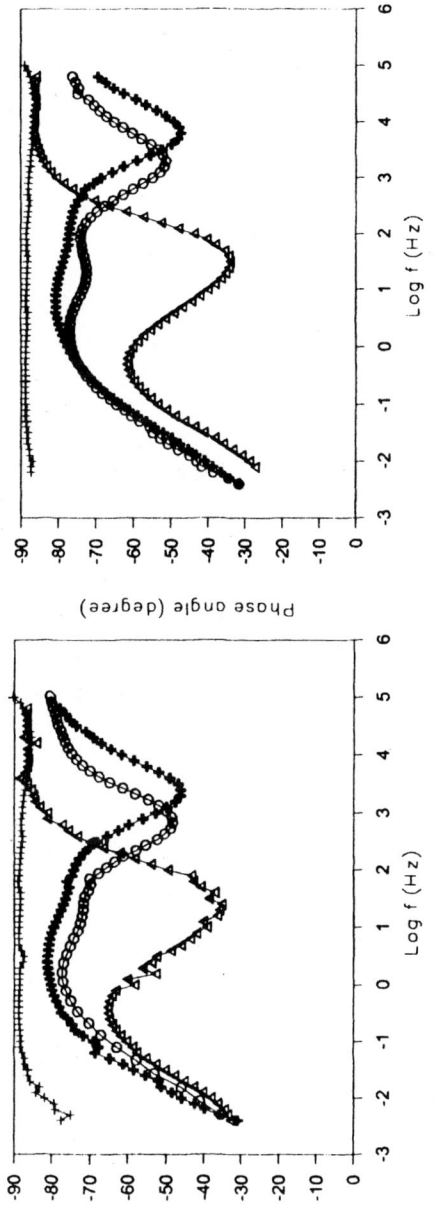

Figure 5. Impedance spectra of samples C, C6, C8, C12 for 7, 21 and 35 days exposure. (Reproduced with permission from reference 8. Copyright 2005 The Electrochemical Society.) Continued on next page.

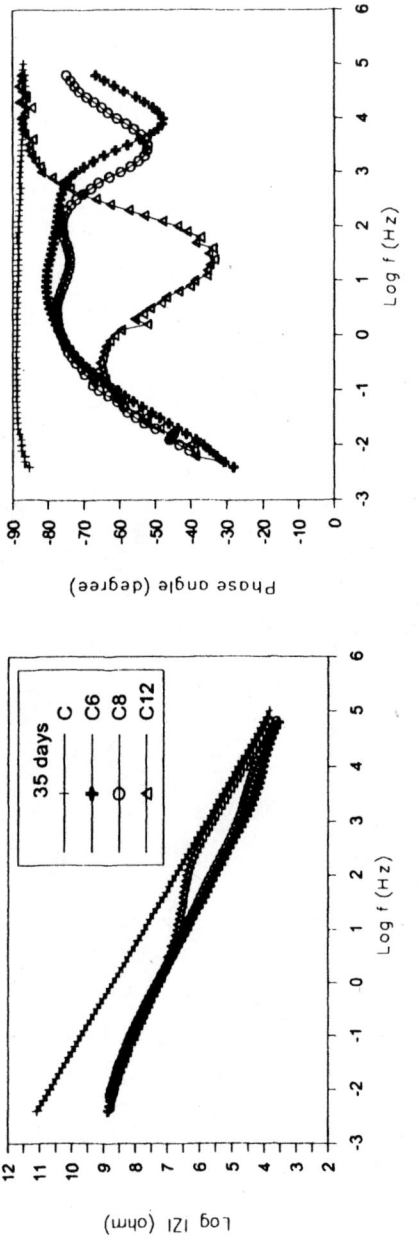

Figure 5. Continued.

areas. C_c is the coating capacitance that depends on its dielectric constant ε and d. R_{po} is the pore resistance due to the formation of ionically conducting paths in the coating. Its changes can be related with degradation of the coating as will be discussed below (4 – 7). R_p is the polarization resistance of the delaminated area at the polymer/coating interface at which corrosion occurs and C_{dl} is the corresponding capacitance. The results of the analysis are given elsewhere (8).

The changes of C_{dl}, R_p and R_{po} with exposure time can be related with the area of the steel surface at which corrosion occurs. The delamination ratio Δ can be defined as $\Delta = A_d/A_t$, where A_d is the delaminated area and A_t is the total exposed area (4–9). As coating porosity increases and A_d increases, R_p and R_{po} decrease, while C_{dl} increases. R_{po}, R_{po} and C_{dl} are related with A_d according to the following equations (7):

$$R_{po} = R°_{po}/A_d \text{ with } R°_{po} = \rho d \text{ (ohm.cm}^2\text{)} \tag{2}$$

$$R_p = R°_p/A_d \tag{3}$$

$$C_{dl} = C°_{dl} A_d \tag{4}$$

$$C_c = (\varepsilon\varepsilon_0/d)A = C°_c (A_t - A_d) \tag{5}$$

In Eq. 2, where ρ is the coating resistivity and d is the coating thickness, it has been assumed that ρ remains constant with exposure time. However as will be shown below ρ decreased with time to an extent that depends on coating formulation. Therefore Eq. 2 can be re-written as:

$$R_{po} = \rho d/A_t \Delta \tag{2a}$$

It has been suggested that the extent of delamination can be determined experimentally from the break-point frequency f_b, which is defined as the frequency at which the phase angle is - 45° in the high-frequency region (4,10). The breakpoint frequency f_b is related to Δ and ρ according to the following relationships (7):

$$f_b = (2\pi R_{po} C_c)^{-1} = f°_b A_d/A = f°_b \Delta \tag{6}$$

$$\text{with } \Delta = A_d/A = f_b/f°_b$$

$$\text{and } f°_b = (2\pi\varepsilon\varepsilon_0\rho)^{-1}$$

The f_b values for samples A and B are given in Figure 7 as a function of exposure time. There is a significant difference between the f_b values of samples A and B during the whole exposure time. While for sample B f_b remained constant at very low frequencies, it increased continuously for sample A indicating increased coating degradation. Figure 8a shows the time dependence of f_b for samples A 12 and C 12 which were the only two samples for which Φ had decreased below $-45°$ at high frequencies. The breakpoint frequencies for these two samples showed a similar initial increase, but did not change much with time after about 200 h.

In addition to f_b, the minimum of the phase angle Φ_{min} and its frequency f_{min} can be used to qualitatively evaluate the delamination process of the coatings and to determine its Δ and ρ values (7). Φ_{min} and f_{min} can be related to Δ and ρ as follows:

$$F_{min} = k_1 \Delta^{1/2}/\rho \quad \text{where} \quad k_1 = (2\pi d)^{-1}(C_c^0 C_{dl}^0)^{-1/2} \quad (7)$$

$$\tan \Phi_{min} = k_2 \Delta^{1/2} \quad \text{where} \quad k_2 = 2(C_c^0/C_{dl}^0)^{1/2} \quad (8)$$

$$f_b/f_{min} = k_3 \Delta^{1/2} \quad \text{where} \quad k_3 = (C_{dl}^0/C_c^0)^{1/2} \quad (9)$$

$$f_b/(f_{min})^2 = k_4 \rho \quad \text{where} \quad k_4 = 2\pi d C_{dl}^0 \quad (10)$$

f_b and f_{min} move to higher frequencies as coating degradation increases, while Φ_{min} changes to less negative values.

Experimental values of Φ_{min} and f_{min} are shown as a function of time in Figure 8b and 8c. For the more protective coatings A12 and C12 Φ_{min} had values close to $-30°$ which did not change much with time, while for the other four samples it changed from about $-40°$ to $-50°$ for longer exposure times (Figure 8b). For f_{min} more or less constant values were found at low frequencies for A12 and C12, while for A6 f_{min} had shifted to higher frequencies, but did not change much with time. For the other three samples f_{min} increased continuously with time (Figure 8c). While the impedance spectra in Figures 4 and 5 in general agree with model in Fig. 6, some inconsistencies are noted. For example in Figure 8b Φ_{min} became more negative with exposure time for all samples except A12 and C12 which according to Eq. 8 implies that coating damage decreased with time for these samples assuming that k_2 did not change with time. On the other hand, f_{min} increased with time for samples A8, C6 and C8 suggesting that coating damage increased with time.

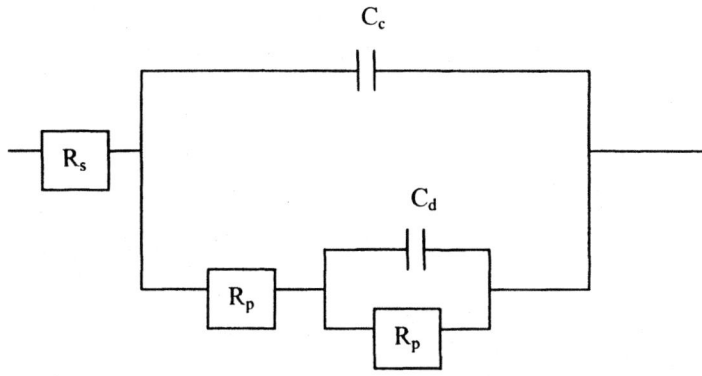

Figure 6. Equivalent circuit for coating model. (Reproduced with permission from reference 8. Copyright 2005 The Electrochemical Society.)

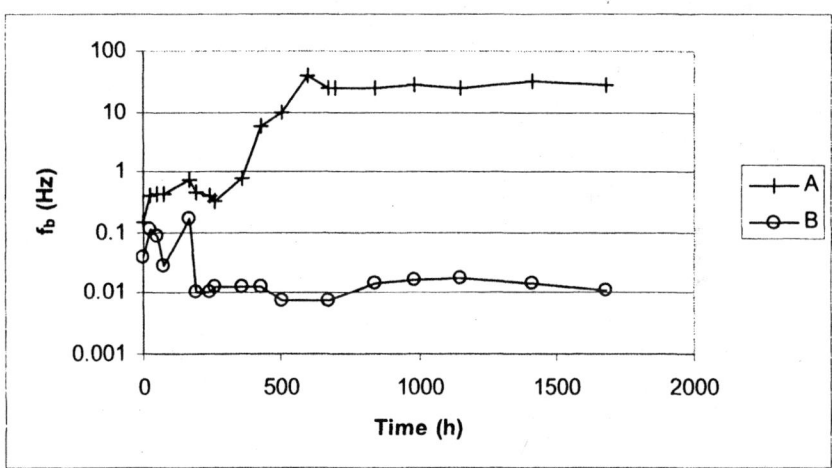

Figure 7. Time dependence of f_b for samples A and B. (Reproduced with permission from reference 8. Copyright 2005 The Electrochemical Society.)

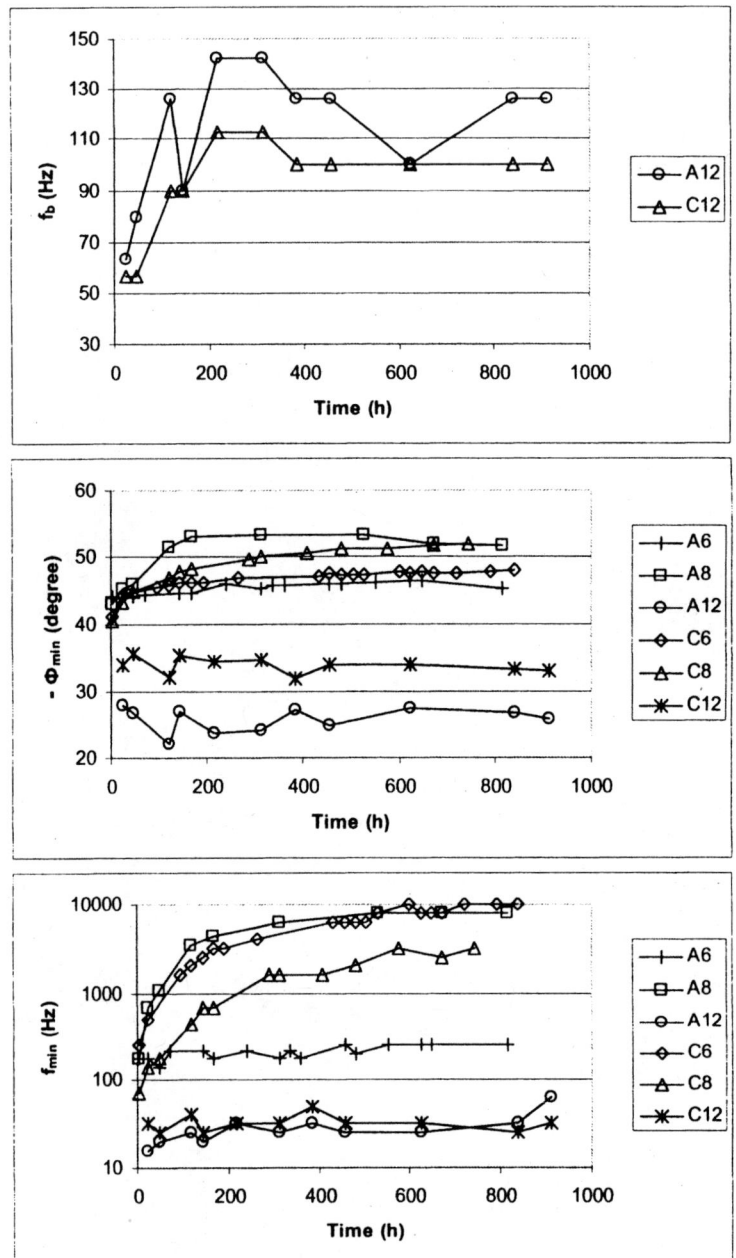

Figure 8. Time dependence of (a) f_b for A12 and C12; (b) Φ_{min} for A6, A8, A12 and C6, C8, C12 and (c) f_{min} for A6, A8, A12 and C6, C8, C12. (Reproduced with permission from reference 8. Copyright 2005 The Electrochemical Society.)

The experimental values of Δ and ρ determined according to Eq. (8) and (7) respectively are plotted as a function of time in Figures 9a and 9b. For these calculations $C°_{dl} = 20$ μF/cm^2 was used and $C°_c$ was taken as the C_c value for t = 0. The observed Δ-values in the range of 0.002 – 0.005 % for all samples except sample A12 (Figure 9a) are very low and did not show significant changes with time. This result suggests that defects that were present in the coating layers at the start of the exposure test did not become larger during exposure to 0.5 N NaCl. The almost constant high values of ρ between 100 and 1000 kohm.cm for samples A12 and C12 (Figure 9b) are typical for polymer coatings and suggest that the coating layers remained more or less pore-free during exposure. For sample A6 ρ was lower by about a factor of 10, but did not change much with time. For samples A8, C6 and C8 ρ decreased continuously to values close to 1 kohm.cm (Figure 9b).

Set II. BAM-PPV Coated Al 2024-T3

A BAM-PPV powder coated Al 2024-T3 sample was exposed to 0.5 N NaCl for 50 days. The EIS data demonstrated that the coating was very protective as evidenced by the fact that the spectra were capacitive in the entire frequency region and did not change with time (Figure 10a). Since the coating acted like an insulating capacitor, valid E_{corr} values could not be obtained, therefore the EIS data were collected at – 0.8 V vs. SCE which is the range of E_{corr} values observed in this study for bare Al 2024-T3. Very similar C_c values were obtained for the 50 days of exposure. Using ε = 4 a coating thickness of 21 μm was calculated based on the following relation between the capacitance and the thickness:

$$C_c = (\varepsilon\varepsilon_o/d) A \qquad (11)$$

Since no significant changes in the impedance spectra were observed after 50 days the sample was scribed with one small line. The spectra in Figure 10b show a decrease of |Z| by a factor of 10^6 at the lowest frequencies due to the coating damage which demonstrates the sensitivity of EIS to very small changes in sample properties. Initially the spectra were similar to those observed for pitting of Al alloys (5). After about one week an additional time constant was observed. After about three weeks an additional small scribe was applied producing the impedance spectra shown in Figure 10c. The impedance values increased with time accompanied by an increase of E_{corr} which suggests that the overall corrosion damage was decreasing with time.

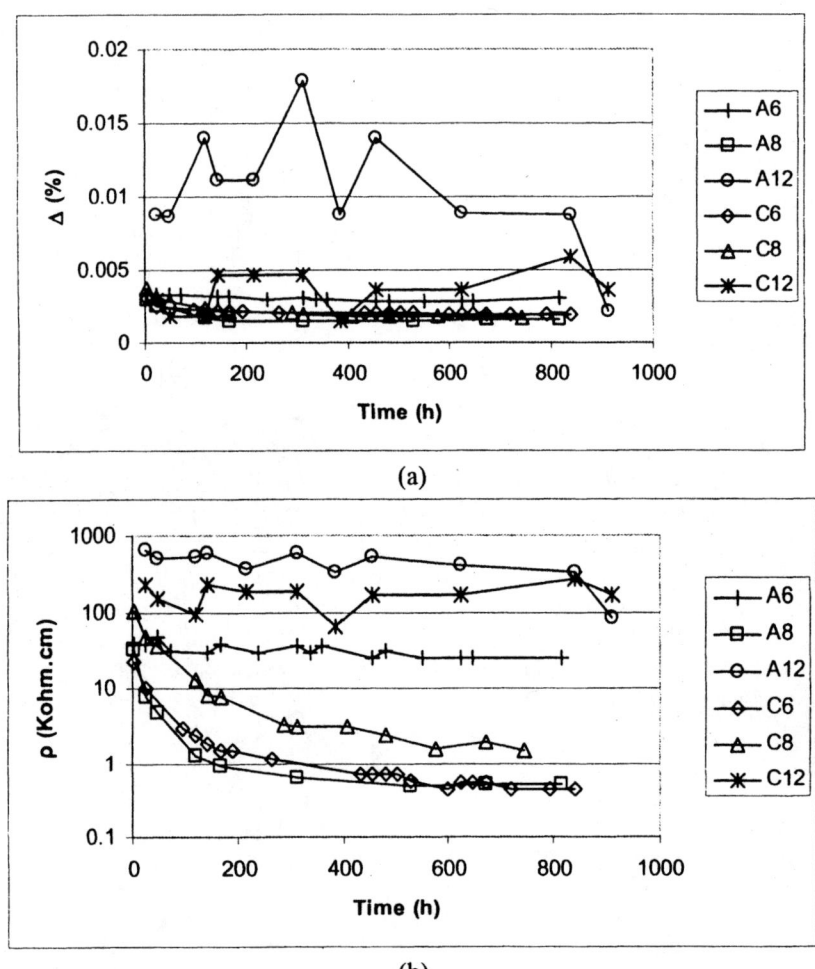

Figure 9. Time dependence of the delamination ratio Δ (a) and specific resistance ρ of the coating (b). (Reproduced with permission from reference 8. Copyright 2005 The Electrochemical Society.)

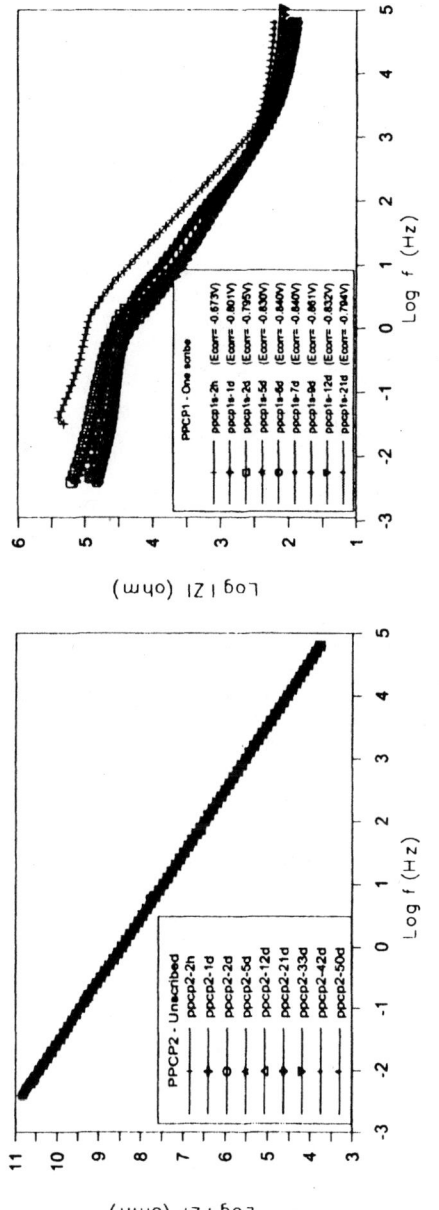

Figure 10. Impedance spectra for Al 2024 with BAMPPV coating exposed to 0.5 N NaCl; a) unscribed; b) one scribe; c) two scribes. (Reproduced with permission from reference 8. Copyright 2005 The Electrochemical Society.) Continued on next page.

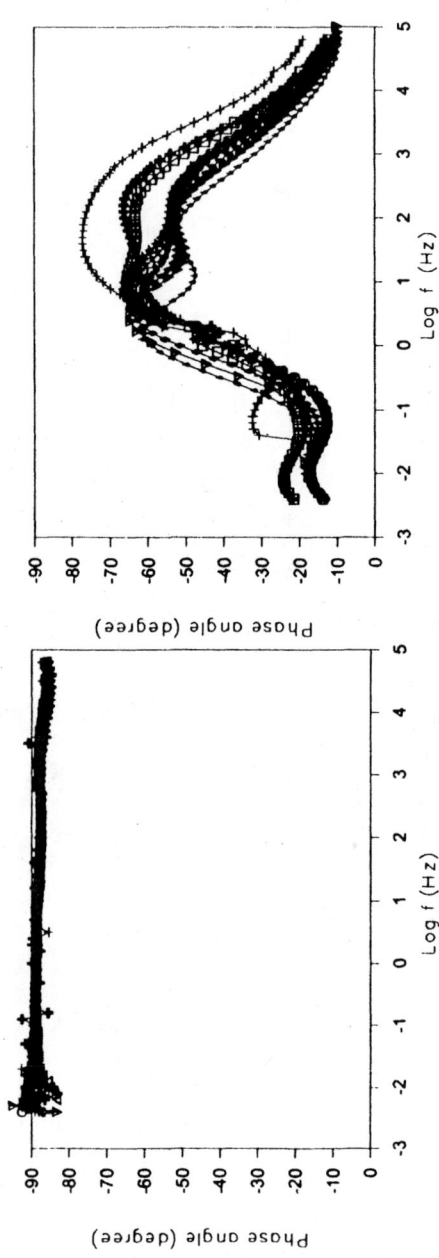

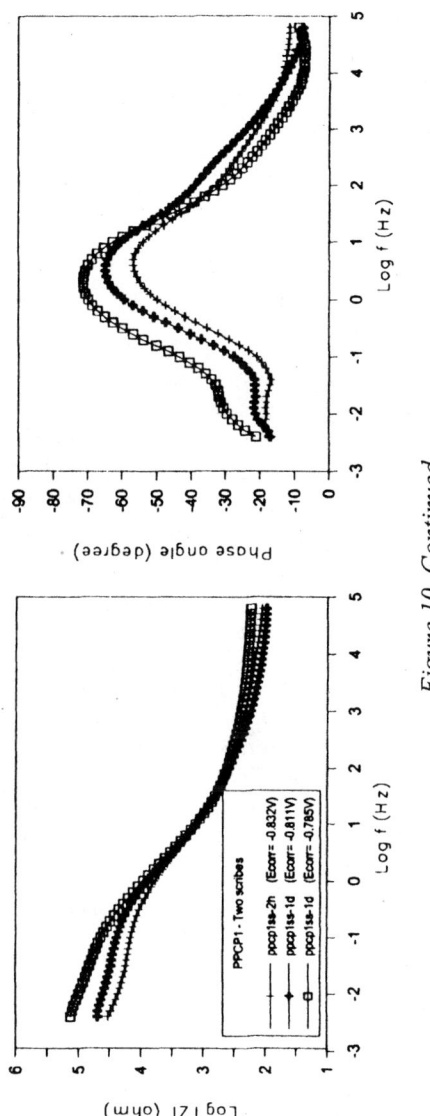

Figure 10. Continued.

The results of the SEM observation of the delaminated are shown in Figure 11. It is interesting to note that hardly any pits were formed in the scribed area. However, there was a round area under the delaminated area where general corrosion and pitting had occurred (Figure 11). A few small pits were also found in other areas. No areas with significant corrosion were found which is surprising for Al 2024-T3 which has poor corrosion resistance in chloride media.

Conclusions

Coating C which was a nonfluorinated pentasiloxane provided the best corrosion protection among all coatings on the steel coupons evaluated here. Its impedance showed capacitive behavior at all frequencies and there were no significant changes of R_p and C_c values with time. The most significant changes in the coating properties were observed for sample A which are assumed to be due to an increase of the delaminated area. The impedance spectra for sample A with coatings prepared by phenol-catalyzed thermal crosslinking (fluorinated) and sample C (nonfluorinated) with α,ω-diaminoalkanes (6, 8 and 12) followed the coating model. The best corrosion protection was provided by coatings with crosslinkings of n=12. The observation that f_b, f_{min} and Φ_{min} did not change significantly with time for samples A12 and C12 suggests that defects in the coatings that were present initially did not grow with increasing exposure time. The calculated coating resistivity was the highest for these two types of samples. The delamination ratio Δ was very small for all samples and did not change much with time.

The BAM-PPV powder coating provided remarkable corrosion protection to the Al 2024-T3 samples. The impedance spectra for the coated sample did not change for several weeks. The impedance spectra recorded after the sample was scribed were much lower reflecting the corrosion activity in the scribed area and demonstrating the great sensitivity of EIS in detecting even very small changes in coating properties. SEM observation showed that pitting did not occur in the scribed areas, but was found mainly in an area under the delaminated coating.

Acknowledgements

This work was supported by the Office of Naval Research (N00014-02-1-0521) and the Naval Air Warfare Center (N68936-03-P-0698). The authors would also like to thank Ron Burr of MacroSonix Inc. and Wade Popham formally of MacroSonix Inc for the powder blending and Mr. Mark Barilla of the NAVAIR Depot, Jacksonville, Florida for e-coat application.

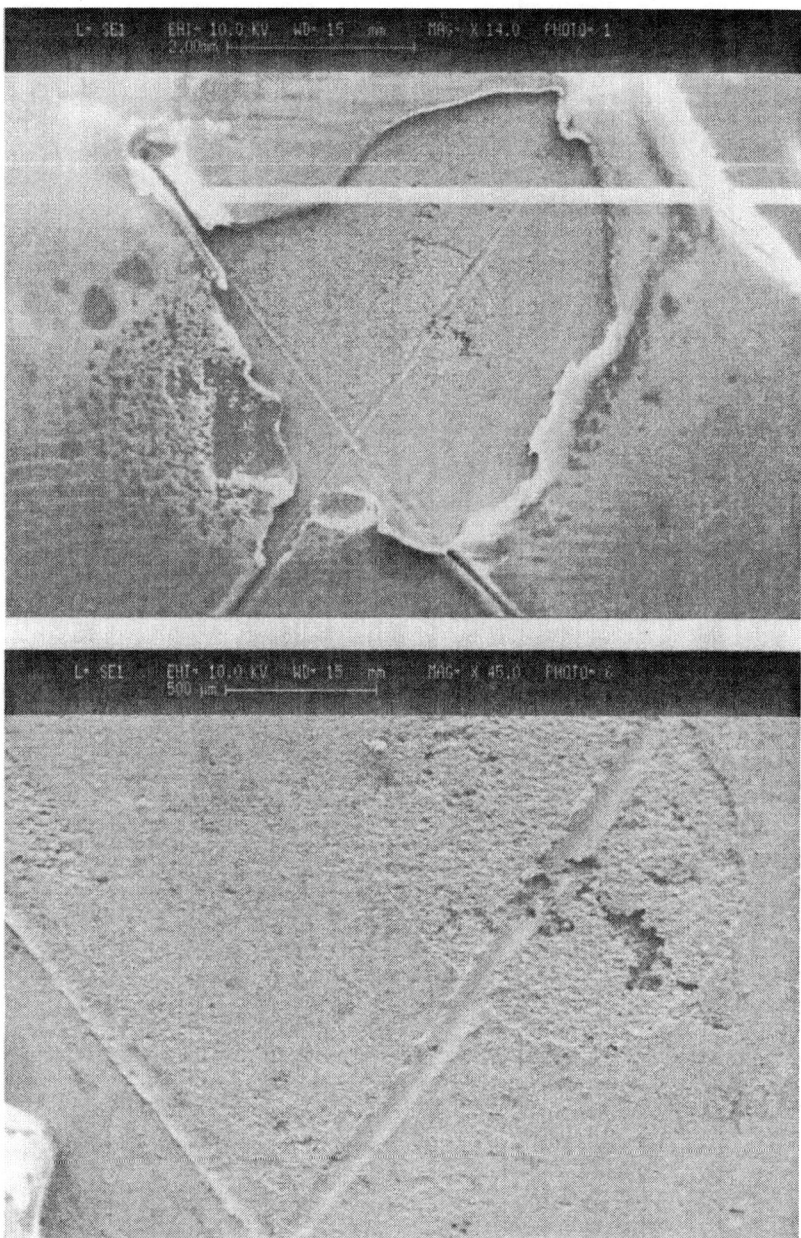

Figure 11. SEM images of the scribed sample with BAMPPV coating. (Reproduced with permission from reference 8. Copyright 2005 The Electrochemical Society.)

References

1. M. A. Grunlan, N. S. Lee, G. Cai, T. Gaedda, J. M. Mabry, F. Mansfeld, E. Kus, D. E. Wendt, G. L. Kowalke, J. A. Finlay, J. A. Callow, M. E. Callow and W. P. Weber, *Chem. Mater.,* 16, 2433 (2004).
2. M. A. Grunlan, N. S. Lee, and W. P. Weber, *J. Appl. Poly. Sci.* 94, 203 (2004).
3. N. Anderson, D. J. Irvin, J. D. Stenger-Smith, A. Guenthner, C. Webber, and P. Zarras; In ACS Symposium Series 843, Electroactive Polymers for Corrosion Control, P. Zarras, J. D. Stenger-Smith and Y. Wei Eds., American Chemical Society, Washington DC, 2003; Chapter 8, p. 140.
4. F. Mansfeld, *J. Appl. Electrochem.* 25, 187 (1995).
5. F. Mansfeld, H. Shih, H. Greene and C. H. Tsai, *ASTM STP* 1188, 37 (1993).
6. F. Mansfeld, L. T. Han, C. C. Lee and G. Zhang, *Electrochim. Acta* 43, 2933 (1998).
7. F. Mansfeld and T. S. Tsai, *Corrosion* 47, 958 (1991).
8. E. Kus, M. A. Grunlan, W. P. Weber and F. Mansfeld, "Evaluation of Non-toxic Polymer Coatings with Potential Biofoul Release Properties", *J. Electrochem. Soc.* 152 (7): B236-B243 (2005).
9. F. Mansfeld, E. Kus, M. A. Grunlan, W. P. Weber "Development of Non-toxic Polymer Coatings with Improved Biofoul Release Properties", *The Electrochem. Soc., Proc.* Vol. 2004-14, p.396 (2004).
10. C. H. Tsai and F. Mansfeld, *Corrosion* 49, 726 (1993).

Indexes

Author Index

Adamsons, Karlis, 279
Anderson, N., 40, 297
Audenaert, Marc, 190
Battocchi, Dante, 8
Benicewicz, Brian C., 1, 54
Bierwagen, Gordon P., 8
Boudjouk, Philip, 69
Buchheit, R. G., 108
Callow, J. A., 91
Callow, M. E., 91
Chen, Ru, 54
Choi, Seok-bong, 69
Davis, K. R., 146
Decker, Christian, 176
Garcia, Dana, 190
Gindin Lyuba, Sr., 225
Gray, H. Neil, 162
Grunlan, M., 297
Guenthner, Andrew J., 146
He, Jie, 8
Jorgensen, Betty, 162
Kus, E., 297
Li, Chunzhao, 54
Lockhart, Aaron, 225
Lunney, Phil, 225
Mahajanam, S. P. V., 108
Mansfeld, F., 297

McAndrew, T. Page, 190
Ni, Hai, 135
Pentony, J. M., 146
Petersheim, Jerry, 190
Potyrailo, Radislav A., 240
Pourreau, Daniel B., 201
Prokopuk, N., 40
Ready, Thomas E., 69
Richards, Thomas, 190
Richey, B., 1
Schmitt, Peter D., 225
Seth, Anuj, 24
Simões, Alda M., 8
Simonsick, William J., 135
Smith, Allan L., 261
Soucek, Mark D., 135
Steinmetz, L., 146
Stenger-Smith, J. D., 40, 297
Tallman, Dennis E., 8
Thomas, Johnson, 69
van Ooij, Wim J., 24
Webber, C., 297
Weber, W. P., 297
Wendt, D. E., 91
Wood, T., 1
Zarras, P., 1, 40, 297

Subject Index

A

Acetone
  calculated Hansen solubility parameters, 212f
  physical properties, 207t
  See also Volatile organic compounds (VOC)-exempt solvents
Acrylic coatings. *See* UV-cured acrylic coatings
Active corrosion protection coatings
  alumina passive film vs. polarized film, 20–21
  anodic and cathodic currents at defect area of scribed/topcoated sample of Mg-rich primer, 16, 17f
  current density map and optical micrograph for Ppy/DBA coated aluminum alloy AA 2024-T3, 13f
  definition, 9
  doping 4,5-dihydroxy-1,3-benzenedisulfonic acid disodium salt (DBA) into polypyrrole (Ppy), 9f
  electrochemical cell and instrumentation, 10–12
  experimental, 10–12
  gas bubble formation on magnesium electrode, 20
  Mg-rich coating, 16–21
  Mg-rich primer, 9
  Mg-rich primer as cathodic protection coating, 18, 20–21
  oxidation process occurring in exposed metal in scribe area, 16
  oxygen concentration profiles comparison, 21
  oxygen concentration profiles using SECM scans, 20
  Ppy/DBA coating on AA 2024-T3, 12, 14, 16
  proposed Al oxidation, 14
  sample preparation, 10
  scanning electrochemical microscope (SECM), 9, 12
  scanning ion electrode technique (SIET), 9, 11–12
  scanning probe techniques, 9
  scanning vibrating electrode technique (SVET), 9, 10–11
  scan over magnesium electrode at constant distance from surface, 21f
  SECM scans over AA 2024-T3 electrode, 20f
  SIET result for Ppy/DBA coated AA 2024-T3, 15f
  SVET scan over Al alloy substrate after disconnecting couple, 19f
  SVET scan over Al alloy substrate after reconnection, 19f
  SVET scan over bare Al alloy substrate immediately after electrical connection with Mg substrate, 18f
  SVET scan over bare Al alloy substrate with couple disconnected, 18f
  two-substrate cell setup for SVET, 11f
Aerospace industry
  chromate conversion coating, 25
  Taguchi's methods, 25
Alkyd coatings, t-butyl acetate (TBAC) in, 221, 222t
Alkyd spray enamel

loss compliance of uncured and
cured Decrolon vs. toluene
content, 274, 275f
monitoring O$_2$-induced curing of
enamel, 274, 276f
short-term drying of uncured
Decrolon film, 268, 270f
sorption of toluene vapor in
uncured Decrolon film, 271,
272f
toluene sorption isotherms for
uncured and cured Decrolon,
271, 273f
toluene uptake and loss in uncured
and air-cured, 271, 274
See also Quartz crystal
microbalance/heat conduction
calorimetry (QCM/HCC)
Aluminum alloys
anodic polarization curves of,
exposed to NaCl with vanadate
additions, 121f
cathodic polarization curves of,
exposed to ZnCl$_2$ and NaCl,
122f
current density map, 13f
doped polypyrrole coating, 12, 14,
16
optical micrograph with
superimposed current density
vectors, 13f
oxidation process in scribe area, 16
scanning ion electrode technique
(SIET) result, 15f
zinc inhibiting oxygen reduction,
121
See also Active corrosion
protection coatings
Amines, incorporation in polymers, 55
Analytical tools. See Service life
predictions
4-Anilinophenyl(meth)acrylates
kinetics of RAFT polymerization,
64, 66–67
monomers, 56f, 57–58

reversible addition-fragmentation
chain transfer (RAFT)
polymerization, 58, 61–62
time-conversion plot for
polymerizations, 62f, 66f
See also Reversible addition-
fragmentation chain transfer
(RAFT) polymerization
Aromatic amines, incorporation in
polymers, 55
Atomic absorption (AA), service life
prediction, 283, 292
Automotive coatings. See Service life
predictions

B

Baking time. See Polyimides
Balanus amphitrite
antifouling settlement assays, 98–
99
assays evaluating foul-release
potential, 100
key test organisms, 94
See also Marine coatings, non-
biocidal
Barnacle cutting
silicone coatings, 83, 87f
See also Marine coatings
Barnacles
antifouling settlement assays, 98–
99
assays evaluating foul-release
potential, 100
key test organisms, 94
See also Marine coatings, non-
biocidal
Bentonite
aluminosilicate-based clays, 110
Ce-exchanged, 116, 117f, 118
Ce-exchanged, pigment and cut
edge corrosion, 115–116
sensing with Ce-bentonite
pigments, 125, 128, 129f

structure, 111f
See also Ion exchange compounds (IECs)
Bioassays
  correlation between screening protocols and oceanside fouling, 76f
  marine coatings, 75, 78
Biocidal coatings
  challenge of fouling, 70–71
  See also Marine coatings
Biofouling, marine environment, 70, 92
Biological fouling, ship hulls, 70, 92
Birefringence, film spinning conditions, 156–157, 159
Bulk modulus of elasticity
  effect of modulus and biocide loading on fouling rating, 83, 88
  marine coatings, 73
  variation of, with crosslinking density for silicone coatings, 83, 84f
  See also Marine coatings

C

Cavitation erosion
  equation for loss, 194
  infrared spectra of polyamide coatings, 198f, 199f
  loss vs. time for coatings and metals, 194, 195f
  material loss during, 197, 199
  measurement, 194
  polyamide-11 performance vs. metals, 194, 197
  shock waves generating collapse, 193
  See also Polyamide-11 powder coatings
Ceramer coatings. See Polyurethane/polysiloxane ceramer coatings

Cerium, characteristics of Ce inhibition within organic coating, 119
Chain transfer agents (CTAs)
  chemical structures, 56f
  α-cyanobenzyl dithiobenzoate, 57
  2-cyanopropyl dithiobenzoate, 57
  selection in reversible addition-fragmentation polymerization, 59–60
  synthesis, 57
  See also Reversible addition-fragmentation chain transfer (RAFT) polymerization
Chemical characterization. See Service life predictions
Chemical resistance test, performance of chromate-free novel primer, 29t, 30t, 31t, 32
5-Chloro-2-(2,4-dichlorophenoxy)phenol). See Triclosan (5-chloro-2-(2,4-dichlorophenoxy)phenol)
Chromate conversion coatings (CCC)
  alternatives to, 41
  electrochemical impedance studies of CCC and alternative coating, 49, 50f
  inhibiting aluminum alloy corrosion, 41
  pretreatment alternatives to CCC, 43t
  See also Electroactive polymers (EAPs)
Chromate-free novel primer system
  analysis of responses to varying parameters, 35f, 36f, 37f
  chemical resistance test, 32
  coating and curing of coated samples, 28
  corrosion performance evaluation tests, 29t, 30t, 31t, 32t
  DI water contact angle, 29, 31
  electrochemical impedance spectroscopy (EIS), 28–29

experimental, 27–28
experimental design for
optimization, 25–27
L9 orthogonal array for
optimization, 25–27
metal substrates, 27
methyl ethyl ketone (MEK) rub
test, 31–32
NaCl aqueous solution immersion
test, 29
optimization and verification using
Taguchi method, 32–34, 38
organofunctional silanes and
crosslinkers, 27
performance evaluation, 28–32
primer formulations, 28
resins for formulation, 27
scanned images exposed to NaCl
solution with and without
topcoat, 38f
tape adhesion test, 31
verification of optimized coatings,
34, 38
Chromate-free polymer coatings
BAM-PPV [poly(2,5-bis(N-methyl-
N-hexylamino)phenylene
vinylene)] coating, 298, 299
BAM-PPV powder coated Al 2024-
T3, 315, 320
breakpoint frequency, 311–312
breakpoint frequency as function of
exposure time, 313f
coating capacitance, 311
coating delamination capacitance,
311
data analysis, 299
delamination ratio, 311–312, 315
diffusion impedance Z_{OFLD}, 300
electrochemical impedance
spectroscopy (EIS) for
evaluating, 298, 299
equivalent circuit for coating
model, 313f
experimental approach, 299
extent of delamination, 311
fluorinated (A and B) and nonfluorinated (C) pentasiloxanes,
299
impedance spectra for Al 2024 with
BAM-PPV coating, 317f, 318f,
319f
impedance spectra for A samples
for 7, 21, and 35 days exposure,
305f, 306f, 307f
impedance spectra for C samples
for 7, 21, and 35 days exposure,
308f, 309f, 310f
impedance spectra of samples for 7,
21, and 35 days exposure, 301f,
302f, 303f
open boundary finite length
diffusion (OFLD) model, 299,
300, 304f
pentasiloxanes coatings on steel,
300, 311–312, 315
polarization resistance, 311
pore resistance, 311
scanning electron microscopy
(SEM) images of scribed sample
with BAM-PPV coating, 321f
time dependence of breakpoint
frequency, phase angle, and
frequency minimum, 312, 314f
time dependence of delamination
ratio and specific resistance of
coating, 315, 316f
time dependence of Warburg
coefficient and fit parameter,
300, 304f
See also Electrochemical
impedance spectroscopy (EIS)
Coating development
combinatorial technology, 71, 72f
prospective scheme for down-
selecting marine coatings, 73,
74f
screening protocols, 71, 72f, 73
See also Marine coatings;
Polyurethane/polysiloxane
ceramer coatings

Coating formulations
  critical role of solvent blend, 212–213
  ozone forming potential, 203, 204f
  solvent usage, 201–202
  TBAC (t-butyl acetate) in alkyd coatings, 221, 222t
  TBAC in nitrocellulose lacquers, 213–215
  TBAC in two-component epoxy amine coatings, 215
  TBAC in two-component epoxy systems, 215–216, 217f
  TBAC in two-component urethane coatings, 219–221
  two-component epoxy reformulation, 216, 218
  volatile organic compound (VOC) regulation, 202–205
  See also Tertiary-butyl acetate (TBAC); Volatile organic compounds (VOC)-exempt solvents
Coatings
  drying and curing, 262
  See also Service life predictions
Colorimetric response, selectivity and sensitivity of sensor coatings, 169
Combinatorial and high-throughput development of polymer sensor coatings
  application of moisture-sensitive sensor films, 254f
  capabilities of combinatorial screening of sensor materials, 252–253
  chlorine model sensor system, 250
  combinatorial cycle for sensor materials development, 241f
  combinatorial screening process of sensor materials, 240–241
  computer modeled response pattern of polymers to three isomers of dichloroethylene and water vapor, 242f
  field testing of GE's polymer coatings using high-throughput screening, 246f
  formulated polymer coatings for optical sensors, 247–253
  high-throughput screening method, 251–252
  linear solvation energy relationships (LSER), 241–243
  materials screening strategy, 244
  optical sensor coatings arrays, 249f
  optimization of formulated materials using gradient sensor films, 252f
  optimization of formulation components, 249–250
  outlook, 253–254
  polymer coatings for resonant sensors, 241–245
  primary screen of gas-sorption properties of sensor coatings, 244, 245f
  screening dynamic responses of formulated optical ionic sensor coatings, 251f
  secondary screen of gas-sorption properties of sensor coatings, 244, 246f
  sensor coatings development at GE from concept to pilot scale, 240f
  spatial gradients in polymers, 249
  tertiary screen for evaluation of long-term stability of two sensor coatings, 245, 246f
  testing of operation of GE resonant sensor system in field, 242f
  thickness-shear mode (TSM) sensor array for high-throughput materials characterization, 243f
  TSM devices, 243
  typical formulated polymeric sensor materials, 248t
Combinatorial screening
  sensor materials, 240–241

See also Combinatorial and high-throughput development of polymer sensor coatings
Combinatorial technology, coating development, 71, 72f
Compliance, transverse shear mode quartz plate resonator, 264
Corrosion
  cut edge, 115–116
  silicone coatings, 83, 87f
  See also Marine coatings
Corrosion inhibition, tetraethyl orthosilicate (TEOS) in polysiloxane, 136
Corrosion performance tests, aerospace industry, 25
Corrosion protection
  release of soluble inhibitors, 109
  solubility and potency constraints, 109–110
  SrCrO$_4$ pigment, 109
  See also Active corrosion protection coatings; Ion exchange compounds (IECs)
Crosslinked films
  vs. non-crosslinked, 227f
  See also Chromate-free polymer coatings
Curing process
  coatings, 262
  monitoring O$_2$-induced curing of enamel, 274, 276f
  toluene uptake and loss in uncured and air-cured enamel, 271, 274
  See also Quartz crystal microbalance/heat conduction calorimetry (QCM/HCC)
Cut edge corrosion, Ce-exchanged bentonite pigment, 115–116
α-Cyanobenzyl dithiobenzoate. See Chain transfer agents (CTAs)
2-Cyanopropyl dithiobenzoate. See Chain transfer agents (CTAs)

D

Decontamination studies, sensor coatings, 167–169
Decrolon film. See Alkyd spray enamel
Deionized water contact angle, performance of chromate-free novel primer, 29, 30t, 31t, 32t
Depth profiling. See Service life predictions
Differential scanning calorimetry (DSC), service life prediction, 282, 286, 287t
Diphenoxybenzophenone (DPBP)
  performance of photoinitiator, 180
  surface and deep-through cures, 180–181
  See also UV-cured acrylic coatings
Dispersions
  production process, 228, 229f
  solvency parameter, 211
  See also UV-curable waterborne polyurethane dispersion (PUD)
Down-selecting coatings, coating development, 73, 74f
Drying process
  coatings, 262
  FTIR spectra of polyimide sample at various points in, 155f, 158f
  gel collapse theory, 154–155
  polyimides, 156–157
  short-term drying of enamel, 268, 270f
  See also Quartz crystal microbalance/heat conduction calorimetry (QCM/HCC)
Dual-curing process, isocyanate and hydroxyl functionalized diacrylate oligomers, 183
Dynamic mechanical thermal analysis, marine coatings, 80

333

E

Electroactive polymers (EAPs)
  alternative to chromate conversion coatings (CCC), 41
  alternative trivalent chromium pretreatment (TCP), 41
  BAM-PPV [poly(2,5-bis(*N*-methyl-*N*-hexylamino)phenylene vinylene)] and MEH-PPV [poly((2-(2-ethylhexyl)oxy-5-methoxy-p-phenylene)vinylene)], 41
  Bode plots of CCC and BAM-PPV coated Al 2024-T3, 50*f*
  electrochemical impedance studies of CCC and BAM-PPV coated Al 2024-T3, 49, 50*f*
  experimental, 41–42
  neutral salt fog exposure of alternative pretreatments on Al 2024-T3, 42–43
  photos of BAM-PPV coated Al 2024-T3, 46*f*, 47*f*
  photos of MEH-PPV coated on Al 2024-T3, 48*f*
  photos of TCP on Al 2024-T3, 44*f*, 45*f*
  pretreatment alternatives to CCC, 43*t*
  total impedance for CCC and BAM-PPV coated Al 2024-T3, 50*f*
Electrochemical impedance spectroscopy (EIS)
  Al 2024 steel with poly(2,5-bis(*N*-methyl-*N*-hexylamino)phenylene vinylene) (BAM-PPV) coating, 317*f*, 318*f*, 319*f*
  Ce-exchanged bentonite as coatings, 116, 117*f*, 118
  coating capacitance, 311
  coating delamination capacitance, 311
  coating model, 300, 311, 313*f*
  delamination ratio, 311, 315, 316*f*
  diffusion impedance Z_{OFLD}, 300
  electroactive polymers, 49, 50*f*
  fluorinated and non-fluorinated pentasiloxanes with photo-crosslinking for 7, 21, and 35 days exposure, 301*f*, 302*f*, 303*f*
  fluorinated pentasiloxanes with thermal crosslinking for 7, 21, and 35 days exposure, 305*f*, 306*f*, 307*f*
  pentasiloxanes with thermal crosslinking for 7, 21, and 35 days exposure, 308*f*, 309*f*, 310*f*
  performance of chromate-free novel primer, 28–29, 30*t*, 31*t*, 32*t*
  polarization resistance, 311
  pore resistance, 311
  Warburg coefficient, 300, 304*f*
  See also Chromate-free polymer coatings
Electron spectroscopy for chemical analysis (ESCA), service life prediction, 283, 292, 293*f*
Enteromorpha
  macroalga, 92
  See also Marine coatings, non-biocidal
Environmental remediation
  development of sensor coatings, 163–164
  hazardous chemical and biological contaminants, 162–163
  See also Sensor coatings
Epoxy amine coatings, two-component, t-butyl acetate (TBAC) in, 215
Epoxy reformulation, solvent blend with t-butyl acetate (TBAC) in, 216, 218
Epoxy system, two-component, kinetics and stability, 217*f*

Epoxy systems, two-component, t-butyl acetate (TBAC) in, 215–216
Esterification reaction, high-solid polyesters, 136–137
Evaporation profiles, solvent blends, 210f
Experimental design
  L9 orthogonal arrays for novel primer system, 25–27
  polyurethane dispersion for soft feel, 231–232
  predicted properties for combinations of factors, 234f
  See also UV-curable waterborne polyurethane dispersion (PUD)

F

Field testing, sensor coating, 171
Film formation, drying and curing, 262
Film geometry, polymer films by spin casting, 149–150
Film thickness. See Polyimides
Fluorinated pentasiloxanes. See Chromate-free polymer coatings
Fouling
  average rating of coatings after 44 and 80 days static immersion (Florida), 85f, 86f
  biological, of ship hulls, 70, 92
  effect of modulus and biocide loading on fouling rating, 83, 88
  fouling-release coatings, 71
  levels, 70
  See also Marine coatings
Free radical polymerization
  poly(acrylate)s and poly(acrylamide)s, 55–56
  reversible addition-fragmentation chain transfer (RAFT), 55
  See also Reversible addition-fragmentation chain transfer (RAFT) polymerization

G

Gas chromatography (GC), service life prediction, 282, 288, 291
Gas-sorption properties, screening sensor materials, 244–245, 246f
Gel collapse
  drying process in terms of, theory, 154–155
  orientation development, 152–153
Green seaweed Ulva. See Ulva

H

Hansen parameters
  solubility, for toluene and VOC–exempt solvents, 212f
  solvency, 211
Hardening profiles, UV-cured coatings, 184, 185f
Hazardous air pollutant (HAP), solvent blend reformulation for, compliance, 209t, 210
Heat, thermodynamics, 262
Heat conduction calorimetry (HCC). See Quartz crystal microbalance/heat conduction calorimetry (QCM/HCC)
High performance coatings. See UV-cured acrylic coatings
High performance liquid chromatography (HPLC), service life prediction, 282, 288, 289f, 290t
High-throughput screening. See Combinatorial and high-throughput development of polymer sensor coatings
Hydrogen bonding, solvency parameter, 211
Hydrotalcites
  aluminum-zinc hydroxide decavanadate, 120
  coating on aluminum alloy substrate, 120f

inverse of cation exchanging clays, 111
pigmented coatings releasing vanadate and Zn^{2+} inhibitors, 122
pigmented epoxy coatings against scribe corrosion, 123–124
pigments and corrosion protection, 119–124
sensing water uptake in coatings, 125, 126f, 127f
sensing with, 129–130, 131f
structure, 112f
See also Ion exchange compounds (IECs)

I

Imides. *See* Polyimides
Infrared (IR) spectroscopy, service life prediction, 282, 286, 289f
Ion exchange compounds (IECs)
aluminum-zinc hydroxide decavanadate hydrotalcite, 120
anodic polarization curves of Al alloy 2024-T3 with vanadate additions, 121f
basics of ion exchange, 110–113
bentonite structure, 111f
Bode magnitude and phase angle plots for epoxy coating with Ce-exchanged bentonite, 117f
cathodic and anodic polarization curves for 2024-T3 in Ce-exchanged bentonite, 118f
cathodic polarization curves of Al alloy 2024-T3 exposed to $ZnCl_2$ and NaCl, 122f
Ce, Ce-exchanged clays and rare-earth inhibitors and pigments, 114–119

Ce-exchanged bentonite in coatings, 116, 117f, 118
Ce inhibition within organic coating, 119
Ce inhibitors and pigments in corrosion resistant organic coatings, 115–116
cerium conversion coatings, 115
commercial ion exchange pigments, 113–114
corrosion inhibiting pigments in paints, 110
cut edge corrosion, 115–116
exchange process, 111–112
hydrotalcite coating on aluminum alloy substrate, 120f
hydrotalcite-pigmented coatings releasing vanadate and Zn^{2+} inhibitors, 122
hydrotalcite-pigmented epoxy coatings protecting against scribe corrosion, 123–124
hydrotalcite pigments and corrosion protection, 119–124
hydrotalcites, 111, 112f
optical photographs of simulated scratch cell coupons after exposure, 122, 123f
oxidation of Ce and Pr to tetravalent state, 115
sensing practicalities, 130–132
sensing water uptake in coatings, 125, 126f, 127f
sensing with Ce-bentonite pigments, 125, 128, 129f
sensing with hydrotalcites, 129–130, 131f
solubility and potency constraints, 109–110
$SrCrO_4$ pigment, 109
synthesis, 112–113
zinc inhibiting oxygen reduction on aluminum, 121

K

Kinetics, reversible addition-fragmentation chain transfer (RAFT) polymerization, 64, 66–67

L

L9 orthogonal arrays
　optimization of novel primer system, 25–27
　Taguchi method, 26t
Light-induced polymerization
　UV-radiation curing, 176–177
　See also UV-cured acrylic coatings
Linear solvation energy relationships (LSER) model, polymer-based sensors, 241–243

M

Magnesium-rich primer
　anodic and cathodic currents at defect area, 16, 17f
　cathodic protection coating, 18, 20–21
　concept of sacrificial primer, 9
　gas bubble formation, 20
　oxygen concentration profiles, 20–21
　sample preparation, 10
　scanning vibrating electrode technique (SVET) scans over bare Al alloy during and after current, 18f, 19f
　scan over Mg electrode, 21f
　See also Active corrosion protection coatings
Marine coatings
　addressing technical challenges, 71, 73
　barnacle cutting and corrosion in coatings, 87f
　biocidal coatings, 70–71
　biocide modification and attachment to siloxane, 75, 78, 79f
　bulk moduli of elasticity depending on polymer backbone and cross link types, 82–83
　bulk modulus of elasticity, 73
　coatings and their characteristics, 81t
　correlations between screening protocols and oceanside fouling, 76f
　dynamic mechanical thermal analysis, 80
　effect of modulus and biocide loading on fouling rating, 83, 88
　emulating ocean environment, 71, 73
　fouling of ship hulls, 70
　fouling rating of coatings after 44 days static immersion (Florida), 85f
　fouling rating of coatings after 80 days static immersion (Florida), 86f
　fouling-release coatings, 71
　generic types of modified silicones, 77f
　materials, 78
　modification of biocide Triclosan by bromoalkene, 78
　multifunctional silicone resin (North Dakota State University), 77f
　organism bioassays, 75, 78
　panel preparation and deployment site, 82
　preparation of coatings, 80, 82
　prospective scheme for down-selecting, 74f
　pseudo-barnacle testing, 73, 75
　simultaneous incorporation of alkene modified Triclosan and

allyl glycidyl ether onto siloxane, 80
surface energy, 73
utilizing combinatorial tools and screening protocols, 72*f*
variation of bulk modulus of elasticity with crosslinking density for NDSU silicones, 84*f*
See also Marine coatings, non-biocidal
Marine coatings, non-biocidal
antifouling settlement assays, 95–99
apparatus, 100–102
assays evaluating foul-release potential, 99–100
assays with algae, 95, 98, 99–100
assays with barnacles, 98–99, 100
barnacles (*Balanus amphitrite*) test, 94
crosslinked hyperbranched fluoropolymers (HBFP) and poly(ethylene glycol) (PEG), 102
false-color image of settled spore of *Ulva*, 96, 97
force gauge test stand, 101–102
fully turbulent flow channel and water jet apparatus, 100–101
green seaweed *Ulva* test, 94
lab-scale evaluation, 102, 105
need for lab-scale evaluations, 93–94
percentage removal of *Ulva* sporelings in flow apparatus, 104*f*
percent settlement of cypris larvae of *B. amphitrite* on xerogel coatings and glass and polystyrene controls, 105*f*
removal of *Ulva* spores from HBFP-PEG networks in flow apparatus, 103*f*
settlement preference of cypris larvae of *B. amphitrite* to proprietary coatings, 104*f*
test sample format, 95
Ulva spores on cross-linked HBFP-PEG network coatings, 103*f*
zoospores of *Ulva*, 96, 97
Mar resistance, UV-cured coatings, 184–185, 186*f*
Mass, thermodynamics, 262
Material screening strategy, sensor materials, 244–245, 246*f*
Mechanical properties, polyurethane/polysiloxane coatings, 138–139, 140*t*
Methyl acetate
calculated Hansen solubility parameters, 212*f*
physical properties, 207*t*
See also Volatile organic compounds (VOC)-exempt solvents
Methyl ethyl ketone rub test, performance of chromate-free novel primer, 29*t*, 30*t*, 31–32
Microtomy, service life prediction, 281, 283
Model
coating, 300, 311
equations for coating, 311
equivalent circuit for coating, 313*f*
one-time-constant (OTC), 300
Moisture resistance, UV-cured coatings, 185, 186*f*
Moisture sensitivity, polymer sensor films, 253, 254*f*
Motional resistance, transverse shear mode quartz plate resonator, 263–265

N

NaCl aqueous immersion test
  performance of chromate-free novel primer, 29, 30*t*, 31*t*, 32*t*
  scanned images of exposed samples with and without topcoat, 38*f*
National security
  removal of chemical and biological weapons, 162–163
  See also Sensor coatings
Neutral salt fog exposure, alternative pretreatment coatings on Al 2024-T3, 42–43
Nitrocellulose lacquers, t-butyl acetate (TBAC) in, 213–215
Non-biocidal coatings. *See* Marine coatings, non-biocidal
Novel primer system. *See* Chromate-free novel primer system
Nuclear magnetic resonance (NMR)
  analysis of polymerization reaction, 59
  in situ, for reversible addition-fragmentation chain transfer (RAFT) polymerization, 64, 66–67
  partial ^{1}H NMR spectra at different reaction times for RAFT polymerizations, 65*f*
  time-conversion plots for RAFT polymerizations, 66*f*
Nylon-11. *See* Polyamide-11 powder coatings

O

One-step coatings. *See* Chromate-free novel primer system
One-time-constant (OTC) model, coating, 300
Optical sensors
  capabilities of combinatorial screening of sensor materials, 252–253
  chlorine as model sensor system, 250
  formulated polymer coatings, 247–253
  gradient sensor films for optimization, 252*f*
  high-throughput screening method, 251–252
  optimization of formulation components, 249–250
  screening dynamic responses of formulated, 251*f*
  spatial gradients in polymers, 249
  typical formulated polymeric materials, 248*t*
Optical waveguides
  fabrication of, device, 147
  polymer-based, 160
  schematic of waveguide formation using polyimides, 147*f*
  See also Polyimides
Organism bioassays
  correlation between screening protocols and oceanside fouling, 76*f*
  fouling organisms, 92
  marine coatings, 75, 78
Organofunctional silanes
  chromate replacement, 25
  novel primer system, 25–27
Orthogonal arrays (L9), optimization of novel primer system, 25–27
Oxygen, photopolymerization inhibitor, 181
Ozone
  formation potential of solvents, 203, 204*f*
  reactions for formation, 203, 205

P

Parachlorobenzotrifluoride (PCBTF)
  calculated Hansen solubility parameters, 212*f*
  physical properties, 207*t*
  See also Volatile organic compounds (VOC)-exempt solvents
Pentasiloxanes. *See* Chromate-free polymer coatings
Photodegradation, resistance of UV-cured coatings, 186, 187*f*
Photoinitiators
  advantages, 180–181
  diphenoxybenzophenone (DPBP), 180
  performance analysis in UV-curing, 181*f*
  See also UV-cured acrylic coatings
Photopolymerization. *See* UV-cured acrylic coatings
Pigments in organic coatings
  Ce inhibitors and pigments, 114–115
  characteristics of Ce inhibition, 119
  commercial ion exchange pigments, 113–114
  $SrCrO_4$-pigmented primers, 109
  See also Ion exchange compounds (IECs)
Plutonium decontamination
  coating compositions for, 166*t*
  potential masking agents, 170*t*
  safety considerations, 171–172
  See also Sensor coatings
Polar, solvency parameter, 211
Poly(amic acid)
  commercial, Pyralin 5878G, 148
  field thickness vs. spin speed, 151*f*
  field thickness vs. spin time, 150*f*
  refractive index profile, 152*t*
  See also Polyimides
Polyamide-11 powder coatings
  cavitation erosion, 193
  comparison of polyamides, 192*f*
  equation for loss, 194
  experimental, 193–194
  impact of cavitation-generated shock waves on, vs. metals, 194, 197
  infrared spectra for coatings, 198*f*, 199*f*
  loss vs. time for coatings and metals, 194, 195*f*
  measurement of cavitation erosion, 194
  nylon-11 [poly(11-aminoundecanoic acid)], 191*f*
  photographs before and after testing, 196*f*, 197*f*
  polyamide-11, 191
  process of material loss during cavitation erosion, 197, 199
  Rilsan® Fine Powders, 191–192
  sample preparation, 193
Poly(ethenyl formamide) (PEF), component of sensor coatings, 164–165
Polyimides
  baking time variables, 157, 159
  behavior during baking, 159–160
  commercial poly(amic acid) Pyralin 5878G, 148
  drying process, 156–157, 158*f*
  drying process in terms of gel collapse theory, 154–155
  effect of film processing on refractive index profile, 154–160
  estimated Tg vs. solids content for Polyimide B/veratrol wet films, 156*f*
  fabrication of optical waveguide device, 147
  field thickness vs. spin speed for Pyralin 5878G and Polyimide A in 1,1,2,2-tetrachloroethane, 151*f*

field thickness vs. spin time for
Pyralin 5878G, 150f
film geometry, 149–150
FTIR spectra of Polyimide C films
after drying process, 158f
gel collapse, 152–153
imidization during baking at high
temperature, 153
materials and methods, 148–149
performance of integrated optical
devices, 147–148
refractive index, 150, 152–160
refractive index profile of
Polyimide B vs. spin time, 154t
refractive index profile of
Polyimide C vs. spin time, 158t
refractive index profiles for
Polyimides A–C and Pyralin
5878G, 152t
rotational mobility of chains, 152–153
schematic of waveguide formation,
147f
spin time, 154
weight fraction Polyimide C in
films at points in baking cycle
vs. baking time, 158t
Polyisobutylene, shear storage
modulus and shear loss modulus,
266, 267f
Polymer sensor coatings. *See*
Combinatorial and high-throughput
development of polymer sensor
coatings
Polymer viscoelasticity, Williams–
Landel–Ferry theory, 264–265
Poly(phenylene vinylene) (PPV)
derivatives. *See* Electroactive
polymers (EAPs)
Polypyrrole
doped with 4,5-dihydroxy-1,3-
benzenesulfonic acid disodium
salt (DBA), 9
sample preparation, 10

See also Active corrosion
protection coatings
Polysiloxane
corrosion inhibition and tetraethyl
orthosilicate (TEOS), 136
See also Polyurethane/polysiloxane
ceramer coatings
Polystyrene
shear loss compliance of
plasticized, 266, 268, 269f
shear storage modulus and shear
loss modulus of plasticized, 266,
267f
Polyurethane-acrylate (PUA) coating
conversion vs. time profiles, 180f
photocuring, 179–184
See also UV-cured acrylic coatings
Polyurethane dispersion (PUD)
waterborne, 226–227
See also UV-curable waterborne
polyurethane dispersion (PUD)
Polyurethane/polysiloxane ceramer
coatings
approach to develop, 136
2-butyl-2-ethyl-1,3-propanediol
(BEPD) structure, 137
chromate pretreatment, 137–138
corrosion photographs, 142f
esterification reaction, 136, 137
experimental, 137–138
formulations, 139t
hardness properties, 138–139
mechanical properties, 140t
mechanical tests, 138
organic/inorganic phases by
scanning electron microscopy
(SEM), 141–142, 143f
primer, 137–138
reverse impact resistance, 139–140
salt spray experiment, 140–141
tensile properties, 140, 141t
typical high-solid polyesters from
diols and diesters, 136, 137

Poly(vinyl alcohol) (PVA), component of sensor coatings, 164, 165, 167
Poly(vinyl pyrrolidone) (PVP), component of sensor coatings, 164, 167
Potassium ionization of desorbed species (K+IDS) mass spectrometry, service life prediction, 283, 293
Powder coatings. *See* Polyamide-11 powder coatings
Predictions. *See* Service life predictions
Primer system. *See* Chromate-free novel primer system
Processing parameters. *See* Polyimides
Protection coatings. *See* Active corrosion protection coatings; Chromate-free polymer coatings
Pseudo-barnacle testing, marine coatings, 73, 75

Q

Quartz crystal microbalance/heat conduction calorimetry (QCM/HCC)
    change in mass per unit area, loss compliance, and thermal power in Decrolon film, 274, 276f
    compliance, 264
    glass transition temperature shift, 266
    loss compliance of uncured and cured Decrolon vs. toluene content, 274, 275f
    measurement technique, 262–263
    monitoring O_2-induced curing of enamel with, 274, 276f
    motional resistance, 263–265
    shear loss compliance, 264
    shear loss compliance of polystyrene plasticized with 2-chlorotoluene, 266, 268, 269f
    shear storage compliance, 264
    shear storage modulus and shear loss modulus for polyisobutylene, 266, 267f
    shear storage modulus and shear loss modulus for polystyrene plasticized with 2-chlorotoluene, 266, 267f
    short-term drying of uncured Decrolon film, 268, 270f
    sorption of toluene vapor in uncured Decrolon film, 271, 272f
    theory, 263–268
    time-temperature superposition principle, 264–265
    toluene sorption isotherms for uncured and cured Decrolon, 271, 273f
    toluene uptake and loss in uncured and air-cured enamel, 271, 274
    transverse shear mode (TSM) quartz plate resonator, 263
    Williams–Landel–Ferry theory of polymer viscoelasticity, 264–265

R

RAFT polymerization. *See* Reversible addition-fragmentation chain transfer (RAFT) polymerization
Refractive index. *See* Polyimides
Resonant sensors
    linear solvation energy relationships (LSER) model, 241–243
    polymer coatings for, 241–245
Reversible addition-fragmentation chain transfer (RAFT) polymerization

4-anilinophenyl acrylate, 58
4-anilinophenyl methacrylate, 57–58
chain transfer agent (CTA) selection, 59–60
characterization methods, 56–57
chemical structures of monomers and CTAs, 56f
evolution of molecular weights and polydispersities, 63f
experimental, 56–59
fragmentation rate, 67
gel permeation chromatography (GPC) elution profiles for poly(4-anilinophenyl methacrylate), 61f
in situ ^{1}H NMR analysis of reaction, 59
kinetic information by in-situ NMR spectrometry, 64, 66
mechanism, 59, 60
monomer synthesis, 60–61
partial ^{1}H NMR spectra at different reaction times, 65f
polymerization rates, 62, 64
RAFT procedure, 58, 61
retardation, 66–67
time-conversion plot, 62f
time-conversion plots by in-situ proton NMR, 64, 66f
Rilsan® Fine Powders thermoplastic, 191–192
See also Polyamide-11 powder coatings

S

Salt spray experiment, polyurethane/polysiloxane coatings, 140–141, 142f
Scanning electrochemical microscopy (SECM)
instrumentation, 10, 12
oxygen concentration over aluminum alloy electrode, 20f
oxygen concentration over magnesium electrode, 21f
scanning probe, 9
Scanning electron microscopy (SEM)
Al 2024 steel with poly(2,5-bis(*N*-methyl-*N*-hexylamino)phenylene vinylene) (BAM-PPV) coating, 321f
polyurethane/polysiloxane coatings, 141–142, 143f
Scanning ion electrode technique (SIET)
doped polypyrrole coated aluminum alloy, 15f
instrumentation, 11–12
mapping pH above doped polypyrrole coating, 14, 15f
scanning probe, 9
Scanning vibrating electrode technique (SVET)
instrumentation, 10–11
scanning probe, 9
scribed/topcoated sample of Mg rich primer, 17f, 18f, 19f
two-substrate cell setup for, 11f
Scratch resistance, UV-cured coatings, 184–185, 186f
Screening protocols, coating development, 71, 72f, 73
Screening strategy, sensor materials, 244–245, 246f
Screening study. *See* UV-curable waterborne polyurethane dispersion (PUD)
Selectivity, colorimetric response in sensor coatings, 169
Sensing
hydrotalcites, 129–130, 131f
practicalities, 130–132
water uptake in coatings, 125, 126f, 127f
with Ce-bentonite pigments, 125, 128, 129f

Sensitivity, colorimetric response in sensor coatings, 169
Sensor coatings
chemical and biological weapons (CBWs) detection and removal, 163
decontamination factors for uranium contaminated planchets, 168t
decontamination studies, 167–169
development at GE from concept to pilot scale, 240f
development with focused remediation utility, 163–164
environmental remediation, 162–163
experimental, 171–172
field testing, 171
general coating preparation, 172
general process for detection and removal of contaminants from surfaces, 165f
masking complexans interfering with colorimetric response, 165f
materials, 171
national security, 162–163
poly(ethenyl formamide) (PEF), 164–165
polymer compositions studied for uranium/plutonium sensing, 164–165
poly(vinyl alcohol) (PVA), 164, 165, 167
poly(vinyl pyrrolidone) (PVP), 165, 167
potential masking agents for uranium/plutonium decontaminating coatings, 170t
safety considerations, 171–172
selectivity and sensitivity of colorimetric response, 169, 170t
strippable coatings, 163
surface decontamination in laboratory, 172
typical compositions, 164

See also Combinatorial and high-throughput development of polymer sensor coatings
Service life predictions
AA (atomic absorption) analyses, 283
acid and microwave digest of microtomed sections, 282
chemical depth profiling using slab microtomed sections, 287t
depth profiling, 283
differential scanning calorimetry (DSC) analyses, 282
direct analysis of slab microtomed sections, 281
DSC analysis: T_g as function of depth, 286, 287t
DSC T_g depth profiles for automotive clearcoat, 287t
electron spectroscopy for chemical analysis (ESCA), 283
elemental analysis by AA, 292
ESCA analysis of catalyst distribution vs. depth, 292, 293f
experimental, 281–283
factors driving research, 280–281
gas chromatography (GC) analyses, 282
GC analysis: solvent trapping vs. depth, 288, 291
high performance liquid chromatography (HPLC) analyses, 282
HPLC analysis: interlayer mixing as function of depth, 288, 289f, 290t
infrared (IR) attenuated total reflectance (ATR) analyses, 282
IR photo-oxidation index (POI) as function of depth, 286, 289f
K+IDS (potassium ionization of desorbed species) mass spectrometry (MS) analyses, 283

K+IDS MS analysis: degradation
products vs. depth, 293
sectioning, 283–284
sliding (slab) microtomy of
automotive finishes, 281
standard microtomy, 281
through-film (T-mode) UV-vis
spectroscopic analysis, 282
time-of-flight secondary-ion MS
(ToF–SIMS) analyses, 283
ToF–SIMS analysis of pigment
distribution vs. depth, 291
tools, 280
typical UVA depth profile for
outdoor exposed
clearcoat/basecoat (CC/BC)
bilayer, 284, 285f
typical UVA depth profile for
unexposed clearcoat (CC) layer,
284, 285f
ultraviolet-visible (UV-vis)
analyses, 282
UV-vis analyses for depth
profiling, 284, 285f
Shear loss compliance
polystyrene plasticized with 2-
chlorotoluene, 266, 268, 269f
transverse shear mode quartz plate
resonator, 264–265
Shear storage compliance, transverse
shear mode quartz plate resonator,
264–265
Ship hulls
biological fouling, 70, 92
See also Marine coatings
Silicone coatings
bulk moduli of elasticity
dependence on polymer
backbone and cross link types,
82–83
coatings and chemical
characteristics, 81t
fouling release, 71

generic types of modified silicones,
77f
North Dakota State University
multifunctional silicone resin,
77f
preparation, 80, 82
See also Marine coatings
Siloxanes. See Chromate-free polymer
coatings
Slab microtomy, service life
prediction, 281, 283, 287t
Sliding microtomy, service life
prediction, 281, 283
Sodium chloride solution immersion
performance of chromate-free
novel primer, 29, 30t, 31t, 32t
scanned images of exposed samples
with and without topcoat, 38f
Soft-touch coatings
analysis of variance for softness,
231t
combined effect of percent chain
extension and NCO content on
softness, 233f
effect of percent chain extension on
softness, 233f
plastic applications, 225–226
preparation and cure of soft-touch
panels, 230
preparation of UV, formulation,
229–230
softness measurement, 230
softness of statistical design
dispersions, 235t
See also UV-curable waterborne
polyurethane dispersion (PUD)
Solvency properties
selecting solvent, 211–212
viscosities of alkyd resins in
various solvents, 213f
Solvent blends
evaporation profiles, 210f
selecting, 208–210

Solvent-cast polyimides. *See* Polyimides
Solvents. *See* Volatile organic compounds (VOC)-exempt solvents
Spin casting
geometry of polymer films, 149–150
refractive index profile of polyimide, 154
See also Polyimides
Spin time
refractive index profiles of polyimides, 154*t*, 158*t*
refractive index profile with film processing, 154–160
weight fraction of polyimide in films during baking cycle, 155*t*
See also Polyimides
Statistical design. *See* UV-curable waterborne polyurethane dispersion (PUD)
Surface energy, marine coatings, 73

T

Taguchi method
aerospace industry, 25
optimization and verification, 32–34, 38
typical L9 orthogonal array, 26*t*
See also Chromate-free novel primer system
Tape adhesion test, performance of chromate-free novel primer, 29*t*, 30*t*, 31, 32*t*
Tensile properties, polyurethane/polysiloxane coatings, 140, 141*t*
Tertiary-butyl acetate (TBAC)
alkyd coatings, 221, 222*t*
benefits, 223
calculated Hansen solubility parameters, 212*f*

nitrocellulose lacquers, 213–215
percent amine retained in commercial polyamide crosslinkers, 216, 217*f*
physical properties, 207*t*
pseudo first-order aminolysis rates, 216, 217*f*
space filling molecular model, 216*f*
stability in two-component epoxy systems, 215–216
two-component epoxy amine coatings, 215
two-component epoxy reformulation, 216, 218
two-component urethane coatings, 219–221
See also Volatile organic compounds (VOC)-exempt solvents
Thermodynamics, heat and mass, 262
Thickness-shear mode (TSM) devices, sensor array for high-throughput material characterization, 243
Time-of-flight secondary-ion mass spectrometry (ToF–SIMS), service life prediction, 283, 291
Time-temperature superposition, 264–265
Toluene
calculated Hansen solubility parameters, 212*f*
See also Alkyd spray enamel
Tools. *See* Service life predictions
Transverse shear mode (TSM), quartz plate resonator, 263
Triclosan (5-chloro-2-(2,4-dichlorophenoxy)phenol)
antimicrobial/antibacterial agent, 75
incorporation onto siloxane, 80
modification and attachment to siloxane, 75, 78, 79*f*
modification by bromoalkene, 78
See also Marine coatings

Trivalent chromium pretreatment (TCP)
  alternative to chromate conversion coating (CCC), 41
  experimental, 41–42
  neutral salt fog exposure on Al 2024-T3, 42–43, 44f, 45f

U

Ultraviolet-visible spectroscopy (UV-vis)
  service life prediction, 282, 284, 285f
  See also UV-curable waterborne polyurethane dispersion (PUD); UV-cured acrylic coatings; UV-radiation curing
Ulva
  antifouling settlement assays, 95, 98
  assays evaluating foul-release potential, 99–100
  false-color image of settled spore of, 96, 97
  lab-scale evaluation, 102, 105
  macroalga, 92
  test organisms, 94
  zoospores of, 96, 97
  See also Marine coatings, non-biocidal
Uranium decontamination
  coating compositions for, 166t
  decontamination factors, 168t
  potential masking agents, 170t
  safety considerations, 171–172
  See also Sensor coatings
Urethane coatings, two-component, t-butyl acetate (TBAC) in, 219–221
UV-curable waterborne polyurethane dispersion (PUD)
  analysis of variance for softness, 231t
  blending study to improve chemical resistance, 235–237
  chemical spot test, 231
  combined effect of percent chain extension and NCO content on softness, 233f
  crosslinked vs. non-crosslinked films, 227f
  dispersion production process, 228, 229f
  effect of percent chain extension on softness, 233f
  experimental, 227–231
  experimental design, 231–232
  generalized structure, 228f
  methyl ethyl ketone (MEK) double rubs, 230
  performance testing of formulation, 237t
  predicted properties for combinations of factors, 234f
  preparation, 227
  preparation and cure of soft-touch panels, 230
  preparation of UV soft-touch coating formulation, 229–230
  screening study, 228
  softness and chemical resistance of statistical design dispersions, 235t
  softness measurement, 230
  softness measurement and MEK double rubs, 232, 235
  statistical design, 229
  suntan lotion resistance, 231
  urethane acrylate blends, 236t, 237t
UV-cured acrylic coatings
  chemical formulas of compounds used, 178
  conversion vs. time by real-time infrared spectroscopy for polyurethane-acrylate (PUA) coating, 180f
  dual-curing process, 183
  experimental, 177, 179

hardening profiles, 184, 185*f*
influence of reactive acrylic diluent on photopolymerization of PUA, 180*f*
infrared spectroscopy (IR) monitoring chemical modifications, 183
inhibitory effect of atmospheric oxygen, 181
IR spectra of stabilized waterborne PUA, before and after UV-aging, 187*f*
mechanical properties, 184, 185*f*
moisture resistance, 185, 186*f*
monomers, 178
oxygen influence on photopolymerization of PUA, 181*f*
performance analysis of radical-type photoinitiators in PUA coating, 181*f*
performance of diphenoxybenzophenone (DPBP) photoinitiator, 180
peroxides initiating polymerization in non-illuminated areas, 183–184
photocuring, 179–184
photoinitiators, 178
photoinitiators comparison, 180*f*
polymerization and hardening of waterborne PUA coating, 182*f*
properties, 184–186
reactivity of acrylate double bond, 179
resistance to photodegradation, 186, 187*f*
scratch resistance, 184–185, 186*f*
solvent and chemical resistance, 185
surface cure and deep-through cure of DPBP, 180–181
telechelic oligomers, 178
thermal and photochemical curing of mixture of NCO and OH functionalized diacrylate oligomers, 182*f*
uncured remote areas, 182
water-based systems, 182
UV-radiation curing
light-induced polymerization, 176–177
See also UV-cured acrylic coatings

V

Volatile organic compounds (VOC)
ozone, 202–203
ozone formation, 203, 205
ozone forming potential of common coating solvents, 204*f*
regulations, 201
regulators, 202
role in coatings and inks, 202
science behind VOC regulation, 202–205
solvent blend reformulation for, compliance, 209*t*
Volatile organic compounds (VOC)-exempt solvents
acetone, 205, 207*t*
benefits of t-butyl acetate (TBAC), 223
calculated Hansen solubility parameters for toluene and, 212*f*
coating formulations, 212–213
comparing, 205–206, 208
effect of temperature on pseudo first-order aminolysis rate constants, 217*f*
epoxy (two-component, 2K) reformulation, 216, 218
evaporation profiles for two solvent blends, 210*f*
Hansen solvency parameters, 211
hazardous air pollutant (HAP) content, 209t, 210
methyl acetate, 205–206, 207*t*

parachlorobenzotrifluoride
(PCBTF), 205–206, 207t
percent amine retained in
commercial polyamide
crosslinkers, 217f
physical properties of exempt
solvents for coatings, 207t
reformulated two-component epoxy
formulation with TBAC, 218t
selecting solvent blend, 208–210
solvency properties, 211–212
solvent blend reformulation for
HAP and VOC compliance,
209t
solvent popping, 205, 206f
space filling molecular model of
TBAC, 216f
stability of TBAC in 2K epoxy
systems, 215–216
TBAC, 205–206, 207t
TBAC in alkyd coatings, 221,
222t
TBAC in nitrocellulose lacquers,
213–215
TBAC in two-component epoxy
amine coatings, 215
TBAC in two-component urethane
coatings, 219–221
urethane reformulations with
TBAC, 220t
viscosity of commercial alkyd resin
solutions, 213f
wood lacquer reformulation for
HAP and VOC compliance, 214t
See also Tertiary-butyl acetate
(TBAC)

W

Waterborne polyurethane-acrylate
coating
chemical formulas, 178
dual-curing process, 183
infrared spectra of stabilized UV-
cured, before and after UV-
aging, 186, 187f
polymerization and hardening on
UV exposure, 182f
UV-curing technology, 182
See also UV-cured acrylic coatings
Waterborne polyurethane dispersion
(PUD)
hydroxyl-functional, 226
technology, 226–227
See also UV-curable waterborne
polyurethane dispersion (PUD)
Water contact angle, performance of
chromate-free novel primer, 29,
30t, 31t, 32t
Water uptake, sensing, in coatings,
125, 126f, 127f
Waveguides. See Optical waveguides;
Polyimides
Williams–Landel–Ferry theory,
polymer viscoelasticity, 264–265

X

X-ray diffraction
Al-Li hydroxide carbonate
hydrotalcite compound, 125,
126f
Al-Zn-chloride and Al-Zn-
decavanadate hydrotalcite
compounds, 130f
Al-Zn-decavanadate hydrotalcite
compounds exposed to Cl
solutions, 131f
epoxy coating on 2024-T3 Al with
Ce-exchanged bentonite, 128f

Z

Zinc, inhibiting oxygen reduction on
aluminum, 121